Communications
in Computer and Information Science
2086

Rationale

The CCIS series is devoted to the publication of proceedings of computer science conferences. Its aim is to efficiently disseminate original research results in informatics in printed and electronic form. While the focus is on publication of peer-reviewed full papers presenting mature work, inclusion of reviewed short papers reporting on work in progress is welcome, too. Besides globally relevant meetings with internationally representative program committees guaranteeing a strict peer-reviewing and paper selection process, conferences run by societies or of high regional or national relevance are also considered for publication.

Topics

The topical scope of CCIS spans the entire spectrum of informatics ranging from foundational topics in the theory of computing to information and communications science and technology and a broad variety of interdisciplinary application fields.

Information for Volume Editors and Authors

Publication in CCIS is free of charge. No royalties are paid, however, we offer registered conference participants temporary free access to the online version of the conference proceedings on SpringerLink (http://link.springer.com) by means of an http referrer from the conference website and/or a number of complimentary printed copies, as specified in the official acceptance email of the event.

CCIS proceedings can be published in time for distribution at conferences or as post-proceedings, and delivered in the form of printed books and/or electronically as USBs and/or e-content licenses for accessing proceedings at SpringerLink. Furthermore, CCIS proceedings are included in the CCIS electronic book series hosted in the SpringerLink digital library at http://link.springer.com/bookseries/7899. Conferences publishing in CCIS are allowed to use Online Conference Service (OCS) for managing the whole proceedings lifecycle (from submission and reviewing to preparing for publication) free of charge.

Publication process

The language of publication is exclusively English. Authors publishing in CCIS have to sign the Springer CCIS copyright transfer form, however, they are free to use their material published in CCIS for substantially changed, more elaborate subsequent publications elsewhere. For the preparation of the camera-ready papers/files, authors have to strictly adhere to the Springer CCIS Authors' Instructions and are strongly encouraged to use the CCIS LaTeX style files or templates.

Abstracting/Indexing

CCIS is abstracted/indexed in DBLP, Google Scholar, EI-Compendex, Mathematical Reviews, SCImago, Scopus. CCIS volumes are also submitted for the inclusion in ISI Proceedings.

How to start

To start the evaluation of your proposal for inclusion in the CCIS series, please send an e-mail to ccis@springer.com.

Jaume Baixeries · Dmitry I. Ignatov ·
Sergei O. Kuznetsov · Sergey Stupnikov
Editors

Data Analytics and Management in Data Intensive Domains

25th International Conference, DAMDID/RCDL 2023
Moscow, Russia, October 24–27, 2023
Revised Selected Papers

 Springer

Editors
Jaume Baixeries ⓘ
Departament Ciències de la Computació
Universitat Politecnica de Catalunya
Barcelona, Spain

Dmitry I. Ignatov ⓘ
National Research University Higher School
of Economics
Moscow, Russia

Sergei O. Kuznetsov ⓘ
National Research University Higher School
of Economics
Moscow, Russia

Sergey Stupnikov ⓘ
Federal Research Center "Computer Science
and Control" of RAS, Russia
Moscow, Russia

ISSN 1865-0929 ISSN 1865-0937 (electronic)
Communications in Computer and Information Science
ISBN 978-3-031-67825-7 ISBN 978-3-031-67826-4 (eBook)
https://doi.org/10.1007/978-3-031-67826-4

This Springer imprint is published by the registered company Springer Nature Switzerland AG
The registered company address is: Gewerbestrasse 11, 6330 Cham, Switzerland

If disposing of this product, please recycle the paper.

Preface

This CCIS volume published by Springer contains the proceedings of the XXV International Conference on Data Analytics and Management in Data Intensive Domains (DAMDID/RCDL 2023), which was held at the HSE University, Moscow, Russia during October 24–27, 2023.

DAMDID is a multidisciplinary forum of researchers and practitioners from various domains of science and research, promoting cooperation and exchange of ideas in the area of data analysis and management in domains driven by data-intensive research. Approaches to data analysis and management being developed in specific data-intensive domains (DID) of X-informatics (where $X \in$ {astro, bio, chemo, geo, medical, neuro, physics, chemistry, material science, social science, ... }), as well as in various other branches of informatics, industry, new technologies, finance, and business, contribute to the conference content.

Previous DAMDID/RCDL conferences were held in St. Petersburg (1999, 2003, 2022), Protvino (2000), Petrozavodsk (2001, 2009), Dubna (2002, 2008, 2014), Pushchino (2004), Yaroslavl (2005, 2013), Pereslavl (2007, 2012), Kazan (2010, 2019), Voronezh (2011, 2020), Obninsk (2016), and Moscow (2017, 2018, 2021).

The program of DAMDID/RCDL 2023 was oriented towards data science and data-intensive analytics as well as data management topics. This year's program included three keynotes and three invited talks with four scientific talks and two industrial talks among them.

Alexander Petrenko (head of the Programming Technologies Department, Ivannikov Institute for System Programming of the Russian Academy of Sciences) gave a Memorial speech on Sergey Dmitrievich Kuznetsov, who passed away in July 2023. Sergey D. Kuznetsov was a Soviet and Russian database scientist, a principal researcher at the Ivannikov Institute for System Programming, a professor at the Lomonosov Moscow State University, Moscow Institute of Physics and Technology, and HSE University, the Deputy Chair of the Moscow ACM SIGMOD Chapter, and a member of the DAMDID/RCDL Coordinating and Program Committees.

The keynote by Bernhard Thalheim (full professor of Computer Science at the Christian-Albrecht University of Kiel, Germany) was devoted to various kinds of models in sciences, engineering, and daily life as universal tools of the art of thinking and acting. Mikhail Zymbler (deputy director of the Scientific and Educational Center "Artificial Intelligence and Quantum Technologies", South Ural State University, Russia) gave a keynote on anomaly discovery in time series with parallel algorithms and high-performance computing. The keynote by Vladislav A. Blatov (head of the General and Inorganic Chemistry Department at Samara State Technical University, Russia) was devoted to topological methods and tools for the analysis of big crystallographic data.

An industrial invited talk by Pavel Velikhov (leading developer of YDB, Yandex, Russia) was devoted to advanced query optimization for modern analytical scenarios. Andrey

Borodin (team lead of open-source DBMS development at Yandex.Cloud, Yandex, Russia) gave an industrial invited talk on collaborative work of the PostgreSQL community and researchers in the field of data management. Ildar Baimuratov (L3S Research Center, Leibniz University Hannover, Germany) gave an invited talk on a semantic approach to scientific communication via the Open Research Knowledge Graph.

The workshop on Data and Computation for Materials Science and Innovation (DACOMSIN) constituted the first day of the conference on October 24. The workshop aimed to address the communication gap across communities in the domains of materials data infrastructures, materials data analysis, and materials in silico experiments. The workshop brought together professionals from across research and innovation to share their experience and perspectives of using information technology and computer science for materials data management, analysis, and simulation.

The conference Program Committee, comprising members from 11 countries, reviewed 75 submissions during a single-blind two-round reviewing process. The average number of reviews per paper was 3.1. In total, 51 submissions were accepted as full papers, 15 as short papers.

According to the conference and workshops program, 63 oral presentations were grouped into 18 sessions. Most of the presentations were dedicated to the results of research conducted in research organizations located in Russia, including Chelyabinsk, Kazan, Moscow, Murom, Novosibirsk, Obninsk, Samara, St. Petersburg, Tomsk, and Tumen. However, the conference also featured talks prepared by foreign researchers from countries such as Armenia, China, Germany, India, Japan, Pakistan, the UK, and the USA.

For the CCIS conference proceedings, 21 peer-reviewed papers including two keynote talks and one invited memorial paper were selected by the Program Committee (acceptance rate of 28%) and organized in five sections: Models and Knowledge Graphs (five papers), Databases in Data Intensive Domains (two papers), Machine Learning Methods and Applications (six papers), Data Analysis in Astronomy (four papers), and Information Extraction from Text (three papers).

We are grateful to the Program Committee members for reviewing the submissions and selecting the papers for presentation, to the authors of the submissions, and to the host organizers from the HSE University.

We are also grateful for the use of the EasyChair Conference Management System, which provided great support during various phases of the paper submission and reviewing process.

The conference was supported by a grant from the Ministry of Science and Higher Education of the Russian Federation, internal number 00600/2020/51896, agreement dated April 21, 2022, no. 075-15-2022-319.

May 2024

Jaume Baixeries
Dmitry I. Ignatov
Sergei O. Kuznetsov
Sergey Stupnikov

Organization

General Chair

Sergei O. Kuznetsov   HSE University, Russia

Program Committee Chairs

Dmitry I. Ignatov   HSE University, Russia
Jaume Baixeries   Universitat Politècnica de Catalunya, Spain
Sergey Stupnikov   Federal Research Center "Computer Science and Control" of RAS, Russia

DACOMSIN Workshop Chairs

Nadezhda Kiselyova   Baikov Institute of Metallurgy and Materials Science, RAS, Russia
Victor Dudarev   Ruhr University Bochum, Germany
Alexandra Khvan   National University of Science and Technology MISIS, Russia

Organizing Committee

Nadezhda Zotova (Secretary)   HSE University, Russia
Larisa Antropova   HSE University, Russia
Pavel Zhidkih   HSE University, Russia
Alexandra Khvan   National University of Science and Technology MISIS, Russia
Natalya Polkovnikova   National University of Science and Technology MISIS, Russia
Majid Sohrabi   HSE University, Russia
Stefan Nikolić   HSE University, Russia
Sergey Lvov   HSE University, Russia

Coordinating Committee

Igor Sokolov (Co-chair)	Federal Research Center "Computer Science and Control" of RAS, Russia
Nikolay Kolchanov (Co-chair)	Institute of Cytology and Genetics, SB RAS, Novosibirsk, Russia
Sergey Stupnikov (Deputy Chair)	Federal Research Center "Computer Science and Control" of RAS, Russia
Arkady Avramenko	Pushchino Radio Astronomy Observatory, RAS, Russia
Pavel Braslavsky	Ural Federal University and SKB Kontur, Russia
Vasily Bunakov	Science and Technology Facilities Council, UK
Alexander Elizarov	Kazan (Volga Region) Federal University, Russia
Alexander Fazliev	Institute of Atmospheric Optics, SB RAS, Russia
Alexei Klimentov	Brookhaven National Laboratory, USA
Mikhail Kogalovsky	Market Economy Institute, RAS, Russia
Vladimir Korenkov	Joint Institute for Nuclear Research, Russia
Sergei D. Kuznetsov	Ivannikov Institute for System Programming, RAS, Russia
Vladimir Litvine	Evogh Inc., California, USA
Archil Maysuradze	Lomonosov Moscow State University, Russia
Oleg Malkov	Institute of Astronomy, RAS, Russia
Alexander Marchuk	Institute of Informatics Systems, SB RAS, Russia
Igor Nekrestjanov	Verizon Corporation, USA
Boris Novikov	Finland
Nikolay Podkolodny	Institute of Cytology and Genetics, SB RAS, Russia
Aleksey Pozanenko	Space Research Institute, RAS and HSE University, Russia
Vladimir Serebryakov	Federal Research Center "Computer Science and Control" of RAS, Russia
Yury Smetanin	Federal Research Center "Computer Science and Control" of RAS, Russia
Vladimir Smirnov	Yaroslavl State University, Russia
Bernhard Thalheim	Kiel University, Germany
Konstantin Vorontsov	Moscow State University, Russia
Viacheslav Wolfengagen	National Research Nuclear University "MEPhI", Russia
Victor Zakharov	Federal Research Center "Computer Science and Control" of RAS, Russia

Program Committee

Hammad Afzal	National University of Sciences and Technology, Pakistan
Ildar Baimuratov	Leibniz University Hannover, Germany
Dmitry Borisenkov	Voronezh State University, Russia
Vasily Bunakov	Science and Technology Facilities Council, UK
George Chernishev	Saint Petersburg State University, Russia
Radhakrishnan Delhibabu	Vellore Institute of Technology, India
Dmitrii Deviatkin	Federal Research Center "Computer Science and Control" of RAS, Russia
Salvatore Distefano	University of Messina, Italy
Boris Dobrov	Lomonosov Moscow State University, Russia
Victor Dudarev	Ruhr University Bochum, Germany
Alexander Elizarov	Kazan (Volga Region) Federal University, Russia
Alexander Fazliev	Institute of Atmospheric Optics, SB RAS, Russia
Irina Filozova	Joint Institute for Nuclear Research, Russia
Fail Gafarov	Kazan (Volga Region) Federal University, Russia
Yury Gapanyuk	Bauman Moscow State Technical University, Russia
Anna Glazkova	University of Tyumen, Russia
Artem Grachev	HSE University, Russia
Anna Grinevich	Institute of Philology, SB RAS, Russia
Vasily Gromov	HSE University, Russia
Md Shahriar Hassan	University of Clermont Auvergne, France
Martin Thomas Horsch	Norwegian University of Life Sciences, Norway
Dmitry Ilvovsky	HSE University, Rusiia
Eugene Ilyushin	Lomonosov Moscow State University, Russia
Robiul Islam	HSE University, Russia
Mirjana Ivanovic	University of Novi Sad, Serbia
Jeyhun Karimov	Technical University of Berlin, Germany
Anton Khritankov	HSE University, Russia
Gennady Khvorykh	Institute of Molecular Genetics, RAS, Russia
Vitaliy Kim	HSE University, Russia
Nadezhda Kiselyova	Baikov Institute of Metallurgy and Materials Science, RAS, Russia
Dmitry Kovalev	Federal Research Center "Computer Science and Control" of RAS, Russia
Dana Kovaleva	Institute of Astronomy, RAS, Russia
Evgeny Lipachev	Kazan (Volga Region) Federal University, Russia
Natalia Loukachevitch	Lomonosov Moscow State University, Russia
Ivan Luković	University of Belgrade, Serbia

Azhar Mahmood   University of Portsmouth, UK
Sergey Makhortov   Voronezh State University, Russia
Muhammad Shahid Iqbal Malik   HSE University, Moscow
Oleg Malkov   Institute of Astronomy, RAS, Russia
Archil Maysuradze   Lomonosov Moscow State University, Russia
Mikhail Melnikov   Federal Research Center of Fundamental and
                                  Translational Medicine, Russia
Francesco Mercuri   National Research Council, Italy
Alexey A. Mitsyuk   HSE University, Russia
Igor Morozov   Joint Institute for High Temperatures, RAS,
                                  Russia
Dmitry Namiot   Lomonosov Moscow State University, Russia
Dmitry Nikitenko   Lomonosov Moscow State University, Russia
Panagote Pardalos   University of Florida, USA
Alexei Pozanenko   Space Research Institute, RAS, Russia
Surya Prasath   Cincinnati Children's Hospital Medical Center,
                                  USA
Aleksei Romanov   ITMO University, Russia
Roman Samarev   Bauman Moscow State Technical University,
                                  Russia
Andrey Savchenko   Sber AI Lab, Russia
Vladimir Serebryakov   Federal Research Center "Computer Science and
                                  Control" of RAS, Russia
Nikolay Skvortsov   Federal Research Center "Computer Science and
                                  Control" of RAS, Russia
Ivan Smirnov   Federal Research Center "Computer Science and
                                  Control" of RAS, Russia
Manfred Sneps-Sneppe   Ventspils University of Applied Sciences, Latvia
Ilya Sochenkov   Federal Research Center "Computer Science and
                                  Control" of RAS, Russia
Majid Sohrabi   HSE University, Russia
Alexander Sychev   Voronezh State University, Russia
Bernhard Thalheim   Christian-Albrechts-Universität zu Kiel, Germany
Irina Uspenskaya   Lomonosov Moscow State University, Russia
Pavel Velikhov   Yandex, Russia
Alexander Veretennikov   Ural Federal University, Russia
Alina Volnova   Space Research Institute, RAS, Russia
Yury Zagorulko   Ershov Institute of Informatics Systems, RAS,
                                  Russia
Victor Zakharov   Federal Research Center "Computer Science and
                                  Control" of RAS, Russia
Nataly Zhukova   St. Petersburg Institute for Informatics and
                                  Automation, RAS, Russia

Sergej Znamenskij   Ailamazyan Pereslavl University, Russia
Mikhail Zymbler   South Ural State University, Russia

DACOMSIN Workshop Program Committee

Karine Abgaryan   Federal Research Center "Computer Science and
                                        Control" of RAS, Russia
Toshihiro Ashino   Toyo University, Japan
Vladislav Blatov   Samara State Technical University, Russia
Vasily Bunakov   Science and Technology Facilities Council, UK
Martin Thomas Horsch   Norwegian University of Life Sciences, Norway
Galina Kuzmicheva   MIREA - Russian Technological University,
                                        Russia
Igor Morozov   Joint Institute for High Temperatures, RAS,
                                        Russia
Irina Uspenskaya   Lomonosov Moscow State University, Russia
Yibin Xu   National Institute for Materials Science, Japan

Additional Reviewers

Vyacheslav Manevich   HSE University, Russia
Vasily Yadrintsev   Federal Research Center "Computer Science and
                                        Control" of RAS, Russia
Denis Zubarev   Federal Research Center "Computer Science and
                                        Control" of RAS, Russia

Sponsors and Partners

HSE University, Russia
Federal Research Center "Computer Science and Control" of the Russian Academy of
    Sciences, Moscow, Russia
Moscow ACM SIGMOD Chapter, Russia
Fund for Infrastructure and Educational Programs, Russia

Memorial Talk

Sergei Dmitrievich Kuznetsov
04/08/1949 – 07/28/2023

Fig. 1. S. D. Kuznetsov. Sophia Antipolis, France, February 2007.

Sergei D. Kuznetsov, a well-known Russian specialist in the field of database and information systems technologies, one of the founders of the Moscow ACM SIGMOD Chapter[1], passed away in July 2023. This short article is a tribute to the memory of a famous scientist and wonderful person, for whom the life of the Moscow ACM SIGMOD Chapter was an important part of his own life. Sergei Dmitrievich Kuznetsov (Sergei) was a major specialist in several areas for System Programming. The main ones are: operating systems, object-oriented approach in the design and development of programs, database management systems, and information systems.

In 1971, he graduated from the Faculty of Mechanics and Mathematics of Lomonosov Moscow State University with a degree in mathematics, then worked at the Institute of Precision Mechanics and Computer Science of the Academy of Sciences of the USSR

---

[1] https://synthesis.frccsc.ru/sigmod/eng/index_html.html.

(now the S. A. Lebedev Institute of Precision Mechanics and Computer Science of the Russian Academy of Sciences). He was one of the main developers of the operating system for the central processor of the AC-6 computer complex, which was actively used in real-time systems for controlling spacecraft flights. The first large-scale use of the AS-6 was the joint Soviet-American Soyuz-Apollo program (1975).

In 1979 Sergei defended a dissertation for the academic degree of Candidate of Physical and Mathematical Sciences on the topic "Organization of multiprogramming in the OS" (under supervision of V. P. Ivannikov), and in 1995 a dissertation for the scientific degree of Doctor of Technical Sciences on the topic "Research and development of a free SQL server"; in 2006 he received the academic title of professor.

In the 1980s Sergei worked at the Delta Research Institute of the USSR Ministry of Electronic Industry and at the Institute of Cybernetics of the Academy of Sciences of the USSR, where he was involved in the creation and implementation of supercomputer software "Electronics SS BIS-1", and the development of the CLOS cluster operating system and a relational database management system based on CLOS.

After the formation of the Institute for System Programming of the Russian Academy of Sciences in 1994, Sergei was the first Scientific Secretary of the Institute, and headed the Department of Data Management and Information Systems Development, whose tasks included the development of system software for data processing and analysis, database management systems, distributed big data processing technologies, and cloud computing technologies. Under the scientific guidance of Sergei development of algorithms for statistical data analysis and machine learning began, as well as software for solving applied problems, including text mining, social network analysis, bioinformatics problems, and multimedia data processing.

Sergei did a lot of public work and scientific and organizational work. In the 1990s, he was the chairman of the council of the Soviet (Russian) Association of Users of the UNIX operating system (SUUG), a member of the European Association EurOpen, Usenix and Uniforum associations, a member of the Association for Computing Machinery, a member of the IEEE Computer Society and its representative in Moscow, deputy chairman of the Moscow section of ACM SIGMOD, a member of program committees of international conferences, and co-chairman of the international conference of young scientists SYRCODIS. He was the editor-in-chief of the Open Systems Journal, the deputy editor-in-chief of the journal Proceedings of the Institute for System Programming of the Russian Academy of Sciences, etc.

Sergei conducted extensive teaching work at the Faculty of Computational Mathematics and Cybernetics of Lomonosov Moscow State University, at the Moscow Institute of Physics and Technology and the National Research University Higher School of Economics. A large team of highly qualified research programmers was trained under the leadership of Sergei.

Sparing words listing the deeds and the results obtained are not enough to tell about a person, especially about a person like Sergei, who had many talents, a gentle soul, and a sympathetic heart. Friends of Sergei know about his unique collection of jazz music, his graphic drawings, and the books he was fond of.

Fig. 2. Drawing by Sergei Kuznetsov. 1960s–1970s.

We cover only three topics in the next sections:

- Scientific publisher and editor;
- Teacher;
- Scientist.

Scientific Publisher and Editor[2]

Sergei D. Kuznetsov apparently had always had a passion for working with texts. He loved to read and write, which, unfortunately, is rare among programmers. Probably, his first experience of professional immersion in working with the "scientific word" took place in the 1970s–80s, when he and Viktor Petrovich Ivannikov actively collaborated with the Abstract Journal of the Academy of Sciences of the USSR, which was published by VINITI. There was a huge flow of information that passed through Sergei and needed to be summarised and presented in a brief but meaningful form. In addition, at the same time, Sergei participated in scientific conferences, where one of his main roles was the drafting and editing of theses and reports. He worked extremely carefully with the authors. I remember one of the conversations with Sergei on the eve of a large conference on databases in Tver (then in Kalinin). I was excited to tell Sergei about the development of an integrated turbo environment that combined editing, compilation, and debugging tools and convince him that I was not involved in any databases, and therefore he should not expect any report from me for this conference. Sergei listened to me with interest and noted that, in a turbo environment, it is probably necessary to have an effective mechanism for working with internal representations of the program, that is, there was an intersection of my topic with the topic of the conference. The result was a good

---

[2] The section was prepared by Alexander K. Petrenko.

article, and the topic of DBMS for supporting integrated development environments is still relevant today.

An important stage in his editorial career was the creation of the Open Systems Journal[3], of which he was the first editor-in-chief. But the longest part of his life as an editor was associated with the Proceedings of the Institute for System Programming, 2000–2023. The editor-in-chief of the journal was Viktor P. Ivannikov, but for all editorial work and correspondence with the authors Sergei took over. In the first year, one issue was published, containing 8 articles. Then 2 issues were published per year, then a process was built: 6 issues per year, each with 10 to 20 articles, sometimes more.

It should be said that Sergei sometimes would formally pull rank if the work was not to his liking, and would immerse himself in it completely if he liked it. There was a happy accident with the Proceedings; he liked this work, and as a result he liked both the articles and the journal itself. He said more than once that the Proceedings had become the best programming journal in Russia in recent years, and he was proud of it. His contribution to the level of the journal primarily consisted in the fact that he not only read articles, but communicated with the authors if it was necessary to bring the article to the required quality. I have never heard of any of the authors being offended by such close attention. At the same time, Sergei always treated all authors very kindly, and the high praise from his lips always inspired and gave new strength.

Fig. 3. Drawing by Sergei Kuznetsov. Calendar. 1960s–1970s.

---

[3] https://www.osp.ru/os/about?lang=en.

Teacher

Academician Israel M. Gelfand, according to the recollections of his students, defined the concept of "good teacher" as follows: "To be a good teacher you need: first, *love your subject*, secondly, *love to teach*, and thirdly, *love those you teach*". Sergei Dmitrievich Kuznetsov fully met this high level.

Supervisor with a Remarkable Breadth of Knowledge[4]

"It has to be either databases or networks! Wouldn't need to study hard in either and there's a lot of potential!" said my best friend in the uni while we were discussing the area to study.

So he ended up doing networking and I went on an open day at the database seminar led by Sergei Dmitrievich. I was really unsure of what relational databases were at the time, but it all sounded very exciting. And it still is, 20 years later!

Sergei was an extraordinary teacher. He was across both hardware and software architectures and could describe a B-Tree index retrieval with a few waves of a cigarette, using its smoke to illustrate operational complexity. After one such impromptu explanation, something clicked, and I finally grasped the underlying connections between a whole lot of subjects I was studying at the time.

Sergei didn't force your hand in choosing the research topic. Instead, he supported and mentored you along the way. This wasn't the more strict study pattern many of us were used to, and it caused a few students to drop out. However, this freedom to choose quite offbeat and trendy topics was exactly what attracted like-minded students under Sergei's wing.

My PhD thesis was based on a niche mathematical concept of lattices and implemented using the MapReduce/Apache Hadoop stack, which was quite novel at the time (2005) [6]. Sergei's breadth of interests, all the reading, and his help with various topics meant that he was already familiar with both concepts and could guide me along the way. That was very impressive.

Sergei was also very interested in the real-life applications of database technologies. For many years, he hosted the "Corporate Databases" conference where his research students and technology professionals from large companies would spend hours in breakout sessions sketching on napkins and debating specific implementations of particular algorithms in different software products. Sergei's interest in both the theoretical and practical aspects of database systems was contagious; we were all buoyed by his enthusiasm.

I followed his advice to start translating some of the articles I found inspiring from English to Russian, as was his long-time habit. This served as both a great writing exercise and also helped me connect with a lot of interesting people once some of these translations were published (with Sergei's help, of course). Later on, I stumbled upon the idea to try out a thesis based on one of the articles.

---

[4] The subsection was prepared by Yuri Kudryavcev.

That writing itch also made me start blogging, connect with more people through the blog, and eventually move to Australia to work with some of them on implementing database technologies. Meeting Sergei definitely defined my life!

Excellent Lecturer and Mentor[5]

I was lucky to be not only a student who attended Sergei D. Kuznetsov's lectures, but also his supervisee.

As a listener to Sergei's lectures, I can say that he was an unsurpassed lecturer who brought to each lecture not just knowledge, but the spirit of researchers and, DBMS developers, and the history of the development of science and the database industry. With full notes available, students actively attended the lectures because they were informative and interesting.

Sergei treated people very warmly, with sincere participation and empathy. He tried to find interesting activities for each of his students, of whom he actually had many. By the way, Sergei had always an "overkill" of scientific students; many wanted to get him as a research advisor. But Sergei was an inexhaustible source of ideas for scientific activity; both Viktor Petrovich Ivannikov and Lyudmila Sergeevna Korukhova, the Scientific Secretary of the System Programming Department, knew this, so they often turned a blind eye to the "overkill". Sergei had a wide circle of acquaintances in the IT industry, thanks to which many of his students, including me, found their first job in IT companies, or on IP projects. Recommendations and requests from Sergei meant a lot. They trusted him and respected his opinion.

I remember how Sergei recommended me to one company to gain my first practical experience. I had interviews and went to communicate in person, but they didn't hire me. Sergei was very worried that it didn't work out. I remember his face, how he silently smoked, sighed, and thought. In the end, he promised to find me an equally good place and, of course, he kept his promise. Sergei was not a demanding leader. He was kind, understanding, and even treated the students like a father. When setting exams, he could challenge a student mercilessly on all topics, fail the student, and in the end give him a good grade. I can't speak for everyone, but Sergei's kind attitude forced me not to let my supervisor down and to respond to condescension with diligence.

As a leader, Sergei was attractive for his aura of openness, on the one hand, and the incredible "coolness" of his personal and joint works with Russian and foreign partners. This was the magic of his figure, sitting in cloudy smoke on the balcony of the hall of the Faculty of Computational Mathematics and Cybernetics of the Moscow State University.

I can't imagine how my supervisor managed to do so much and at the same time communicate with people without rushing. By the way, he always had many pages open at the same time in his browser. But when I came to him, he almost never kept me waiting and said: "Let's go have a smoke."

Subsequently, already during my work at the Faculty of Computational Mathematics and Cybernetics, I gave a lecture course on the theory of databases based on the notes of

---

[5] The subsection was prepared by Andrey Suslov.

Sergei during my business trip to Kazakhstan, at the Moscow State University branch. It was already clear to me then that students' interest was determined not so much by the content of the course as by the presentation of the material and its connection to the life stories of interesting people. I didn't have my own baggage of stories, of course, but I gladly (and probably with a smart look) retold stories from Sergei

We continued to have very warm relations even after the University. We once met in the mid-2000s on a winter evening at a hotel in Kazakhstan, where Sergei came to give lectures, and I was already working in another company, by the way, related to the research work of Pavel Velikhov. Sergei Dmitrievich was with his wife Lena. It was cold in the room, but we drank hot tea all evening, discussed world trends, and just warmly talked about life. Sergei was always sincerely happy about my successes, as if I was his own son.

As an indicator of the demand for textbooks by Sergei, statistics on the number of downloads can serve as a guide. Over the past three months it has been like this:

– Databases. Introductory course

  - November 2023–1615 downloads
  - December 2023–1922 downloads
  - January 2024–1938 downloads

– Fundamentals of Modern Databases

  - November 2023–1851 downloads
  - December 2023–2044 downloads
  - January 2024–2239 downloads

– Summary statistics by article

  - 600–700 downloads per month.

Further unique evidence of the students' gratitude is the course of lectures "Databases" [6].

In the header of each page there is the following inscription: "These lecture notes were prepared by students, have not been professionally edited and may contain errors. Please follow https://vk.com/teachinmsu for updates."

Not everyone gets such recognition and respect from students.

We didn't have time to meet before the pandemic, but we regularly called each other. I was looking forward to our meeting, but Sergei, as it turned out later, was already ill and did not want to be seen like that.

I think the most important thing that has always been in my supervisor is love. For people, for business, for the world in general. To love your subject, to love to teach and to love those you teach – this is 100% about our dear Sergei Dmitrievich.

---

[6] https://teach-in.ru/file/synopsis/pdf/database-M.pdf.

Fig. 4. The course of lectures on Databases

Fig. 5. Drawing by Sergei Kuznetsov. 1997

Scientist[7]

Sergei D. Kuznetsov was a leading Russian scientist in the field of Database Management Systems. His activities in this area included not only academic and educational activities, but also deep practical activities: Sergei was the scientific director of the development of the GNU SQL Server DBMS and Sedna XML Database. Sergei delved into completely different aspects of creating a DBMS, from complex scientific issues to purely technical ones. And besides these two systems in which Sergei was directly involved, he also maintained connections with the entire Russian DBMS development community and helped with advice and his expertise on other DBMS projects.

I was lucky enough to work with Sergei on the Sedna project, and subsequently Sergei and I worked together at Huawei, where Sergei advised the company's management on the development of its product line. Sergei quickly immersed himself in the topic of massively parallel analytical DBMSs and wrote fundamental articles about the prospects and limitations of real-time analytics systems.

Also Sergei asked me to review his latest work in the field of DBMS. This fundamental work by Sergei proposed an elegant solution to a very old problem of relational DBMSs and the SQL standard – working with unknown values. And although some solutions to the problem have already been proposed in the industry, Sergei described many years of development of thoughts and initiatives in this direction and justified his solution to this problem in very clear terms. At the same time, Sergei worked hard on the article in the last months of his life, despite already serious health problems. The work turned out to be of very high quality and is one of the most valuable scientific contributions of Sergei

I would especially like to note the deep human interest of Sergei in researchers in the field of DBMS. Sergei was always interested not only in the scientific results of his colleagues, but also always showed deep human interest in their activities and destinies.

Acknowledgements. We would like to thank Nadezda Zotova from HSE University for her help with typing this text.

Selected works. In the References section below a list of selected papers by Sergei Dmitrievich is provided. The list of papers that he himself compiled consists of 219 items. Here we have selected the very first and most recent papers. For the most part, these are brilliant analytical reviews, as well as landmark works that determined entire areas of research and development of the teams in which he worked. A more complete list of works, including textbooks and monographs written by Sergei Kuznetsov can be found on his website https://kuz.me/.

---

[7] The section was prepared by Pavel Velikhov.

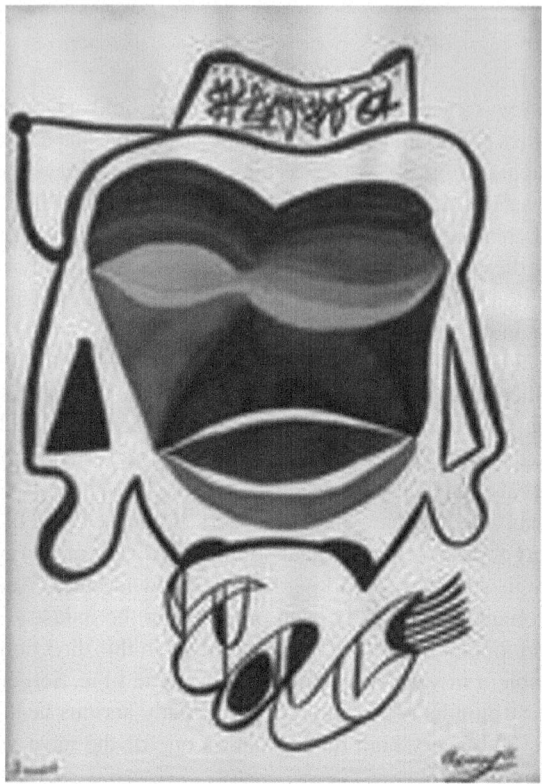

Fig. 6. 9 myself. 1960s–1970s.

References

1. Kuznecov, S., Kudryavcev, Y.: Applying map-reduce paradigm for parallel closed cube computation. In: Chen, Q., Cuzzocrea, A., Hara, T., Hunt, E., Popescu, M. (eds.): The First International Conference on Advances in Databases, Knowledge, and Data Applications, DBKDS 2009, Gosier, Guadeloupe, France, 1–6 March 2009, pp. 62–67. IEEE Computer Society (2009)
2. Kuznetsov, S.D., Velikhov, P.E., Fu, Q.: Real-time analytics: benefits, limitations, and tradeoffs. Prog. Comput. Softw. **49**(1), 1–25 (2023)
3. Kuznetsov, S.D.: Typed unknown values: a step towards solving the problem of representing missing information in relational databases. Trudy ISP RAN/Proc. ISP RAS 35 (2) (2023) 73–100
4. Chistyakova, A., Cherepnina, M., Arkhipenko, K., Kuznetsov, S.D., Oh, C.S., Park, S.: Evaluation of interpretability methods for adversarial robustness on real-world datasets. In: 2021 Ivannikov Memorial Workshop (IVMEM), pp. 6–10. IEEE (2021)

5. Ryndin, M.A., Turdakov, D.Y., Kuznetsov, S.D.: Catalyst: Combining co-training and active learning for lifelong classification. In: 2020 Ivannikov Ispras Open Conference (ISPRAS), pp. 96–101. IEEE (2020)
6. Kuznetsov, S.D.: In anticipation of native DBMS architectures based on non-volatile main memory. Proc. Inst. Syst. Prog. RAS **32**(1), 153–180 (2020)
7. Nguyen, V.A., Duong, N.S., Nguyen, T.T.H., Kuznetsov, S., Nguyen, T.Q.V.: A method for determining information diffusion cascades on social networks. **6**(2), 61–69 (2018)
8. Kuznetsov, S.D.: New storage devices and the future of database management. Baltic J. Modern Comput. **6**(1), 1–12 (2018)
9. Drobyshevskiy, M., Turdakov, D., Kuznetsov, S.: Reproducing network structure: a comparative study of random graph generators. In: 2017 Ivannikov ISPRAS Open Conference (ISPRAS), pp. 83–89. IEEE (2017)
10. Kuznetsov, S.D.: Perspectives and problems of nonvolatile memory usage. In: Proceedings of the 5th International Conference on Actual Problems of System and Software Engineering, pp. 7–21 (1989)
11. Kuznetsov, S.D.: Data management: 25 years of forecasts. Proc. Inst. Syst. Prog. RAS **29**(2), 117–160 (2017)
12. Kuznetsov, S.D.: ODMG and SQL object models ten years later: no inconsistencies. Proc. Inst. Syst. Prog. RAS **27**(1), 173–192 (2015)
13. Kuznetsov, S.D., Poskonin, A.V.: NoSQL data management systems. Prog. Comput. Softw. **40**, 323–332 (2014)
14. Turdakov, D.Y., et al.: Texterra: A framework for text analysis. Prog. Comput. Softw. **40**, 288–295 (2014)
15. Bourdonov, I.B., Ivannikov, V., Kossatchev, A.S., Kuznetsov, S.D., Tomilin, A.: Operating system of the multi-machine computer AS-6. In: Impagliazzo, J., Proydakov, E. (eds.) Perspectives on Soviet and Russian Computing. SoRuCom 2006. IFIP Advances in Information and Communication Technology, vol. 357, pp. 31–35. Springer, Heidelberg (2011). https://doi.org/10.1007/978-3-642-22816-2_5
16. Kuznetsov, S.D.: Transactional parallel DBMSs: a new wave. Proc. Inst. Syst. Prog. RAS **20**, 189–251 (2011)
17. Taranov, I., et al.: Sedna: native XML database management system (internals overview). In: Proceedings of the 2010 ACM SIGMOD International Conference on Management of Data, pp. 1037–1046 (2010)
18. Kuznetsov, S.D.: Object-relational databases: former stage or depreciated capabilities? Proc. Inst. Syst. Prog. RAS **13**(2), 115–140 (2007)
19. Kostenko, B., Kuznetsov, S.D.: History and urgent problems of temporal databases. Proc. Inst. Syst. Prog. RAS **13**(2), 77–114 (2007)
20. Novak, L., Kuznetsov, S.D.: Canonical forms of XML schemas. Prog. Comput. Softw. **29**, 283–293 (2003)
21. Grinev, M., Kuznetsov, S.D.: UQL: A UML-based query language for integrated data. Prog. Comput. Softw. **28**, 189–196 (2002)

22. Kuznetsov, S.D., Ivannikov, V., Burdonov, I., Kosachev, A., Kopytov, G.: The CLOS project: Towards an object-oriented environment for the development of applied systems. Control Syst. Comput. 61–65 (1992)
23. Kuznetsov, S.D.: State-of-the-art and perspectives of database theory and practice: An analytical survey. Appl. Inf. **4**, 153–172 (1989)
24. Kuznetsov, S.D., Byakov, A.Y., Ivannikov, V.: Standard operational system for the central processor of a computational system. Preprint (1974)
25. Kuznetsov, S.D., Byakov, A.Y.: Standard operational system for the central processor of a computational system. Proc. Young Spec. **1**, 79–93 (1973)

Alexander K. Petrenko
Yuri Kudryavcev
Andrey Suslov
Pavel Velikhov

Contents

Machine Learning Methods and Applications

Data Analysis in Astronomy

Information Extraction from Text

Keynote Talks

Kinds of Models in Sciences, Engineering, and Daily Life

Bernhard Thalheim[(✉)]

Christian-Albrechts University at Kiel, 24098 Kiel, Germany
bernhard.thalheim@email.uni-kiel.de
http://bernhard-thalheim.de/

Abstract. Models are universal, partly silent and wonderful tools of the art of thinking and acting. Computer science is particularly affected here and has developed hundreds of different approaches for model conceptions and modelling approaches. One wonders, therefore, whether there should not be an art and science of models with a common conception of the term "model" for the whole science. So far, this has seemed unrealistic. We use the generic conception of model and develop on its basis derived particular conceptions depending on the use case, depending on the CoP, depending on the requirements, and satisfying concerns. We can also use it to transfer knowledge and findings for one kind of model to the other kind, thus developing modeling as one of the four guiding paradigms of Computer Science in coexistence with three other paradigms: small and large scale structuring, small and large scale dynamics, and collaboration with all its facets.

Keywords: modelology · science and art of modelling · models · conceptual model · modelling theory · modelling practices

1 Introduction

Models accompany us since time immemorial in all spheres of life, support the thinking and also every activity and are also in the transfer of knowledge and insights no longer imaginable without. Computer science uses models in all matters; indeed, it is downright model-loving and model-infested. It is astonishing, however, that this part of computer science is only neglected in research. It is time to develop (1) a science and art of modelling ($\_\_M$), (2) modelling and model usage activities ($\_M\_$)and (3) models (MMM) ($M\_\_$), i.e. a science and art of MMM what can be called *modelology*.

Modelology also brings together modelling branches within other disciplines, remedying a problematic specialization and fragmentation (see, for instance, the compendium [40]). Modelology is at the same time an art and especially an engineering art, a science with a theory building, and an action doctrine for MMM. We want to devote special attention here to the question of the extent to which one can arrive at a systematization at all, given the rather large variety of model concepts, quite different model usage, and the abundance of approaches to model development.

1.1 400+ Model Notions and Conceptions

Computer Science uses far more than 50 different particular kinds of models in all its sub-disciplines [42]. There are also well over 400 different model conceptions to date. Even for conceptual modeling, there is more than threescore of conceptions of conceptual model. Mostly these conceptions are also so different that a harmonization and systematization seem to be beyond the possibilities of a science. But if you look more closely, you discover that these conceptions are centrally focused on a specific use in a specific application domain, for a specific CoP (community of practice), for a limited area of responsibility, and also a dominant language support.

If this variety and diversity is continued in such a way, then one develops again and again quite similar solutions, approaches and also justifications without a chance of taking over existing experiences and the rich body of MMM knowledge. We found a general model definition after an extensive bottom-up analysis for science and engineering [40]. This concept seems very abstract at first. To make it usable we use a generic parameter-based form. This can also be used to obtain corresponding special forms by concretization and coherent assignment of the entire parameter ensemble. This path is followed in this paper to show how insights and success stories can be developed and exploited for many of the particular model types simultaneously.

1.2 Modelology Inherits Success Stories of Other Developments in Computer Science

However, the incredulous question arises whether such harmonization and systematization can be possible at all, or whether this would be unique for the whole of Computer Science. An almost hidden success story is algorithmics [4] that treats algorithms as general solution pattern which have parameters for their instantiation, handling mechanisms for their specialisation to a given environment, and enhancers for context injection. So, an algorithm can be derived based on explicit selectors and control rules [3] if we neglect context injection.

We can use this approach for data mining design (DMD) [14]. For instance, an algorithm pattern such as regression uses a generic model of parameter dependence, is based on blind search, has parameters for similarity and model quality, and selection support for specific treatment of the given data set. In this case, the controller is based on enablers that specify applicability of the approach, on error rules, on data evaluation rules that detect dependencies among control parameters and derive data quality measures, and on quality rules for confidence statements.

1.3   Typical Particular and Specific Models and Their Deployment

The easiest and most convincing way to determine the types of models is to use them. This separation into particular models is common and allows us to consider particular models further on. This has also led to the different kinds of models and the many conceptions of model.

Modeling can be done routinely and quickly or after thorough preparation and thoughtfulness. One does not need the optimal solution, but only a sufficient one. An approximate model can also be sufficient in other application scenarios. We can observe different thorough approaches to modeling. For the description we use the metaphor the *cooking princes*:

Quick and somehow successful use – *[cookery with ready-to-eat food]*: A new model is created according to a routine procedure based on a prefabricated mold and used schematically. Mostly tools are used directly.

Use following existing experience –*[recipe book cookery]*: Analog use cases are selected as models and defaults. Then analogous procedures are followed without further adaptation and critical reappraisal. Many of the modeling approaches in Mathematics, for example, follow this approach.

Assemble, configure, compose for thoughtful use – *[source-driven cookery]*: It aligns modeling with available capabilities and resources so that an approach to a configuration can be pursued roughly at first, then refined steadily, and then, with a little more effort, a model calibrated to the task is developed and deployed.

Professional journeyman use – *[craftsmanship cookery]*: A professional analyzes the use case, knows about applicable resources and their limitations, selects best possible resources from them and then acts professionally, especially based on the already successfully practiced cases as well as the lessons learned from less successful ones.

Modelology – *[master class 1star cookery]*: A deeper understanding of the art and practice of modeling exists and, after thorough preparation, is used professionally at the master level so that adequate models that are fully aligned with the concern are created and used.

Within this understanding of modeling as an auxiliary act, particular models have emerged which are then developed and used with the background of experience. This has resulted in only a small contribution to modelology. To what extent then all the knowledge on particular models could be suitable and also integrated for other particular models remains unknown. We therefore set ourselves the task of opening up the body of knowledge for all particular models.

1.4   Research Issues and Questions

Each branch and each special lining of the branch in a thought community led so far to a large body of knowledge of this community for particular model kinds. It seems that all this is so different that one cannot learn and participate from each other. This insufficient state-of-art of modelling can be overcome if we can answer the following research questions in sufficient depth.

RQ₁ – Knowledge and prowess transfer: *Can approaches, techniques, insights, findings, and practice of one particular kind of model be transferred to another kind of model? Can this be learned systematically? What can be rethought in one kind if one knows the other well?*

RQ₂ – Generalising experience to a body of knowledge: *What are commonalities, differences, compatibilities, and interchangeable specifics of individual kinds? Can this be brought together in such a way that a general model science can emerge?*

RQ₃ – Modelology as art and engineering: *Is it possible to prepare a systematics for particular model conceptions in such a way that a doctrine and lore for model kinds or a modelology can be distilled from it?*

RQ₄ – Inspired by others: *If there is a generalisable view on the systematics of the models, then one could also observe in a dedicated way in other modelings what one should also learn for one's own categories. Can the particular model kinds cross-fertilize and inspire each other?*

To answer these research questions, we use a bitter lesson that interoperability research has brought us. Integration at the same level of abstraction – of whatever kind – is impossible with mangled diversity. However, unification and harmonisation of very different approaches is usually possible if we abstract and generalize in a clever way, while also professionally underlaying the mapping from particular to general. This leads directly to a consideration of *generic model conceptions* together with corresponding particular conceptions.

2 From Generic Model Conception to Particular Conceptions and Back

Let us start with the generic model conception, although particular and specific notions exist in an amazing variety. Since this diversity prevents an order and systematization, we want to develop here an order scheme and an approach to systematization. This will also allow particular and specific properties to be transferred from one model category to another. Notice, however, that "anything can be a model of anything ... [I]t is being taken as a model which makes an actual out of a potential model; and every case of being taken as a model involves a restriction with respect to relevant properties." [45] The model-being or the

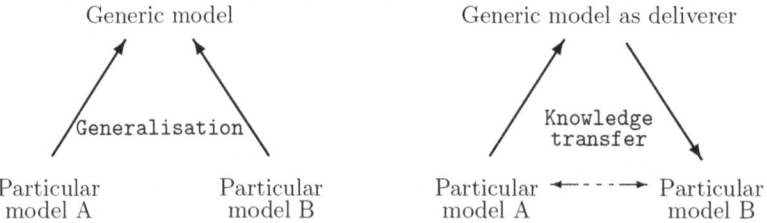

Fig. 1. Resolving RQ₄: Particular models inspired by another particular model

instrument-being is based on a judgement whether something could be used as a model or instrument.

Figure 1 shows the way to learn from one kind of model for another. It maps to a generic model for one kind of model using a generalization mechanism. This generalization mechanism can also be invertible if necessary. In this case, one could transfer knowledge and experience of model kind A to model kind B. For the inverse, we use mappings from generic to particular models right away as a trick.

2.1   The Conception of Model

Let us briefly remember notions to modelling in [32] (see also [35–39])[1]):

"A **model** *is a well-formed, adequate, and dependable instrument that represents origins and that functions in utilisation scenarios."* [8, 32]

"Its criteria of well-formedness, adequacy, and dependability must be commonly accepted by its ... CoP within some context and correspond to the functions that a model fulfills in utilisation scenarios.

The model should be well-formed according to some well-formedness criterion. As an instrument or more specifically an artifact, a model comes with its *background*, e.g. paradigms, assumptions, postulates, language, thought community, etc. The background its often given only in an implicit form and hidden.

A well-formed instrument is *adequate* for a collection of origins if it is *analogous* to the origins to be represented according to some analogy criterion, it is more *focused* (e.g. simpler, truncated, more abstract or reduced) than the origins being modelled, and it sufficiently satisfies its *purpose*.

Well-formedness enables an instrument to be *justified* by an empirical corroboration according to its objectives, by rational coherence and conformity explicitly stated through conformity formulas or statements, by falsifiability or validation, and by stability and plasticity within a collection of origins.

The instrument is *sufficient* by its *quality* characterisation for internal quality, external quality and quality in use or through quality characteristics [32] such as correctness, generality, usefulness, comprehensibility, parsimony, robustness, novelty etc. Sufficiency is typically combined with some assurance evaluation (tolerance, modality, confidence, and restrictions).

A well-formed instrument is called *dependable* " [and thus reliable and trustable] "if it is sufficient and is justified for some of the justification properties and some of the sufficiency characteristics." [32]

2.2   The Tension Between Particular Model Conceptions

As an example, let us consider a specific particular model first in its highly abbreviated form.

---

[1] We omit here a detailed bibliography of the more than 4000 works on models and further on the history of modeling already considered by us and refer especially to [7, 21, 23, 25, 30, 40].

A database model is a conceptual model of a database for an application. [32]
This form hides as its "secret" many details, which can also be considered explicitly. Just try to unfold these secrets.

A conceptual description database structure model *is used as*

(a) an instrument in a database system construction scenario (its nature)

(b) for understanding, communicating, negotiating, agreeing, and specifying by consolidation

(c) of structuring of potential origins in dependence of predicted ways of usage, and

(d) functions well as description in analysis, design, explanation, and realisation of database systems.

The conceptual description database structure model

(α) inherits the platform landscape with its mission and brand and additionally its postulates,

(β) incorporates the matrix with its paradigms,

(γ) uses supporting and enabling means from its workshop with its specific principles within the settings, and (δ) implicitly integrates the specific understanding, meaning and sense of notions in given application.

It represents the conceptual understanding (i.e. the conceptualisation) of some database application domain for a particular purpose in all its viewpoints.

This characterisations of a model kind follows a scheme: (a) defines the application landscape, (b) concern of the model, (c) the role to play by the model, (d) the functions within an application scenario, while (α) promotes the embedding in the landscape, (β) annotates the strategy frame, and (γ) relates to the workshop frame.

Example 1. *Consider, for instance, a special conceptual description database structure model that serves analysis, specification, clarification of a common understanding, and paraphasing for a development of an information system in a banking application with special orientation to the support of the bankers' work (different than for the bank employees).*

The model is canonically formed based on well-known business data structures common in the application as a data structure description model suite for various data usages and equipped with the usual functionality for object-relational analysis, development and design for a coordinated and appropriate specification of the banker's work. The model incorporates the application background (e.g., the way models are thought about, represented, handled), the approach to development (e.g., the database life cycle), the way they are implemented (e.g., concrete system environments), and the world of banking (e.g., organizational theory, specific banking doctrines, bankers' thinking). A banker's information system thrives on the naturalized understanding of how a bank (such as the German bank) works, what processes are performed and in what form (e.g., specific balance sheets), concepts and terms, accepted theories, and also common practices. For a realization as an information system we need as functionality also linguistic forms (e.g. of the banker's everyday life), means of representation (e.g. languages for conceptual modeling such as ER [30]), and also the language of the systems (e.g. database systems within a selected DBMS paradigm).

If one describes a particular model in this way, then it becomes apparent that such types of models are closely connected with the generic model by an explicit concretization of the parameters in the generic conception. We postulate that at least structure-oriented model kinds follow this scheme and that thus also the mapping rules for the connection between generic and particular models are given. If our postulate can be maintained for many model kinds, then we also have the way to let particular model kinds learn from each other. We therefore turn first to these mappings and then discuss in Sect. 4 how even knowledge, prowess, and insight can be transferred between model conceptions.

As an example, we will now examine how knowledge and techniques for other types of models can be derived from the body of knowledge and skills for conceptual models. The direct question here is whether the knowledge, techniques and approaches for conceptual models can also be applied to other types of models. On the one hand, conceptualization is a language-based semiotic extension in which the terminology of an application-oriented language is used to secure a conceptual content. On the other hand, an implicit understanding of the application context is also injected with a conceptualization, i.e. a context abstraction is resolved and linked to the model so that the model elements carry a directly accessible meaning and a sense from the application.

We want to investigate this for a completely different type of particular model: technical models in engineering. We will then discover that these models are also placed in an application context and thus shortened by a context abstraction, whereby this context also resonates implicitly in the model. We want to return to this question in Table 1.

2.3   Meta-Models and Frames for Particular Model Configuration

Meta-models set requirements for models and modelling [43]

(1) by establishing the framework and guidelines for models and the process of representation and design and
(2) by deciding what methods and frames should be used to obtain and to use models.

Particular models are models that are considered everywhere and, if necessary, further adapted to the circumstance and concerns. We recognize that such models are actually no different than framed generic models. *Framing* is an old technique for focusing and concentrating on the essentials. However, if frames are not made explicit, then the way back to generalization and generic conceptions like our model conception is blocked. Frames are therefore nothing more than covert conditioning or strategic and possibly tactical meta-models used to adapt and lead to the normalising "right path".

Frames can also be understood as an arrangement of patterns. A pattern is then a special meta-description of a reusable solution for a problem. A pattern is profound, recurring, and consolidated problem-solution pair in a given context where it is useful whereas the pattern does not only reflect the best fitting solution but provides its background and potential. Pattern summarize regularities to a composite and can be assigned to model kinds.

We summarize this view as follows: *A particular model conception is derived from the generic model conception by applying patterns as instantiation of parameters in the generic model term, i.e. by framing[2] the generic conception. Pattern can be primal and thus explicitly stated, controlling and governing and thus mostly in the background, silent and implicitly assumed, or slaved ones.*

Patterns should focus on a class of problems so that one can run an arrangement of patterns as a join of the individual patterns. We then distinguish the following patterns in such an approach due to the peculiarities of the generic models:

Landscape pattern: Each application area is marked by its specific "pedigree" and postulates.

Scenario pattern: Models have their nature according to concern and matter and are applied in the context of application "games".

Function pattern: Models should function effectively in a scenario.

Instrument pattern: Instrument patterns answer questions such as: who uses what, when, where, why, in what way, and by what means.

Analogy pattern: Analogies expand the understanding of a given situation by simplifying borrowings from a donor domain. They can be simple partial mappings as well as complex weavings from the donor domain.

Focus pattern: Focusing allows concentration and abstraction on essential and at the same time central elements while excluding all side elements.

Purpose and utility pattern: A model must be good to use, suitable, capable, practicable, practical, and fruitful.

Justification pattern: Justification patterns are mostly silent and are already taken from the context of the origins and applications.

Quality pattern: A quality pattern describes central, interwoven characteristics with their methods of evaluation and excludes other, less central ones.

Example 2. *In the application example above, these patterns have already been introduced. We use analogy by homomorphic mapping from business structures, focus by strict reduction to essential elements for the application, and purpose pattern as specification and consolidation of a database structure. The function pattern is definable as a structured bunch of two alternatives either understanding and then comprehending in a communication act or multiple negotiation followed by agreement in a discourse, i.e.*

---

[2] Frames are convenient and directed shortcuts to the world view in the CoP and the known knowledge horizons depending on the perspective based on the concern. In modeling, they intuitively and recurrently guide the development and use of models based on consolidated positive and negative experiences. Foundational abstraction thrives on frames, strategic paradigmatic approaches, tactical principles, and transferable metaphors as a framework for intuitive interpretation and use.

((understand ⊕ comprehend) ⋈ communicate)
  ⩑ ((negotiate* ; agree) ⋈ discourse) [3].

Silent pattern are database language pattern, platform orientation, graphical representability, language restrictions to platform's data type system, specific canonical instrument pattern based on canonical database structure description, justification by tight integration of business understanding, and especially the quality sufficiency (e.g. quality in use as understandability by commonly accepted vocabulary, external quality as exactness, and internal quality especially inner coherence, toleratable redundancy, and simplicity). Specification is, thus, the guiding standard of this model kind.

3 Modelology for Particular Models

The meta-model within a particular model allows us to develop a systematic and doctrine for the development and use of such models. To this end, we examine the utility orientation of these models by aspects, the systematic construction using a worksheet, and the architecture of such models as a conceptualization for their components.

3.1 Aspects for Particular Models

Until now, the variety of uses of models seemed to prevent a model science. We now use an approach to the configuration and categorisation of particular models that allows derivation of particular and specific model conceptions from the generic model conception depending on the application scenario.

Categorizations follow a classification as a division of a totality on the basis of characteristics. Model types can be classified according to the type of origins and, if applicable, the associated context distinguished e.g. by the Universe-of-Concern or whether origins are ideas, models and objects. Another classification follows the nature of the outcome of model use and the effectiveness of models (e.g., problem solving, mental space, usable artifacts as blue pasue or material objects, extension of knowledge or disciplinary insights, behavioral orientation or regulation). One can also combine the two classification approaches using an origin-result relationship. A further classification can be based on the type of instruments and thus models.

Models in science, engineering, and daily life will be categorised based on the four aspects of model use:

(1) presentation aspect for reflection,
(2) realisation, activity, and context aspect for activities,
(3) foundations and intelligible aspect for reasoning, and
(4) interaction and social life aspect for social life models.

---

[3] ⊕ means combined conjunction. ⋈ denotes join of activities. ⩑ means exclusive disjunction. ; denotes sequential order. * denotes recursive repetition (also zero ones).

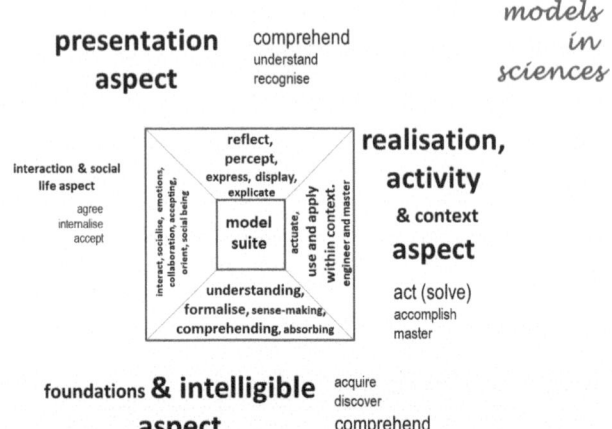

Fig. 2. Aspects and functions of particular models in sciences

These aspects do not always occur separately, but can be combined at will. It should be noted that "something" which is evaluated as a model of a certain type in a certain situation can also go on a "journey" and as a traveler change its type, open it up anew or even give it up.

These aspects can also individually determine model-being. Depending on the discipline, depending on the application, depending on the CoP and depending on the form of use of the models, other aspects are central. For example, simple physical models are more suitable for the presentation of essential properties from the world of origins. Social models of daily life are oriented towards ideas, social interaction and tasks for orientation. Models in engineering are more likely to serve the construction of artifacts, depending on the background of the engineers. Models in the sciences serve at least serve three aspects while to fourth (interaction and social ones) is neglected:

- o that of making foundations from the sciences intelligible,
- o that of a presentation that is sufficiently easy to understand, and
- o that of model-based acting based on the task, purpose, and function.

Figure 2 represents the different aspects of science models according to the intended purposes and the functions they support in scientific discourse, where a model is usually a suite of associated models, which can be quite different depending on the member of the CoP, the discipline, and the orientation of the model.
To reflect the functions, we use verbs here. The size of verbs in Fig. 2 is meant to visualize the meaning for this category of models.

Typical particular model conceptions are therefore

- presentation models, e.g. imagination models, description models, representation models, physical models, communication models, visualisation models, description models, declaration models, illustration models, intuition models, and documentation models,

- activity models, e.g. instruction models, practitioner models, action instructions, usage models, blueprints, prescription models, experience models, and moulds (master as well patrix),
- foundational and intelligible models, e.g. cogitation models, explanation models, exploration models, learning models, rationale models, conceptual models, investigative models, corroboration models, and reasoning models, and/or
- interaction and social life models, e.g. society models, identity models, personality and personae models, manipulation models, memory models, and exchange models.

This classification is based on the nature of the models used, their mission in a given situation, their promoted community, and their potential and capacity to contribute in an application. The respective specific designation brings at the same time the frames that guide to the configuration. For example, description models serve to represent an issue in a form that is understood by the CoP and contributes to solving the problem and satisfying the concern. The focus is then always on the task, the concern and matter. The formation and the functionalization are to take into account with the model, which supports as instrument this concern, for the really interesting origins.

Particular models are not oriented to just one aspect. Thus, mental models usually sweep over all four aspects. For instance, conceptual models are usually also visualized and supported by a background.

Even particular models can be further specialized depending on the application, on the recipient, on the environment, and on the context. *Specialised model conceptions* are then derived from them and, for a specific application, also *specific conceptions.*

Example 3. *A typical specific database model conception is derived from some conceptual description database structure model conception by a coherent refinement of the specialization of the parameters, e.g. as a system for the German Bank within their style of behaviour, for the second-level bankers in Frankfurt according to current policies, object-relational DBMS, and a system operating as a secondary system equipped with data from the outside. This specific model conception is then used to develop a concrete model for the given specific application.*
It should be noted that up to now one also started immediately with the concrete model and carried along in the background all the "secrets" of the particular or specific model conceptions.

3.2 The Worksheet for Particular and Specific Model Conceptions

Let us now turn to the practicality of our approach. The question arises how to proceed systematically within modelology. Here we use the approach that has proven itself in all crafts and also experimental sciences: the creation of a detailed task sheet or a structured general worksheet.

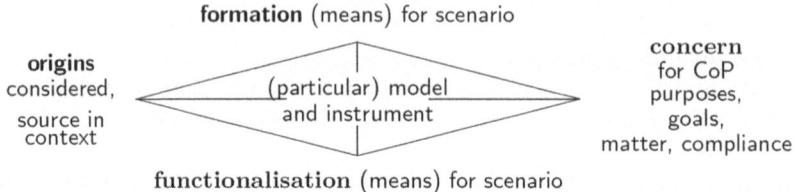

Fig. 3. The Worksheet of a Particular Model

The worksheet presents an intentional model (model-what-is, model-as-should-be, what-how-why-when-where configuration, model-used) and allows for narrowing down for the identity of a model. The situation and scenario in which "something" is considered to be a model are highlighted by the worksheet through answers to questions such as the following, which are analogous for systems theory, e.g. [1,10]: What is the primary problem and what are the secondary possibilities? What models and model kinds allow these problems to be addressed? What factors characterize the model? What are the limitations of such a model kind? Which functions are supported by which approaches by the instrument in the given scenarios? What reliability can be expected from it? In what way is the game of sufficiently accurate analogy and parsimonious focus resolved? Which CoP acts in which role, endeavor, and execution? What support and equipment should the model have as an instrument and what does it currently have? Is the chosen orientation to the given approach and implicit model sufficient? What contribution can "something" have as a model in the given situation? To what extent can the model continue to be used as the situation changes? What potential and capacity does such a model have and lack? What shape and form should the model have as a well-formed and usable tool?

The diamond in Fig. 3 summarizes the configuration for a model-being. This worksheet describes as an initial document the kind of model or instrument, the intended origins (model "whereof") within the given context as a limitation against the full set of all only possible origins, and the concern ("wherefor") for the respective matter with a compliance. A life case of a CoP in a situation triggers a concern in a matter. This means that a target state is set as a goal for the CoP based on the current state. With appropriate means from a user environment, the purpose can also be derived from the goal. The model can then function in a usage scenario. In addition, patterns are given as the means to be used for composing the model (formation) and for supporting the use of the model (functionalisation).

Model construction uses the *complementarity principle* inherent in the model on the one hand through the constructive formation and on the other hand through the integrated usage and functioning enabling as functionalisation that is based on a range of functions provided together with the model. Models have to be appropriate in use, i.e. have to have their utility. The form provides a meaning in the way how models are used. Meaning is tightly associated with meaning and sense in the application area. The function is oriented towards the use of the model both for representation ("whereof") and use in a scenario ("wherefore").

Conceptual models live from the explicit presentation of the formation [32](paper 44) and the implicit functionalisation through instrumentation by a language beside natural language semiotics for terms. Mathematical models are designed for problem solution and typically follow a five-phases procedure [2,23] by starting with an investigation, reformulation, understanding, idealisation, and description of the problem situation. Table 1 also refers to other types of formation and functionalization for other types of models.

That means that models have to have two supporting and enabling means: *formation* and *functionalisation*. The first one enables proper reflection according to the mission or nature of the model. Functionalisation support the appropriateness in use and determines which form has the right utility. Formation and functionalisation do not have to be given in an explicit form. Science, engineering, and daily life use a large variety of supporting and enabling means, thus arriving at different kinds of models. Table 1 illustrates some of these means.

Table 1. Formation and functionalisation for different kinds of models

Formation using supporting means	**Model** kind	**Functionalisation** using enabling means
conceptualisation through encyclopedic notions, application world, supporting models from domain, context, insight	conceptual DB model *(descriptive, prescriptive, exploratory scoping)*	conceptual modelling language as domain-specific language with referencing terms in term spaces
known mathematical approaches for problem solving, problem-solution classes, solution pattern, implicit background	mathematical model *for problem-solution-evaluation-application-revision frame*	Mathematical apparatus for problem solving, techniques, solution moulds, methods, representation frames
typical scientific discipline, statements according actual issue	visual model *as presentation of insight*	visualisation languages, visualisation pattern
art and practice of engineering in application area	engineering model	engineering drawing or formalisation
art of building, facility management, statics, building conceptions, beauty	architectural design *as material artefact or blueprint*	material artefact techniques
art styles and practices, symbolism, stories, narrations, aesthetics	fine arts artefact	languages, symbolics, visual communication forms
ideological ideas, religious cultures, habits, beliefs, ceremony conception etc.	ideological **monstrance**, *shown in public*	ceremonial acts in contexts
book of life (BoL), culture	ethics model	narratives

Example 4. *Technical thinking is the intrinsic professional background for all engineering solutions. This includes the fundamentals, laws and rules of the specific engineering discipline such as physical and chemical theories in depth. The positive and negative experiences are also part of it. In such applications, the full knowledge of engineering, the driving forces, the capacity, the potential, and the entire treasure trove of skills of the discipline are used in a clever way. In addition, there are also deep insights into the possibilities and limitations of the material depending on the requirements for the technical work. And here also*

comes the full professional background of engineering work. This is not specifically anchored in the technical model, but we abstract from it because it can be assumed.

If we remember the context abstraction that works in the conceptual models of the respective application, then we discover a surprising analogy for supporting means, which we can now also integrate for engineering models. In this case, we use the conceptual world of the application, which is only partially reflected directly in the language. The application not only contributes its glossary and thesaurus, but the entire "stable smell" or pedigree, in this case also implicitly and intrinsically. Without this background, a conceptual model is just a nice reflection and a deep context abstraction.

Can these special particular models learn from each other? The answer is positive. Let us compare supporting means for formation of models in conceptual modelling and in engineering in Table 1. In a similar form we can compare functionalisations and learn from each other.

The background of conceptual modeling is still implicit. The driving forces, the knowledge and the prowess are often carried along in other models such as view ensembles, business cases, performance requirements or technical realization options. This means that conceptual modeling can learn from technical modeling and its context abstraction. In context abstraction, for example, the methodological framework is adopted from other particular models.

In the opposite direction, the technical models have already learned the conceptualization, the conceptual worlds, the recipes and the composition mechanisms despite the inadequacy of linguistic representability. For the last problem area, however, conceptual models offer a learning space for engineering that still needs to be developed.

We note that experience can also be adopted for the other particular and special models in Table 1 by taking a closer look at the different forms of abstraction. The simplest adoption is certainly possible for the construction abstraction, where we can start from a small set of constructors and thus compose a model and a solution. We have looked at context abstraction here. The *foundational abstraction*, the *refinement abstraction*, the *usage abstraction*, and the *toolbox abstraction* can also be considered.

3.3   Conceptualisation of the Architecture of a Model

The question now arises as to whether every specific model and thus every particular model follows this approach, so that the specific and particular model can also be derived from the generic model From our point of view, this even applies to monolithic models as well as to most of the systems we know. To understand this more precisely, let us consider the composition of models, i.e. the architecture of models. Such a mostly stereotypical model composition can even be conceptualized with individual components and their roles within the model. This then underpins the worksheet.

Example 5. *To illustrate this, let's take a look at mathematical or formal models based on the forms of abstraction used [2]. Mathematical modeling is characterized by the choice of a particular approach, whereby the potential and capacity are at least implicitly well understood in addition to the justification of this approach. The mathematician can thus intuitively or consciously determine whether this approach with all its specific aspects is suitable at all. Such approaches have been worked out methodically so that the individual components of a model can be obtained as a configuration. This background represents the disciplinary matrix for mathematical problem-solving procedures. The justification of such an approach and also the implicit models are given a priori for the specialised mathematician. This means that the mathematician can also draw on a body of knowledge and a body of prowess to solve the problem.*

Mathematicians are interested in solving problems and evaluating this solution for a specific life case in an application sphere and scenario, whereby a model is assumed to describe the application. The mathematical solution can then be evaluated with the application specialist. This approach has become established in many disciplines that use mathematical models to solve application problems.

Each model has its inner, silent and hidden components and its directly visible and outer components, whereas the explicit components are still adapted for the concrete application. If we return to the metaphor of the cooking prince, we find for the quick and somehow successful use that only the external components are of interest. With the use on the recipe level at least already the visible formation and formalization are interesting. May be also something from the Body of Knowledge and also the usable tools. If, on the other hand, one comes to assemble and thoughtful use, then the internal structure will also be important (esp. the matrix as strategic background) and also the adaptation to the landscape will have to be mastered. A professional also needs background and insight as well as good knowledge of the existing Body of Knowledge and Body of Prowess for the supporting and enabling means. In the master class you should be able to put all these inner elements together as well and thus develop a model that you can also reassemble depending on new challenges, changing fundamentals or even other concerns.

The model conceptions does not yet allow to derive the structure and functionality of models directly. Therefore, a construction doctrine is also needed for models that will be definable as a *conceptual model of model*. The conceptual model in Fig. 4 includes the nature of the model in the application landscape (range of portfolio tasks of the model in scenarios of the community of practice (CoP)), the fundamentals (especially the strategic ones for the inner model behind the model as a matrix), and further both the supporting means and the enabling instrumentation (i.e. the formation and functionalisation) of the model based on tactical support and enablers.

This also provides a partial answer to our RQ_1. We can systematically configure and create models from which we can also extract and transfer the approaches, techniques, formation, functionalization, insights, and findings for other kind of models.

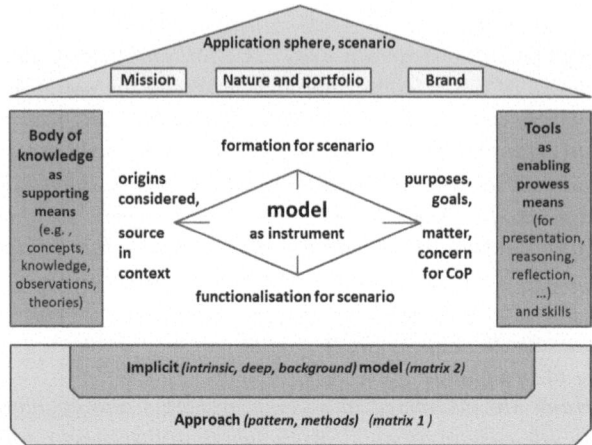

Fig. 4. The conceptual model of the architecture and composition of a model

4 Appropriation of Experience and Inner Accomplishment Among Particular Model

Let us now turn to an answer of all four research questions. We will now show how to transfer knowledge, skills, insights, and practices from one kind of model to another one. We focus here on potential and also partly hypothetical possibilities of transfer rather than on a detailed description of success stories. Already in [40] it was discovered that models of solar wind are at the same time models of epidemic propagation, respectively that models of black holes have a lot to do with models of financial markets[4]. Differential equation systems with a synergetic approach are condensed into simultaneous evolution models, whereby results of the form "the big eats the small" can be derived into dependable and robust estimates of crises.

The transfer of experience to other models of the same type has slowly been incorporated into data mining as inductive or incremental transfer learning [44]. It has so far been limited to the same type. In the first rounds of MMM[5] we have learned that learning of approaches, methods, and mathematical background is possible on a much broader scale. For example, many disciplines use tightly coupled vertical and horizontal model families. A general coupling theory can be derived from this. Mathematical control theory can also be extended to an economic control theory. Approximative algorithms for computer science models can also be extended to an optimization theory of electrical engineering. Many scientific and engineering disciplines have developed competing and complementary frameworks for addressing insufficient data quality. Models have a power and a

---

[4] See the theory of additive component models in [5,6,11].

[5] See [41]and its compilation into the survey compendium on models in mathematics, sciences, and engineering [40].

scope. Very different methodologies have emerged. Model organisms in medicine lead to analogous solutions as substitution models in electrical and mechanical engineering. In contrast, inverse models have already been extensively used in data mining.

4.1 Case Study of Knowledge and Prowess Transfer Among Kinds of Models

We first consider some particular and special kinds of models, which at present are not generally understood in research as one type, but which allow to transfer all the knowledge and skills from one type to the other. It is not our goal here to present the direct transfer of knowledge and skills. Instead, we want to outline the path.

Data Mining Models and Inverse Modelling. Data mining, like the development of algorithms before algorithmics, has led to an unmanageable variety of methods. A first attempt at systematization was presented in [14]. There, a systematization is presented for the six central classes (anomaly detection, association rule learning, clustering, regression, classification, summarisation). You start with data, where this data already has its own background and has a certain quality, and combine this with suppositions.

Inverse modeling starts with a generic or particular model class in which governing parameters can be determined backwards using data. For example, from observations the corresponding instantiated parameters are considered as the cause of behaviour of observations. We start with the effects and then calculate the causes because of the parameters cannot be directly observe.

Essentially, these rather similar approaches start with data as sources in some context and are oriented on detecting some causal relationship and deriving an insight. Based on a problem situation such as the derivation of an appropriate explanation, a specific model class is selected with which one can solve this problem with models. This approach requires a feeling for which model class can possibly solve the problem with the available means such as algorithms and reasoning mechanisms, within which strategic matrix and which knowledge-skill background. From this, a model configuration for the model class is derived, with which one can derive a well-suited model for problem solving based on the available data, the usable concepts due to the chosen school of thought and methodology.

Conceptual Models and Mathematical Models. It seems (see Table 1) that conceptual and mathematical models have little in common. However, they can cross-fertilize each other very well and also resolve their problematic sides by learning the experiences of the other kind.

Mathematical modeling is well aware of the language dependency, especially on the existing as well as non-existing potential of the respective languages. It has also systematically developed a modeling cycle [2]. The modeling cycle paves the way for systematic evolution of models based on the insufficiency of the just completed models. Both can be directly applied to conceptual models – of whatever particular kind – and thus solve problems of conceptual modeling such as: language dependence, solution apparatuses, model evolution and modernisation, patterns inside models, combination of models, succession of models.

In contrast, for all their sophisticated structure, mathematical models explain quite little about the individual constructs, their interrelationship, their interaction, and their mutual constraints. Mostly, this is only done narratively in rudimentary form as a reference to the theories in applications. Conceptual models have a special strength just here. One could also use them to develop integrity constraints for the interaction of the individual components of mathematical models in a systematic form.

Mathematics knows a definition culture[6] which allows to provide different but equivalent definitions depending on context and user. There are several representations for one term or concept. The database practitioners know very well that different viewpoints have to be provided depending on the business user group. A systematics from mathematics could bring here also a systematics of the view ensemble.

Conceptual Modelling and the Meaning of Concepts in Applications. Conceptual modeling of database structures uses names from a controlled vocabulary. With this name control, one also hopes to understand the meaning and significance of the concepts, although in each application these concepts may have a completely different meaning and sense (e.g. *"Bank"* versus *"German Bank"* in the above example). Actually, however, one should know the entire culture of the application and put down at least essential parts in the model. This is precisely the strength of ideological models, which have exactly this culture of their components, the usage ceremonies, the behavior of the users, etc. in mind.

Language Dependency and Language Power. Computer Science knows its "universals controversy". New languages are invented over and over again, especially because some of the older language constructs could not be properly represented, could not be mastered, or could not yet be theoretically underpinned. The same is true for modeling languages. And mostly the same problem is solved, only represented differently. Just look at the different visualizations. Here, much can be learned from the visual models, which focus first on the potential and capacity of the language and then on its suitability.

Natural languages and thus models of daily life have had a long period of maturation, so that too hard a language binding is no longer an obstacle. This is still to come for Computer Science. Just consider the foundation with the Turing machine paradigm and try to represent classical analog computers or the mechanism of Antikythera with this paradigm. At best, a really confusing and

---

[6] See, for instance, the notion of a set defined in explicit form by enumeration of its elements, in algorithmic form by construction rules, in iterative or recursive form by stepwise construction of next elements for already given ones, in axiomatic form by axioms, or in descriptive form using defining expressions.

not sufficiently complete emulation of analog computers succeeds, i.e. we realise a Turing incompleteness as also for many of the human mechanisms of thinking and acting by means of models.

Engineering – like modeling or even speaking in a language – does not assume that a mechanism must be represented and realised in the most exact way. A certain error and thus a tolerance is accepted from the outset. Here all models of computer science can still learn from those of engineering.

Model Suites as Coherent and Well-Associated Model Sets. Models rarely stand isolated and alone. Models of daily life, just like enterprise models, consist of a multitude of models, whereby these models are also closely interwoven and coordinated with each other in a coherent form, i.e. we use a model suite consisting of well associated models[7]. This interweaving can also be explicitly represented in model-like form as a meta-model of the model suite. Instead of an integration on the same abstraction level as with the interoperability one can produce a stepped and coordinated coherence for this [16,40], which is nothing else than also learning the models of the everyday life for model suites of computer science. In daily life, models of thought and action are flexibly linked with each other. Enterprise models also do not need general integration, but rather follow the approach in the enterprise as right "pedigree" of the application within the given enterprise with its style and model of handling things.

The association can occur as a spider web or in a layered form as a model pipe as is common in mathematical models. The data analysis or data mining story was reduced to one step above. In reality, we consider a sequence of modeling steps: First, an approach is selected such as regression analysis. Second, such a procedure is applied. Third, this can also be used to develop explanatory models, from which simplifying models of the driving master and the dependent slaved parameters are created. In this case we use models layered on top of each other.

Global-as-Design versus Local-as-Design. Database development still essentially follows a global-as-design strategy, where local user schemata must be derivable from the global schema while being bound to a specific platform. Since not all views of business users have to be relevant for some other views, one can also follow a generalised global-as-design instead of the classic pratfall[8] and at least keep a second local user schema separate from the global schema as long as user schemata are combinable.

We can learn from the architecture design, on the other hand, how a more flexible strategy and handling can be developed. One does not have to aim for the full flexibility of fine art artefacts, but one can certainly develop a more flexible and above all multi-platform modeling. The extreme form of model-free and scheme-free handling can thus be avoided. It has not worked so far and has

---

[7] See https://citeseerx.ist.psu.edu/, search phrase "Model Engineering: Model Suites".

[8] The classical global-as-design strategy assumes that all local data are derivable from the global data and all local data are of the same granularity and quality compared to the global data.

only led to unmaintainable IT solutions. In contrast, it makes sense to adopt a solution from facility management, e.g. as a model suite with several central schemata.

4.2   Some Particular and Specific Model Conceptions

In introducing the aspects for models, we briefly named very different kinds of models. All these kinds follow our scheme of deriving particular models from the generic model conception. We can also systematize the almost unmanageable variety of specific models in this way and thus underpin them theoretically. From the worksheet, one can derive a model constellation and thus derive particular and specific model conceptions.

Typical Particular Models. As an example, the following particular model types are given here:

- *Imagination models* are specific presentation models. They characterize, harmonize, and explain the imaginations, ideas, and concepts about the origins and their essential properties as well as peculiarities explicitly ordered together into an agreed view in such a way as to provide a sufficient description of the origins depending on their use in a scenario.
- *Representation models* are used to represent perceived or imagined contents in the form of an idea, concepts, signs, narratives, symbols, figures or pictures in an abstracting way or to present them in a concretising way by characteristic and typical instantiation. Thereby they become visual or tangible images of 'something' e.g. objects and thought buildings.
- *Manufacturing models* (also called MODEL) or stencils are to be used for manufacturing, production, creation, making, creating, reconstructing, restoring, etc. of objects starting from objects, models and ideas. They contain the essential and central details and instructions for the creation of objects. Models are intended to achieve a specific end result. Models represent what will be. Models serve as templates, blueprints, instructions, etc. for objects to be created. They are used on the basis of instructional models.
- *Didactic models* explicate, focus, and present meanings of origins in a form that is easy for the recipient to grasp and well-structured. They facilitate the understanding of the origin with a representation, justification, definition, clarification, explanation, embedding in what is known and accepted, etc. Depending on the function of the model in the scenario, didactic models direct the consideration of the origins to their specific important details.
- *Orientation models* mentally enable a recipient on the basis of origins to select alternatives in the space of possibilities, to find his or her way according to the situation and scenario, and to concentrate in space, time and scenario, to orientate in space and time and to concentrate on specific options. The possibility space possesses a structuring as well as permits a preference and guidance according to different points of view.

- *Cosmological models* are mostly complex models for presenting and conveying ideas as well as for orientation on the basis of a focus on a world view. They are often also model systems, are time and process oriented, represent a recourse to a narrative to the past, and are based on a chiseled composition or constitution. The recipients belong to a determining culture and receive such models for their orientation, mostly also unconsciously or preconsciously.

All of these particular model types were derived from the Sect. 3 scheme.

Specific Model Conceptions. One can still agree on a particular model conception – at least for the detailed variant. But this becomes hardly possible, if one considers the many concepts for the respective application areas. In [32](paper 27) over sixty different definitions for the notion of a conceptual model are laid side by side. It is noticeable that almost all of these conceptions are contractions of the actual conception due to an orientation to the concrete landscape, in particular also to the creeds in the CoP and to a specific technology platform. Thus, *specific model* conceptualizations are preferred due to the focus on the functions that a model has to perform in a particular scenario. We consider this once for the particular notion of conceptual model.

A conceptual model is defined as a model [32](paper 29)

- that is a coherent and consolidated presentation model,
- that is supported by abstract concepts from an encyclopedic background space,
- that is formulated by referencing terms in a language that allows well-structured formulations,
- that is based on mental/perceptual/situational models of the origins with their embedded abstract term(s), thus inheriting its reliability and other canons from origins and the applications, and
- that is oriented to a commonly accepted strategic matrix.

We can compare this detailed version with more specific ones, e.g. from [32](paper 27):

- "A conceptual is a model which represents a conceptual understanding (i.e. conceptualisation) of some domain for a particular purpose. A model is an artefact acknowledged by the observer as representing some domain for a particular purpose."
- "A conceptual model is a descriptive model of a system based on qualitative assumptions about its elements, their interrelationships, and system boundaries."

If we compare all such versions, we arrive at the following *specific model conception* of a conceptual model in Computer Science and Informatics:
"A conceptual model is a concise and purposeful consolidation of a set of concepts[9] that are presented by means of terms in a predefined linguistic format.

---

[9] An introduction to the theory of concepts is a task for the future. There are more than two dozen different types of concepts [34]: individual and in particular mental

As such it establishes a view of a given notion space." [32] (paper 44)

Thus, all the specific models simply become truncations and specializations of the particular models with a specific orientation towards model properties of central interest. All other model properties are implicitly assumed without directly representing this.

5   Finally

We now want to respond positively to RQ_1. A transfer is possible if the meta-models behind the actual models correspond sufficiently well. Then one can even use meta-meta-models to systematize the transfer (see [32](paper 41)). An answer to RQ_2 still needs a great deal of systematic study of the many types of models. It is noticeable that a body of knowledge for many types of models – except for some classes of mathematical models – does not exist today. For example, there is only a rudimentary approach to such a BoK for conceptual models of information systems.

Modelology can be developed as a cross-aspect discipline among science, engineering, and systematic support for daily life. We have given here a first initial response to RQ_3. We are currently trying to develop this MODELOLOGY systematically and to present it in a monograph. Individual elements of such a modelology have been explored in our MMM discussion groups for several years. We invite anyone interested to participate here[10].

To answer RQ_4, let us come back to Fig. 1. In Sect. 4 we have illustrated how a transfer can take place. In addition, there are other success stories such as [36], in which the potential of electrical circuits is explored for modeling periodic systems such as the human heart.

References

1. Alter, S.: Work system theory: overview of core concepts, extensions, and challenges for the future. J. Assoc. Inf. Syst. **14**, 72–121 (2013)
2. Berghammer, R., Thalheim, B.: Wissenschaft und Kunst der Modellierung: Modelle, Modellieren, Modellierung. Methodenbasierte mathematische Modellierung mit Relationenalgebren, pp. 67–106. De Gryuter, Boston (2015)

---

concepts [19], unconscious or insufficiently precise thought concepts [13], conventional concepts of a community of practice or a culture [12,27], application concepts such as in database or web applications [20,26], generated or empirical concepts of formal concept analysis [9,24], mathematical or logical concepts [17,28], or property-based concept lattices [18]. A concept can be defined in a mathematical or logical form with definiendum and definiens [33], in an intention/extension form [31], in a graphical form or generalized as a meta-hypergraph [15,29] or even just in a narrative form. For a general introduction to concepts see details in [22].

[10] Please send an email if you are interested in being invited. Prior MMM discussions are summarised in [40].

3. Bienemann, A., Schewe, K.-D., Thalheim, B.: Towards a theory of genericity based on government and binding. In: Embley, D.W., Olivé, A., Ram, S. (eds.) ER 2006. LNCS, vol. 4215, pp. 311–324. Springer, Heidelberg (2006). https://doi.org/10.1007/11901181_24

4. Brassard, G., Bratley, P.: Algorithmics - Theory and Practice. Prentice Hall, London (1988)

5. Colander, D., et al.: The financial crisis and the systemic failure of academic economics. Univ. of Copenhagen Dept. of Economics Discussion Paper (09-03) (2009)

6. Duschl, W.J., Strittmatter, P.A.: The cosmogony of supermassive black holes. Mon. Not. R. Astron. Soc. **413**(2), 1495–1504 (2011)

7. Eck, C., Garcke, H., Knabner, P.: Mathematische Modellierung. Springer (2008). https://doi.org/10.1007/978-3-658-00535-1

8. Embley, D., Thalheim, B.: The Handbook of Conceptual Modeling: Its Usage and Its Challenges. Springer (2011). https://doi.org/10.1007/978-3-642-15865-0

9. Ganter, B., Wille, R.: Formal Concept Analysis – Mathematical Foundations. Springer (1999). https://doi.org/10.1007/978-3-642-59830-2

10. Gregor, S.: The nature of theory in information systems. MIS Q. **30**, 611–642 (2006)

11. Henning, C., Henningsen, G., Henningsen, A.: Networks and transaction costs. Am. J. Agric. Econ. **94**(2), 377–385 (2012)

12. Jaakkola, H., Thalheim, B.: Culture-adaptable web information systems. In: Information Modelling and Knowledge Bases XXVII, Frontiers in Artificial Intelligence and Applications, vol. 280, pp. 77–94. IOS Press (2016)

13. Jackendoff, R.: A User's Guide to Thought and Meaning. Oxford University Press (2012)

14. Jannaschk, K.: Infrastruktur für ein data mining design framework. Ph. D. thesis, Christian-Albrechts University, Kiel (2017)

15. Johnson, J.: Hypernetworks in the Science of Complex Systems. Imperial College Press, Singapore (2013)

16. Kalinichenko, L.A., Stupnikov, S.A.: Constructing of mappings of heterogeneous information models into the canonical models of integrated information systems. In: ADBIS 2008, Local Proceedings, pp. 106–122. Tampere University of Technology, Pori (2008)

17. Kamlah, W., Lorenzen, P.: Logische Propädeutik: Vorschule des vernünftigen Redens. Springer-Verlag (2016). https://doi.org/10.1007/978-3-476-05434-0

18. Kauppi, R.: Einführung in die Theorie der Begriffssysteme. Acta Universitatis Tamperensis, Ser. A, Vol. 15, Tampereen yliopisto, Tampere (1967)

19. Langacker, R.W.: Foundations of Cognitive Grammar: Volume I: Theoretical Prerequisites, vol. 1. Stanford University Press (1987)

20. Mayr, H.C., Thalheim, B.: The triptych of conceptual modeling - a framework for a better understanding of conceptual modeling. Softw. Syst. Model. **20**(1), 7–24 (2021)

21. Müller, R.: Model history is culture history. From early man to cyberspace (2016). http://www.muellerscience.com/ENGLISH/model.htm. Assessed 29 Oct 2017

22. Murphy, G.L.: The Big Book of Concepts. MIT Press (2001)

23. Ortlieb, C.P., von Dresky, C., Gasser, I., Günzel, S.: Eine Einführung in zwölf Fallstudien. Vieweg, Mathematische Modellierung (2009)

24. Poelmans, J., Ignatov, D.I., Kuznetsov, S.O., Dedene, G.: Formal concept analysis in knowledge processing: a survey on applications. Expert Syst. Appl. **40**(16), 6538–6560 (2013)

25. Samarskii, A.A., Mikhailov, A.P.: Principles of Mathematical Modelling: Ideas, Methods, Examples (Translated from Russian, 1997). CRC Press (2001)
26. Schewe, K.-D., Thalheim, B.: Design and Development of Web Information Systems. Springer, Chur (2019). https://doi.org/10.1007/978-3-662-58824-6
27. Seiler, T.B., Wannenmacher, W.: Concept Development and the Development of Word Meaning, vol. 12. Springer Science & Business Media (2012). https://doi.org/10.1007/978-3-642-69000-6
28. Simon, M.A.: Explicating mathematical concept and mathematical conception as theoretical constructs for mathematics education research. Educ. Stud. Math. **94**, 117–137 (2017)
29. Terekhov, V., Gapanyuk, Y., Kanev, A.: Metagraph representation for overcoming limitations of existing knowledge bases. In: FRUCT 2021, pp. 458–464. IEEE (2021)
30. Thalheim, B.: Entity-Relationship Modeling - Foundations of Database Technology. Springer, Berlin (2000). https://doi.org/10.1007/978-3-662-04058-4
31. Thalheim, B.: The conceptual framework to user-oriented content management. In: Information Modelling and Knowledge Bases, Volume XVIII of Frontiers in Artificial Intelligence and Applications. IOS Press (2007)
32. Thalheim, B.: Models, to model, and modelling. collections of papers.https://www.researchgate.net (search keyphrase "Towards a theory of models, especially conceptual models and modelling"), also academia.edu (2009-2021)
33. Thalheim, B.: Towards a theory of conceptual modelling. J. Univ. Comput. Sci. **16**(20), 3102–3137 (2010). http://www.jucs.org/jucs_16_20/towards_a_theory_of
34. Thalheim, B.: Conceptual models and their foundations. In: Schewe, K.-D., Singh, N.K. (eds.) MEDI 2019. LNCS, vol. 11815, pp. 123–139. Springer, Cham (2019). https://doi.org/10.1007/978-3-030-32065-2_9
35. Thalheim, B.: Artificial intelligence enhanced by modelling. Intellect. Syst. Theory Appl. **26**(1), 360–366 (2022)
36. Thalheim, B.: Model-based reasoning for investigating the heart capability. In: Lohff, B., Schaefer, J. (eds.) Cardio-Physiology Challenging Empirical Philosophy, vol. II, pp. 162–199. BoD, Norderstedt (2022)
37. Thalheim, B.: Models: the fourth dimension of computer science - towards studies of models and modelling. Softw. Syst. Model. **21**(1), 9–18 (2022)
38. Thalheim, B.: Auf dem Wege zur Modellkunde. In: Modellierung 2022, volume P-324 of LNI, pp. 11–32. Gesellschaft für Informatik e.V, Bonn (2023)
39. Thalheim, B.: Modellkunde: kurz & knapp. In: Loeben, C.L. (ed.) Modellkunde und Ägyptologie im Dialog. Essays. Kulturverlag Kadmos, Berlin (2023)
40. Thalheim, B., Nissen, I. (eds.): Wissenschaft und Kunst der Modellierung: Modelle, Modellieren. Modellierung. De Gruyter, Boston (2015)
41. Thalheim, B., Slawik, T., MMM CAU team: Science and Art of Modelling SAM^3, p. 137. Christian-Albrechts University Kiel, MMM programme CRC, German (2014)
42. Thomas, M.: Modelle in der Fachsprache der Informatik. Untersuchung von Vorlesungsskripten aus der Kerninformatik. In: DDI, volume 22 of LNI, pp. 99–108. GI (2002)
43. Van Gigch, J.P.: System Design Modeling and Metamodeling. Springer Science & Business Media (1991). https://doi.org/10.1007/978-1-4899-0676-2

44. Vilalta, R., Giraud-Carrier, C., Brazdil, P., Soares, C.: Inductive Transfer. In: Sammut, C., Webb, G.I. (eds.) Encyclopedia of Machine Learning, pp. 666–671. Springer US, Boston, MA, (2017). https://doi.org/10.1007/978-0-387-30164-8_401
45. Wartofsky, M.W.: The model muddle: proposals for an immodest realism. In: Models: Representation and the Scientific Understanding, pp. 1–11. Springer, Dordrecht (1979). https://doi.org/10.1007/978-94-009-9357-0_1

Topological Methods and Tools for the Analysis of Big Crystallographic Data

Vladislav A. Blatov(✉) (iD)

Samara State Technical University, Samara 443100, Russian Federation
blatov@topospro.com

Abstract. We briefly overview mathematical models, methods and computer tools for the topological description, analysis and classification of crystal structures. We present *ToposPro*, a program package for comprehensive geometrical and topological analysis of periodic architectures of any composition and complexity. *ToposPro* is designed for the automated analysis of big crystallographic data, which are either collected in the world-wide electronic databases or generated by theoretical methods. *ToposPro* was used to create a system of topological databases, which are available both in the local version and as a number of interactive web-services integrated into the *TopCryst* system. All described topological tools are considered as applied to solving typical tasks of crystal chemistry and materials science.

Keywords: Topological Analysis · Crystal Structure · Big Data · Software · Web-Service

1 Introduction

Experimental data on crystal structures form one of the biggest and well-organized sets of information in natural sciences. At present, about two million records are collected in the world-wide crystallographic databases such as Cambridge Structural database (CSD), Inorganic Crystal Structure Database (ICSD), Pearson's Crystal Database (PCD), Crystallography Open Database (COD). This dataset definitely contains a lot of correlations 'chemical composition – chemical structure', but has not yet been explored in detail. For a long time, the main reason for that was the absence of automated tools for processing such huge and extremely diverse information with the same universal algorithms. Another important problem was that the initial crystallographic data contained no knowledge about chemical bonding, which is crucial for the usage of these data to solve the tasks of crystal chemistry and materials science.

In this paper, we present a number of computer tools for the description, analysis and classification of crystal structures of any chemical composition and complexity. An important feature of all mathematical models and algorithms behind these tools is that they are based on the topological representation of a crystal structure, where the overall connectivity between the atoms is restored hence providing naturally chemical treatment of the experimental crystallographic data. Such approach enables one to solve the problem of processing the crystallographic datasets mentioned above.

© The Author(s), under exclusive license to Springer Nature Switzerland AG 2024
J. Baixeries et al. (Eds.): DAMDID/RCDL 2023, CCIS 2086, pp. 28–34, 2024.
https://doi.org/10.1007/978-3-031-67826-4_2

2   Models and Algorithms

2.1   Voronoi Partition

The initial experimental crystallographic data provide only geometrical information about positions of atoms in the crystal space and symmetry operations between them. To transform these data into a chemical object we need to establish interatomic contacts. For this purpose we represent the atoms by their *Voronoi polyhedra*, which confine parts of the space belonging to particular atoms (Fig. 1). The Voronoi polyhedra form a partition of the crystal space (*Voronoi partition*) thus assigning volume and shape to the atoms. The faces of the Voronoi polyhedra correspond to possible interatomic contacts; to treat them as chemical bonds or weaker interactions additional chemical and geometrical criteria can be applied. As a result, a rigorous algorithm *Domains* was developed [1], which enables one to determine chemical bonding in crystal structures of any nature, from metals to proteins.

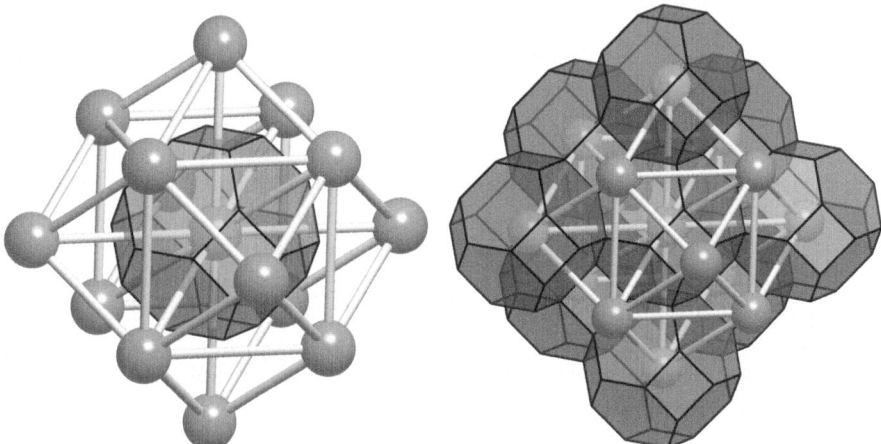

Fig. 1. Body-centered cubic packing: (left) a fragment of atomic net with the Voronoi polyhedron of an atom; each of 14 faces of the polyhedron corresponds to an interatomic contact; (right) a part of the Voronoi partition; the net of the vertices and edges of the Voronoi polyhedra represents the Voronoi net.

However, Voronoi partition describes not only atoms and their interactions, but also the free space in the crystal. Indeed, the vertices of the Voronoi polyhedra are geometrically the most distant points from the surrounding atoms and hence can be considered as centers of voids. Similarly, the edges of the Voronoi polyhedra mimic channels between the voids. This approach is especially important when studying absorption, catalytic, or diffusion properties of the crystal [2].

2.2 Atomic Net

After determining interatomic interactions the crystal structure is represented as an infinite periodic graph (*atomic net*), which nodes and edges correspond to atoms and interactions between the atoms (Fig. 1). This is the *complete* topological representation of the crystal structure as it provides the comprehensive information about the structure connectivity. The topology of the net is completely described by its *labeled quotient graph* [3], which can be obtained by wrapping the net to the unit cell using the periodic conditions and keeping the information about bonding of the atoms inside and outside the unit cell. This approach enables one to transform the infinite net to a finite object and hence store the topological information in a database. The topology of the labeled quotient graph and the corresponding net can be characterized by a set of topological indices (invariants), which can be used for matching periodic nets and uniting the nets with the same topology into the same *topological type*. As a result, a rigorous topological classification of crystals structures becomes possible [4].

2.3 Underlying Net

The complete atomic net can then be represented at several levels of detailing by *simplification* algorithms [5]. Such algorithms enable one to separate atomic groups (building units), which assemble the whole crystal structure. Both chemical and topological criteria are used for this purpose and as a result the atomic net is transformed into a net of the centroids of the building units connected according to the connections of the atoms of these units in the atomic net (Fig. 2). This simplified net is called *underlying* [4] as it encodes the method of constructing the crystal from atomic units and hence underlies the chemical model of the crystal. Since building units can be chosen in different ways depending on the treatment of the chemical nature of the crystalline substance, several underlying nets can match the same crystal structure. This is a powerful tool for determining structural correlations between chemical substances of different composition, crystal structure parameters and bonding [6].

2.4 Voronoi Net

The net of vertices and edges of all atomic Voronoi polyhedra is called *Voronoi net* (Fig. 1). This kind of periodic net describes the topology of the free space between the atoms and can be analyzed by the same methods and tools as the atomic net. This allowed us to talk about *dual crystal chemistry* [2], which inverts the consideration of a crystal from material objects (atoms and bonds as clouds of electron density) to virtual objects (voids and channels), which however admit a similar vision.

2.5 Natural Tiling

Another method of the analysis of the crystal free space is based on the tiling model. A *tile* is a generalized polyhedron, which vertices and edges are vertices and edges of the atomic or underlying net. Unlike a convex Voronoi polyhedron, the tile faces are not necessarily flat and vertices can be bivalent. However, similar to Voronoi polyhedra

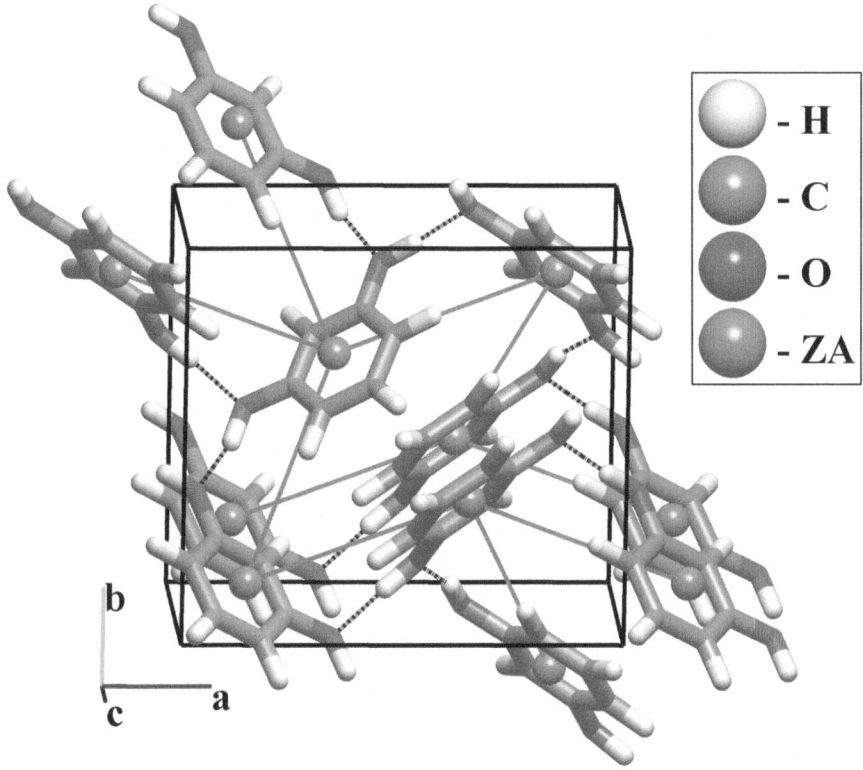

Fig. 2. Crystal structure of resorcinol. The underlying net is shown by green balls ZA (centroids of the resorcinol molecules) and green edges, which mimic the H-bonding between the resorcinol molecules (H-bonds are shown by dotted blue lines). (Color figure online)

tiles form a face-to-face partition (*tiling*) of the crystal space, but the centers of tiles are centers of cages, not atoms (Fig. 3). In general, the number of possible tilings, which can be constructed for a given net, is infinite but we have proposed e set of rigorous conditions [7] that enabled us to algorithmize the construction of a unique *natural tiling* with the smallest possible tiles, the so-called *natural tiles*. The natural tiles have clear physical meaning of the smallest cages in the crystal structure while their faces mimic the sections of the channels between the cages. The net of the tile centers and edges between the centers of adjacent tiles is called *dual net* (Fig. 3); similar to the Voronoi net it indicates the topology of the system of cages and channels between them. Thus the Voronoi net and natural tiling models supplement each other altogether characterizing the free (porous) space of the crystal.

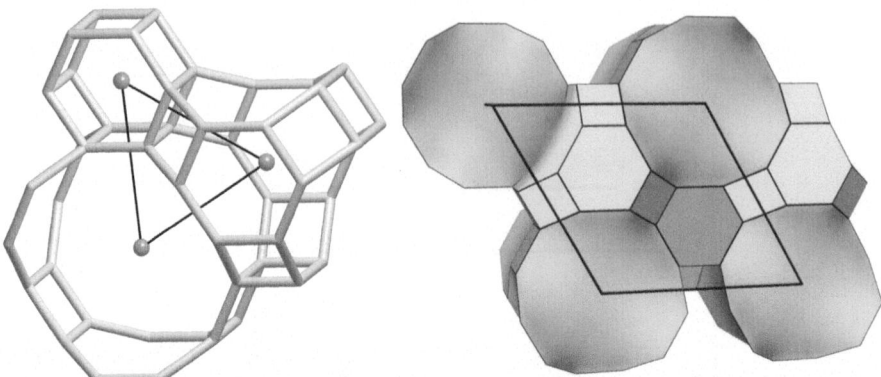

Fig. 3. Silicon framework in the crystal structure of the zeolite gmelinite (GME): (left) three different natural tiles and the corresponding fragment of the dual net (green balls); (right) a fragment of the natural tiling. (Color figure online)

3   Software

We have implemented all mathematical models and algorithms mentioned above into the free program package *ToposPro* [4, 8], which currently (May 2023) has more than 6,800 registered users from 102 countries. *ToposPro* is designed for the Windows© operation systems and allows the users to create their own crystallographic databases in a unique binary format. An important advantage of the *ToposPro* format compared with other crystallographic databases is that it enables one to store the complete information on periodic nets, *i.e.* about topological properties of crystal structures. *ToposPro* includes a database management system, which is supplied with a user-friendly interface and integrates a number of applied programs for the comprehensive geometrical and topological analysis of crystal structures. Each applied program is designed for processing the crystallographic data in a batch mode that enables the user to work with large samples of information on hundreds of thousands of crystalline substances. At present, the researchers over the world use *ToposPro* for solving many tasks of crystal chemistry and materials science and for various classes of crystalline substances [9].

4   Databases

We have used *ToposPro* to extract topological information from the world-wide crystallographic databases and to develop the first *topological* databases. These databases form a set of the *ToposPro* topological collections [4], which contain millions of records with the geometrical and topological parameters characterizing crystal structures at different levels of their organization. The following structural units are characterized: atoms, molecules, ligands, nanoclusters, natural tiles, and periodic nets. For all these objects their occurrences in crystal structures are given, and the classification into topological types is provided. The collections are available at the *ToposPro* website and are distributed in a demo version together with *ToposPro*.

5 Web-Services

We have recently started the implementation of the *ToposPro* tools and databases into free online web-services. All actual *ToposPro* collections are available *via* the *TopCryst* service [10, 11]. The users can upload their crystallographic data as a CIF file and get a detailed topological analysis of the crystal structure in a fully automated manner. *TopCryst* provides the information on all possible structure representations and for each representation it gives a list of building units together with the corresponding underlying net, which is related to one of more than 800,000 topological types from the *ToposPro* collection. The occurrences of all revealed topological objects are output and the links to the website of the CSD and ICSD are provided to let the user see the initial crystallographic information.

6 Conclusion

The models, algorithms, methods and computer tools described above form the first automated system of the topological analysis of crystal structures. This approach is naturally chemical as it introduces the concept of chemical interaction into purely geometrical crystallographic model. Long experience of the applications of this approach to various classes of crystal structures and to different crystallochemical tasks has proved its effectiveness. The system is in progress: the general trend for its further development is the transformation of the topological databases into the knowledge bases by the application of machine-learning methods [6] and the creation of artificial-intelligence systems for materials science.

Acknowledgments. The work was supported by the Russian Science Foundation (grant No. 22-13-00062).

References

1. Blatov, V.A.: A method for topological analysis of rod packings. Struct. Chem. **27**(6), 1605–1611 (2016)
2. Blatov, V.A., Shevchenko, A.P.: Analysis of voids in crystal structures: the methods of 'dual' crystal chemistry. Acta Cryst. **A59**(1), 34–44 (2003)
3. Chung, S.J., Hahn, T., Klee, W.E.: Nomenclature and generation of three-periodic nets: the vector method. Acta Cryst. **A40**(1), 42–50 (1984)
4. Blatov, V.A., Alexandrov, E.V., Shevchenko, A.P.: Topology: ToposPro. In: Comprehensive Coordination Chemistry III. Elsevier, Oxford (2021)
5. Shevchenko, A.P., Blatov, V.A.: Simplify to understand: how to elucidate crystal structures? Struct. Chem. **32**(2), 507–519 (2021)
6. Shevchenko, A.P., Smolkov, M.I., Wang, J., Blatov, V.A.: Mining knowledge from crystal structures: oxidation states of oxygen-coordinated metal atoms in ionic and coordination compounds. J. Chem. Inf. Model. **62**(10), 2332–2340 (2022)
7. Blatov, V.A., Delgado-Friedrichs, O., O'Keeffe, M., Proserpio, D.M.: Three-periodic nets and tilings: natural tilings for nets. Acta Cryst. **A63**(5), 418–425 (2007)

8. ToposPro Homepage. https://topospro.com, last accessed 2023/05/17
9. Blatov, V.A., Shevchenko, A.P., Proserpio, D.M.: Applied topological analysis of crystal structures with the program package *ToposPro*. Cryst. Growth Des. **14**(7), 3576–3586 (2014)
10. Shevchenko, A.P., Shabalin, A.A., Karpukhin, I.Y., Blatov, V.A.: Topological representations of crystal structures: generation, analysis and implementation in the *TopCryst* system. Sci. Technol. Adv. Mater. Methods **2**(1), 250–265 (2022)
11. TopCryst Homepage. https://topcryst.com, last accessed 2023/05/17

Data Models and Knowledge Graphs

Analysis of the Metagraph Data Model in Terms of Metagraph Operations and Category Theory

Stepan Vinnikov, Anatoly Nardid, and Yuriy Gapanyuk$^{(\boxtimes)}$

Bauman Moscow State Technical University, Moscow, Russia
vinnikovss@student.bmstu.ru, {nardid,gapyu}@bmstu.ru

Abstract. A metagraph data model is a type of complex graph model designed to describe complex subject areas. Category theory is a well-known and well-researched mathematical apparatus that studies the properties of relations between various mathematical objects that do not depend on the internal structure of objects. In this article, we make an attempt to analyze the metagraph data model in terms of category theory. Operations over the metagraph are introduced and the properties of operations are considered in detail. The category of metagraphs "MetGr" is proposed, which is a category in which the object class is the set of metagraphs. Morphisms in such a category are based on a union operation. The proposed operations over the metagraph and the category of metagraphs "MetGr" is a necessary basis for further development of a rewriting system on metagraphs.

Keywords: Complex graph · Metagraph · Metavertex · Category theory · Category · Morphism

1 Introduction

There are countless different systems of various degrees of complexity in the world, the implementation and maintenance of which would be impossible or extremely difficult without their preliminary modeling.

Modeling allows, on the one hand, to analyze various characteristics and properties of the object of study without significant resource costs, and on the other hand, to predict the behavior of the object with changes in both its parameters and structure.

One of the most commonly used tools for modeling systems and subject areas is graph theory. This type of modeling is called graph structural. A special case of such modeling is modeling with metagraph data model.

The metagraph model is a type of complex graph model designed to describe complex subject areas. The model was originally proposed by A. Bazu and R. Blanning in the monograph [1] and subsequently received a number of extensions independently proposed by different groups of researchers. In this article,

J. Baixeries et al. (Eds.): DAMDID/RCDL 2023, CCIS 2086, pp. 37–50, 2024.
https://doi.org/10.1007/978-3-031-67826-4_3

the metagraph model is used in the form of annotated metagraph model proposed in [2].

The structure of the absolute majority of existing systems and subject areas is dynamic, that is, it changes over time – new components and connections are added, old ones are changed or removed. This process of changing the system is called its evolution.

The evolution of the system leads to the need for the evolution and models of this system, since they must reflect its current state. This statement is true, in particular, for the case when the model is described using metagraphs.

There are techniques that allow to work with graph transformations formally. One such technique is graph rewriting [3]. This approach is based on category theory – mathematical theory that studies the properties of relations between various mathematical objects. The same approach can be developed for metagraphs, but for this, first of all, it is necessary to consider operations over metagraph and describe metagraph as a category.

Currently, knowledge-based systems often use a Semantic Web approach. In particular, the RDF triple model and the OWL ontology description language are used [4]. Descriptive logics [5] are used as a basis for the formal-logical description of Semantic Web technologies. At the same time, paper [6] shows the fundamental difference between the RDF data structure and the metagraph model. Because of this, the results obtained in the logical study of Semantic Web technologies cannot be directly applied to the metagraph model.

Thus, the main goal this article is the creation of a formal apparatus that can later be used as a basis for development of a rewriting system on metagraphs. As such a formalization, we propose a way of defining the operations over metagraph and representing metagraph data model as a category.

The article is organized as follows. In section two, a formal definition of the metagraph model is given, operations on the metagraph are introduced and their properties are considered. In section three the category of metagraphs "MetGr" is introduced and its properties are considered. In section four the example of the application of the described category is given.

2   Operations Over Metagraphs

2.1   Definitions of the Metagraph Structures

According to [2] metagraphs and their internal structures are described as follows:

Definition 1. Metagraph *is a set* $MG = \{MV, E\}$ *where* MG *is a metagraph,* MV *is a set of metavertices of the metagraph,* E *is a set of edges of the metagraph.*

Definition 2. Empty metagraph *is a metagraph that has both empty* MV *and* E *sets. Empty metagraph is denoted as* $\varnothing = \{\varnothing, \varnothing\}$

Definition 3. Metavertex *is a set* $MV = \{id, mg\}$ *where* MV *is a metavertex, id is the identifier of the metavertex, mg is a nested metagraph of the metavertex, mg can be an empty metagraph.*

Definition 4. Edge *of the metagraph is a set* $E = \{id, (id_{begin}, id_{end})\}$ *where* E *is an edge, id is the identifier of the edge,* id_{begin} *is the identifier of the initial metavertex (the beginning of the edge),* id_{end} *is the identifier of the terminal metavertex (the ending of the edge).*

2.2   Theory of Categories Definitions

In this section and beyond, the definitions and properties of the elements of category theory are given, based both on the classical books of MacLane and Awodey ([7,8]) and on the works of contemporary authors ([9–11]).

Definition 5. *Category* $\mathbb{C}$ *consists of:*

1. *Class of objects* $Ob_{\mathbb{C}}$. *The fact that* X *is an object of* C *is written as* $X \in \mathbb{C}$;
2. *Sets* $hom(X, Y)$ *of morphisms from* X *to* Y *for each* $X \in \mathbb{C}$, $Y \in \mathbb{C}$.

The following axioms hold for any category:

1. *For* $\forall X \in Ob_{\mathbb{C}}$ *there exists an identity morphism* $1_X : X \to X$
2. *Morphisms can be composed: if there is* $f : X \to Y \in$ *and* $g : Y \to Z$ *then there exists their composition* $g \circ f : X \to Z$.
3. *Identity morphism is a neutral element for the composition operation:* $f \circ 1_X = f = 1_Y \circ f$ *where* $f \in hom(X, Y)$.
4. *Composition operation is associative:* $(h \circ g) \circ f = h \circ (g \circ f)$ *for arbitrary* f, g, h.

To apply the apparatus of category theory to the metagraph data model, it is necessary to determine the category carrier class and morphisms in this category. One of the possible categories is a category in which the class will be the set of all possible metagraphs, and the morphisms will be based on the union operation on the set of metagraphs. It is necessary to prove that sets with such morphisms are indeed categories.

2.3   Definition of the Operations

Let's describe operations and relations of metagraphs and metavertices:

1. Metagraph union:

$$mg_1 \cup mg_2 = mg_3 = \{MV_{mg_1} \cup MV_{mg_2}, E_{mg_1} \cup E_{mg_2}\}, \qquad (1)$$

for $\forall\, mg_1, mg_2 \in MG$.

2. Metavertex union:

$$mv_1 \cup mv_2 = \begin{bmatrix} \{id, mg_{mv_1} \cup mg_{mv_2}\} \ if \ id_{MV_1} = id_{MV_2} \\ \{mv_1, mv_2\} \ if \ id_{mv_1} \neq id_{mv_2} \end{bmatrix}, \tag{2}$$

for $\forall \, mv_1, mv_2 \in MV$.

3. Binary relation [12] $\subseteq$:

$$mg_1 \subseteq mg_2 \iff (mg_1 = \varnothing) \vee (\forall mv_i \in MV_{mg_1} \Rightarrow \\ \exists! mv_j \in MV_{mg_2} : mv_i \subseteq mv_j \wedge (E_{mg_1} \subseteq E_{mg_1})), \tag{3}$$

for $\forall mg_1, mg_2 \in MG$.

$$mv_1 \subseteq mv_2 \iff (id_{mv_1} = id_{mv_2}) \wedge (mg_{mv_1} \subseteq mg_{mv_2}), \tag{4}$$

for $\forall mv_1, mv_2 \in MV$.

Let's note the special properties of metagraphs and metavertices:

1. Empty metagraph is a neutral element of the union operation:

$$\varnothing \cup mg = \{\varnothing \cup MV_{mg}, \varnothing \cup E_{mg}\} = \{MV_{mg}, E_{mg}\} =$$

$$\{MV_{mg} \cup \varnothing, E_{mg} \cup \varnothing\} = mg \cup \varnothing = mg$$

2. Each metavertex can be represented as a metagraph:

$$mv = \{\{mv\}, \varnothing\} \in MG, \forall mv \in MV$$

3. The MV structure can be represented as a MV' structure:

$$MV' = \{id, \{mv\}, \{e\}\},$$

and there is a bijection [12] between MV and MV':

$$f : MV \leftrightarrow MV'$$

$$f(mv) = \{id, \{mv\}_{mg_{mv}}, \{e\}_{mg_{mv}}\} \in MV'$$

$$f^{-1}(mv') = \{id, \{\{mv\}_{mv'}, \{e\}_{mv'}\}\} \in MV$$

Note that the MG and MV structures can be expanded with one or several additional structures, such as the string:

$$MG_{str} = \{MV, \, E, \, s\},$$

$$MV_{str} = \{id, \, mg, \, s\},$$

where s is the string.

Then it will be necessary to describe $\cup$ operation and $\subseteq$ relation for these additional structures. For example, $str_1 \cup str_2 = concatinate(str_1, str_2)$, and $str_1 \subseteq str_2 = str_1 \preccurlyeq str_2$ where $\preccurlyeq$ is a lexicographic order.

2.4   Property of the Union Operation

Operation $\cup$ is associative on MG and MV, if all the additional structures of MV and MG also have this property:

$$(mg_1 \cup mg_2) \cup mg_3 = mg_1 \cup (mg_2 \cup mg_3), \forall mg_1, mg_2, mg_3 \in MG \quad (5)$$

$$(mv_1 \cup mv_2) \cup mv_3 = mv_1 \cup (mv_2 \cup mv_3), \forall mv_1, mv_2, mv_3 \in MV \quad (6)$$

For the proof of the 5–6 properties, we introduce a special algebraic structure "simple recursive structure" for which we will prove the above properties.

2.5   Simple Recursive Structure Definition and Properties

One of the remarkable structural properties of metagraphs and metavertices is their nesting property, and such nesting is recursive in nature.

The simplest algebraic structure with such property is a simple recursive structure.

Let ID be a set of identifiers. ID can be any set, for example, the set of natural numbers $\mathbb{N}$.

Definition 6. Simple recursive structure S is a set $S = \{id, \{s\}\}$ where $id \in ID$ is an identifier of the structure instance, $\{s\}$ is a nested set of the structure instances $s \in S$.

Let's note the special properties of simple recursive structures:

1. Each element of a set of simple recursive structures has a unique identifier:

$$s_i \in \{s\} \Rightarrow \nexists s_j \in \{s\} : s_j \neq s_i \wedge id_{s_j} = id_{s_i}$$

2. There is an empty set of simple recursive structures which is neutral to the union operation and is a subset of any set:

$$\varnothing = \{\}$$
$$\{s\} \cup \varnothing = \varnothing \cup \{s\} = \{s\}$$
$$\varnothing \subseteq \forall \{s\}$$

3. If an instance of the simple recursive structure is an element of the set then a set, containing only this element is a subset of this set:

$$s_1 \in \{s_1, s_2\} \Rightarrow \{s_1\} \subseteq \{s_1, s_2\} \Rightarrow s_1 \cup \{s_1, s_2\} = \{s_1, s_2\}$$

Now we will describe operations and relations of simple recursive structures:

1. Binary relation "=":

$$s_1 = s_2 \iff (id_{s_1} = id_{s_2}) \wedge (|\{s\}_{s_1}| = |\{s\}_{s_2}|) \wedge \\ ((\{s\}_{s_1} = \{s\}_{s_2} = \varnothing) \vee (\forall s_i \in \{s\}_{s_1} \Rightarrow \exists! s_j \in \{s\}_{s_2} : s_i = s_j)), \tag{7}$$

for $\forall s_1, s_2 \in S$.

2. Binary relation "$\subseteq$":

$$s_1 \subseteq s_2 \iff (id_{s_1} = id_{s_2}) \wedge \\ ((\{s\}_{s_1} = \varnothing) \vee (\forall s_i \in \{s\}_{s_1} \Rightarrow \exists! s_j \in \{s\}_{s_2} : s_i \subseteq s_j)), \tag{8}$$

for $\forall s_1, s_2 \in S$.

Remark 1. By definition, a binary relation R over sets X and Y is a subset R of the Cartesian product $X \times Y$. We will denote $xRy = True$ if $(x, y) \in R$.

Note the property $s \subseteq s, \forall s \in S$ without proof

3. Binary operation "$\cup$":

$$s_1 \cup s_2 = \begin{bmatrix} \{s_1, s_2\} \ if \ id_{s_1} \neq id_{s_2} \\ \{id, \{s\}_{s_1} \cup \{s\}_{s_2}\} \ if \ id_{s_1} = id_{s_2} \end{bmatrix}, \forall s_1, s_2 \in S, \tag{9}$$

where $\{s\}_{s_1} \cup \{s\}_{s_2} = \{s_i \cup s_j\}, \forall s_i \in \{s\}_{s_1}, \forall s_j \in \{s\}_{s_2}$.

4. Function $h(s) \in \mathbb{N} \cup \{0\}$, that describes the nesting depth $s \in S$ and defined as follows:

$$h(s) = \begin{bmatrix} 0 \ if \ \{s\}_s = \varnothing \\ 1 + h(\{s\}) \end{bmatrix}, \tag{10}$$

$\forall s \in S$, where $h(\{s\}) = max(\{h(s_i)\}), \forall s_i \in \{s\}_s$.

Before proving properties of the operations on S we distinguish a special predicate [13] class $P_r(S_{P_r})$ where $S_{P_r} = \{s_1, s_2, \dots\}$ is a set of variables of the P_r predicate.

Each predicate $p \in P_r$ is defined as follows:

$$(\exists s_i, s_j \in S_p : id_{s_i}! = id_{s_j}) \Rightarrow p(S_p) = True$$

$$(\exists s_i \in S_p : \{s\}_{s_i} = \varnothing) \Rightarrow p(S_p) = True$$

$$(\nexists s_i \in S_p : \{s\}_{s_i} = \varnothing) \Rightarrow p(S_p) = \bigwedge p(s_{1i}, s_{2j}, \dots),$$

for $\forall s_{1i} \in \{s\}_{s_1}, s_{2j} \in \{s\}_{s_2}, \dots$, where $s_1, s_2, \cdots \in S_p$

The following lemma holds for this class of predicates

Lemma 1. $p(S_p) = True \ for \ \forall p \in P_r, \ \forall S_p$

Proof. Take an arbitrary predicate $p \in P_r$ and let $P(h)$ be the statement $p = True$, $\forall S_p$. We induct on $h(S_p)$.

1. Base case:
   $P(0)$ is true by the definition of p.
   Let's show that $P(h)$ is true for $h(S_p) = 1$:

   1) if $(\exists s_i, s_j \in S_p : id_{s_i}! = id_{s_j})$, then $p(S_p) = True$.
   2) if $(\exists s_i \in S_p : \{s\}_{s_i} = \varnothing)$, then $p(S_p) = True$.
   3) if $(\nexists s_i \in S_p : \{s\}_{s_i} = \varnothing)$, then $p(S_p) = \bigwedge p(s_{1i}, s_{2j}, \dots)$, for $\forall s_{1i} \in \{s\}_{s_1}, s_{2j} \in \{s\}_{s_2}, \dots$, where $s_1, s_2, \dots \in S_p$.
   $h(s_{kj}) = 0, \forall s_{kj}, k = \overline{1, |S_p|}, j = \overline{1, |\{s\}_{s_k}|}$, hence $h(s_{1i}, s_{2j}, \dots) = 0$, hence $p(s_{1i}, s_{2j}, \dots) = True$ for $\forall s_{1i} \in \{s\}_{s_1}, s_{2j} \in \{s\}_{s_2}, \dots$, where $s_1, s_2, \dots \in S_p$, hence $\bigwedge p(s_{1i}, s_{2j}, \dots) = True$, hence $p(S_p) = True$.

2. Induction step:
   Let $p(S_p) = True$ for $S_p : h(S_p) = n \in \mathbb{N}$.
   then $p(S_p') = True$ for $S_p' : h(S_p') = n', n' \leq n$ by the definition pf p.
   Let's show that $p(S_p) = True$ for $S_p : h(S_p) = n + 1$:
   1) if $(\exists s_i, s_j \in S_p : id_{s_i}! = id_{s_j})$, then $p(S_p) = True$.
   2) if $(\exists s_i \in S_p : \{s\}_{s_i} = \varnothing)$, then $p(S_p) = True$.
   3) if $(\nexists s_i \in S_p : \{s\}_{s_i} = \varnothing)$, then $p(S_p) = \bigwedge p(s_{1i}, s_{2j}, \dots)$, for $\forall s_{1i} \in \{s\}_{s_1}, s_{2j} \in \{s\}_{s_2}, \dots$, where $s_1, s_2, \dots \in S_p$.
   $h(s_{kj}) \leq n, \forall s_{kj}, k = \overline{1, |S_p|}, j = \overline{1, |\{s\}_{s_k}|}$, hence, $h(s_{1i}, s_{2j}, \dots) = n$, hence, $p(s_{1i}, s_{2j}, \dots) = True$ for $\forall s_{1i} \in \{s\}_{s_1}, s_{2j} \in \{s\}_{s_2}, \dots$, where $s_1, s_2, \dots \in S_p$, hence, $\bigwedge p(s_{1i}, s_{2j}, \dots) = True$, hence, $p(S_p) = True$.

Conclusion: Since both the base case and the induction step have been proved as true for the arbitrary $p \in P_r$, hence by mathematical induction the statement $P(h)$ holds for every $h(S_p)$, hence $p(S_p) = True$ for $\forall p \in P_r$, $\forall S_p$.

The following lemmas are valid for the S structure:

Lemma 2. $s_1 \subseteq s_1 \cup s_2, \forall s_1, s_2 \in S$

Proof. 1. Consider the expression $s_1 \subseteq s_1 \cup s_2$:
   If $id_{s_1} \neq id_{s_2}$, then $s_1 \cup s_2 = \{s_1; s_2\}$. $s_1 \in \{s_1; s_2\} = s_1 \cup s_2 \Rightarrow s_1 \subseteq s_1 \cup s_2$.
   If $id_{s_1} = id_{s_2} = id$, then $s_1 \cup s_2 = \{id, \{s\}_{s_1} \cup \{s\}_{s_2}\}$, hence $s_1 \subseteq s_1 \cup s_2 \iff (id = id) \wedge (\{s\}_{s_1} = \varnothing \vee (\forall s_i \in \{s\}_{s_1} \Rightarrow \exists! s_j \in \{s\}_{s_1 \cup s_2} : s_i \subseteq s_j))$.
2. Consider the expression $(id = id) \wedge (\{s\}_{s_1} = \varnothing \vee (\forall s_i \in \{s\}_{s_1} \Rightarrow \exists! s_j \in \{s\}_{s_1 \cup s_2} : s_i \subseteq s_j))$:
   $id = id$ for $\forall id \in ID$.
   If $\{s\}_{s_1} = \varnothing$, then $s_1 \subseteq s_1 \cup s_2 \overset{def}{=} True$.
3. Consider the case $\{s\}_{s_1} \neq \varnothing$:
   First we prove that s_j exists.
   If $\{s\}_{s_2} = \varnothing$, then $\{s\}_{s_1} \cup \{s\}_{s_2} = \{s\}_{s_1} \Rightarrow (\forall s_i \in \{s\}_{s_1} \Rightarrow \exists! s_j \in \{s\}_{s_1 \cup s_2} = s_i : s_i \subseteq s_j)$. $s_j = s_i \Rightarrow s_i \subseteq s_i, \forall s_i \in \{s\}_{s_1} \Rightarrow s_1 \subseteq s_1 \cup s_2$.

If $\{s\}_{s_2} \neq \varnothing$, then $\{s\}_{s_1} \cup \{s\}_{s_2} = \{s_k \cup s_l\}, \forall s_k \in \{s\}_{s_1}, \forall s_l \in \{s\}_{s_2}$

$$s_k \cup s_l \stackrel{def}{=} \begin{bmatrix} \{s_k, s_l\} \ if \ id_{s_k} \neq id_{s_l} \\ \{id, \{s\}_{s_k} \cup \{s\}_{s_l}\} \ if \ id_{s_1} = id_{s_2} \end{bmatrix} \Rightarrow$$

$\Rightarrow \forall s_i \in \{s\}_{s_1} \exists! s_j \in \{s\}_{s_1 \cup s_2} : id_{s_j} = id_{s_i}$

4. Consider the expression $s_i \subseteq s_j$ where $s_i \in \{s\}_{s_1}, s_j \in \{s\}_{s_1 \cup s_2}: s_j = s_i \cup s_l, \forall s_i \in \{s\}_{s_1}, \forall s_l \in \{s\}_{s_2}$, then $s_i \subseteq s_j \iff s_i \subseteq s_i \cup s_l, \forall s_i \in \{s\}_{s_1}, \forall s_l \in \{s\}_{s_2}$, which is equivalent to the formulation of the lemma.
So,

$$s_1 \subseteq s_1 \cup s_2 \iff \begin{bmatrix} id_{s_1} \neq id_{s_2} \\ \{s\}_{s_1} = \varnothing \vee \{s\}_{s_2} = \varnothing \\ s_i \subseteq s_i \cup s_l, \forall s_i \in \{s\}_{s_1}, \forall s_l \in \{s\}_{s_2} \end{bmatrix} \Rightarrow$$

$$\Rightarrow (s_1 \subseteq s_1 \cup s_2) \in P_r \xrightarrow{Lemma 1} s_1 \subseteq s_1 \cup s_2, \forall s_1, s_2 \in S$$

Lemma 3. *Operation $\cup$ is associative on S.*

$$(s_1 \cup s_2) \cup s_3 = s_1 \cup (s_2 \cup s_3), \forall s_1, s_2, s_3 \in S \tag{11}$$

Proof. Let $id_1, id_2, id_3 \in ID$. We will write down in the table all possible combinations of indexes, while if they match, we will denote them by id:

No	s_1	s_2	s_3
1	$\{id_1, \{s\}\}$	$\{id_2, \{s\}\}$	$\{id_3, \{s\}\}$
2	$\{id, \{s\}\}$	$\{id, \{s\}\}$	$\{id_3, \{s\}\}$
3	$\{id, \{s\}\}$	$\{id_2, \{s\}\}$	$\{id, \{s\}\}$
4	$\{id_1, \{s\}\}$	$\{id, \{s\}\}$	$\{id, \{s\}\}$
5	$\{id, \{s\}\}$	$\{id, \{s\}\}$	$\{id, \{s\}\}$

Consider the expression (11) for all of the five cases:

1. $(s_1 \cup s_2) \cup s_3 = \{s_1, s_2\} \cup s_3 = \{s_1, s_2, s_3\} = s_1 \cup \{s_2, s_3\} = s_1 \cup (s_2 \cup s_3)$
2. $(s_1 \cup s_2) \cup s_3 = \{s_1 \cup s_2\} \cup s_3 = \{s_1 \cup s_2, s_3\} \stackrel{Lemma\ 2}{=} \{s_1 \cup s_2, s_1, s_3\} = s_1 \cup \{s_2, s_3\} = s_1 \cup (s_2 \cup s_3)$
3. $(s_1 \cup s_2) \cup s_3 = \{s_1, s_2\} \cup s_3 = \{s_1 \cup s_3, s_2\} \stackrel{Lemma\ 2}{=} \{s_1 \cup s_3, s_1, s_2\} = s_1 \cup \{s_3, s_2\} = s_1 \cup (s_2 \cup s_3)$
4. $(s_1 \cup s_2) \cup s_3 = \{s_1, s_2\} \cup s_3 = \{s_1, s_2 \cup s_3\} = s_1 \cup \{s_2 \cup s_3\} = s_1 \cup (s_2 \cup s_3)$
5. $(s_1 \cup s_2) \cup s_3 = \{id, (\{s\}_{s_1} \cup \{s\}_{s_2}) \cup \{s\}_{s_3}\}$. On the other hand $s_1 \cup (s_2 \cup s_3) = \{id, \{s\}_{s_1} \cup (\{s\}_{s_2} \cup \{s\}_{s_3})\}$.
   In this case $(s_1 \cup s_2) \cup s_3 = s_1 \cup (s_2 \cup s_3) \iff (\{s\}_{s_1} \cup \{s\}_{s_2}) \cup \{s\}_{s_3} = \{s\}_{s_1} \cup (\{s\}_{s_2} \cup \{s\}_{s_3})$.

Consider the expression $(\{s\}_{s_1} \cup \{s\}_{s_2}) \cup \{s\}_{s_3} = \{s\}_{s_1} \cup (\{s\}_{s_2} \cup \{s\}_{s_3})$ for various $\{s\}_{s_1}, \{s\}_{s_2}, \{s\}_{s_3}$:

1. If $\{s\}_{s_1} = \varnothing$, then $(\{s\}_{s_1} \cup \{s\}_{s_2}) \cup \{s\}_{s_3} = (\varnothing \cup \{s\}_{s_2}) \cup \{s\}_{s_3} = \{s\}_{s_2} \cup \{s\}_{s_3} = \varnothing \cup (\{s\}_{s_2} \cup \{s\}_{s_3}) = \{s\}_{s_1} \cup (\{s\}_{s_2} \cup \{s\}_{s_3})$.
2. If $\{s\}_{s_2} = \varnothing$, then $(\{s\}_{s_1} \cup \{s\}_{s_2}) \cup \{s\}_{s_3} = (\{s\}_{s_1} \cup \varnothing) \cup \{s\}_{s_3} = \{s\}_{s_1} \cup \{s\}_{s_3} = \{s\}_{s_1} \cup (\varnothing \cup \{s\}_{s_3}) = \{s\}_{s_1} \cup (\{s\}_{s_2} \cup \{s\}_{s_3})$.
3. If $\{s\}_{s_3} = \varnothing$, then $(\{s\}_{s_1} \cup \{s\}_{s_2}) \cup \{s\}_{s_3} = (\{s\}_{s_1} \cup \cup \{s\}_{s_2}) \cup \varnothing = \{s\}_{s_1} \cup \{s\}_{s_2} = \{s\}_{s_1} \cup (\{s\}_{s_2} \cup \varnothing) = \{s\}_{s_1} \cup (\{s\}_{s_2} \cup \{s\}_{s_3})$.
4. If $\{s\}_{s_1} \neq \varnothing \wedge \{s\}_{s_2} \neq \varnothing \wedge \{s\}_{s_3} \neq \varnothing$, then:
   1) $(\{s\}_{s_1} \cup \{s\}_{s_2}) \cup \{s\}_{s_3} = \{\{s_i \cup s_j\} \cup s_k\}, \forall s_i \in \{s\}_{s_1}, \forall s_j \in \{s\}_{s_2}, \forall s_k \in \{s\}_{s_3}$
   2) $\{s\}_{s_1} \cup (\{s\}_{s_2} \cup \{s\}_{s_3}) = \{s_i \cup \{s_j \cup s_k\}\}, \forall s_i \in \{s\}_{s_1}, \forall s_j \in \{s\}_{s_2}, \forall s_k \in \{s\}_{s_3}$

Then the expression (11) is true if

$$\{\{s_i \cup s_j\} \cup s_k\} = \{s_i \cup \{s_j \cup s_k\}\}, \forall s_i \in \{s\}_{s_1}, \forall s_j \in \{s\}_{s_2}, \forall s_k \in \{s\}_{s_3}$$

$$(12)$$

We can fix s_i, s_j, s_k because if (12) is true for the fixed arbitrary combination, then it will be true for all other combinations s_l, s_m, s_n.
Then (12) can be written as:

$$((s_i \cup s_j) \cup s_k) = (s_i \cup (s_j \cup s_k)), \forall s_i \in \{s\}_{s_1}, \forall s_j \in \{s\}_{s_2}, \forall s_k \in \{s\}_{s_3},$$

which is equivalent to the formulation of the lemma.

So, $(s_1 \cup s_2) \cup s_3 = s_1 \cup (s_2 \cup s_3)$, if:

$$\begin{bmatrix} id_{s_1} \neq id_{s_2} \vee id_{s_1} \neq id_{s_3} \vee id_{s_2} \neq id_{s_3} \\ \{s\}.s_1 = \varnothing \vee \{s\}.s_2 = \varnothing \vee \{s\}.s_3 = \varnothing \Rightarrow \\ ((s_i \cup s_j) \cup s_k) = (s_i \cup (s_j \cup s_k)) \end{bmatrix}$$

$$\Rightarrow ((s_1 \cup s_2) \cup s_3 = s_1 \cup (s_2 \cup s_3)) \in P_r \xrightarrow{\text{Lemma 1}} (s_1 \cup s_2) \cup s_3 = s_1 \cup (s_2 \cup s_3),$$

for $\forall s_1, s_2, s_3 \in S$

Remark 2. If we expand S structure with an additional structure $p \in \mathbb{P}$ for which a binary relation $\subseteq$ and binary operation $\cup : \mathbb{P} \times \mathbb{P} \to \mathbb{P}$ are defined, and the statement $p_1 \subseteq p_1 \cup p_2, \forall p1, p2 \in \mathbb{P}$ is true, then lemma No2 will still hold true for S because:

$$s_1 \cup s_2 = \{id, p_{s_1} \cup p_{s_2}, \{s\}_{s_1} \cup \{s\}_{s_2}\}$$

Similarly, if lemmas No2,3 hold true for $\forall p_1, p_2 \in \mathbb{P}$, then they will hold for S.

The same conclusions can be done for S expanded with several additional structures $p_i \in \mathbb{P}_i$.

The MV' structure corresponds to the S structure expanded with a set of edges. Set of edges is the usual set, hence, lemmas No2,3 hold true for MV'. Since $MV \leftrightarrow MV'$, lemmas No2,3 hold true for MV. Each metagraph can be represented as a metavertex with the same $id = id_{mg}$, hence each metagraph holds all the properties of the S structure.

It follows from this that the statements 5–6 are proven for $\forall mg \in MG$ and $\forall mv \in MV$.

3   Category of Metagraphs

Having proved the associativity property of the union operation of metagraphs, we can describe a category of metagraphs.

Definition 7. *Category of metagraphs $MetGr_\cup$ is a category in which class of objects is a set of metagraphs MG and morphisms are expressions of the form $\cup mg$, that perform a union operation of the argument mg_1 (the beginning of the arrow) with mg and get metagraph mg_2 (the end of the arrow) as a result:*

$$mg_1 \xrightarrow{\ \cup mg\ } mg_2 \tag{13}$$

Proof of the category axioms for $MetGr_\cup$:

1. Identity morphism: $id_{MetGr_\cup} = \cup\varnothing$.
   Let $f = \cup mg_f$, then:

   $$f \circ id = \cup\varnothing \cup mg_f = \cup mg_f = f$$
   $$id \circ f = \cup mg_f \cup \varnothing = \cup mg_f = f$$

2. Composition of morphisms:
   Let $f = \cup mg_f$, $g = \cup mg_g$, then:

   $$g \circ f = \cup mg_f \cup mg_g = \cup\{MV_{mg_f} \cup MV_{mg_g},\ E_{mg_f} \cup E_{mg_g}\}$$

3. Associativity of the composition operation:
   Let $f = \cup mg_f$, $g = \cup mg_g$, $h = \cup mg_h$, then:

   $$(h \circ g) \circ f = \cup mg_f \cup (mg_g \cup mg_h) =$$
   $$\cup mg_f \cup \{MV_{mg_g} \cup MV_{mg_h},\ E_{mg_g} \cup E_{mg_h}\} =$$
   $$\cup\{MV_{mg_f} \cup (MV_{mg_g} \cup MV_{mg_h}),\ E_{mg_f} \cup (E_{mg_g} \cup E_{mg_h})\} =$$
   $$\cup\{(MV_{mg_f} \cup MV_{mg_g}) \cup MV_{mg_h},\ (E_{mg_f} \cup E_{mg_g}) \cup E_{mg_h}\} =$$
   $$\cup\{MV_{mg_f} \cup MV_{mg_g},\ E_{mg_f} \cup E_{mg_g}\} \cup mg_h =$$
   $$(mg_f \cup mg_g) \cup mg_h = h \circ (g \circ f)$$

Category axioms are proven, hence $MetGr_\cup$ is a category.

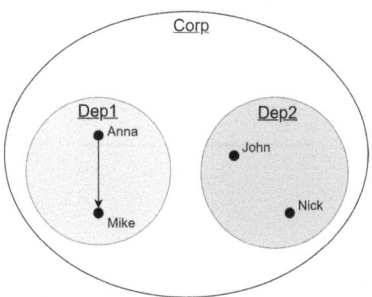

Fig. 1. initial state of the "Corp"

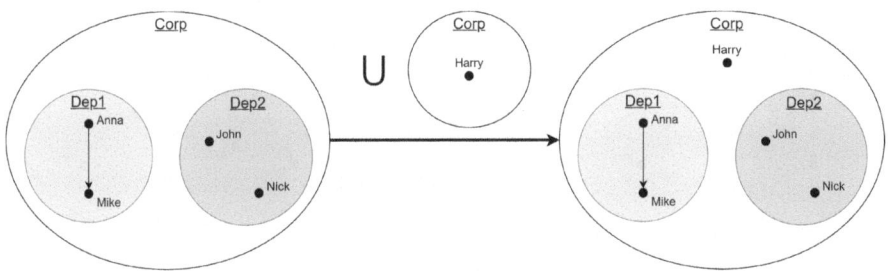

Fig. 2. Harry comes to the company

4   Example

Let 's give an example of using operations over metagraphs.

Let's say there are two departments in some corporation called "Corp". The first department is called "Dep1", and Mike and Anna work in it, Anna is Mike's supervisor. The second department is called "Dep2", John and Nick work in it. John came to work recently, so he is not Nick's supervisor yet.

After the new year, the following changes are planned in "Corp":

1. A manager Harry comes to the corporation, and he will manage the "Dep1" and "Dep2" departments.
2. A new employee Alex comes to the "Dep1", and Anna will be his supervisor.
3. John becomes Nick's supervisor.

Let's simulate this situation using a $MetGr_\cup$ category. Let the metavertices denote departments, the vertices denote employees, and the edges denote the subordination relationship.

Metagraph in the Fig. 1 corresponds to the initial state of the "Corp". Figure 2 corresponds to the moment at which Harry comes to the company. In Fig. 3 Harry starts managing "Dep1" and "Dep2" departments. Alex comes to "Dep1" in Fig. 4. Anna becomes Alex's supervisor in Fig. 5. John becomes Nick's supervisor in Fig. 6.

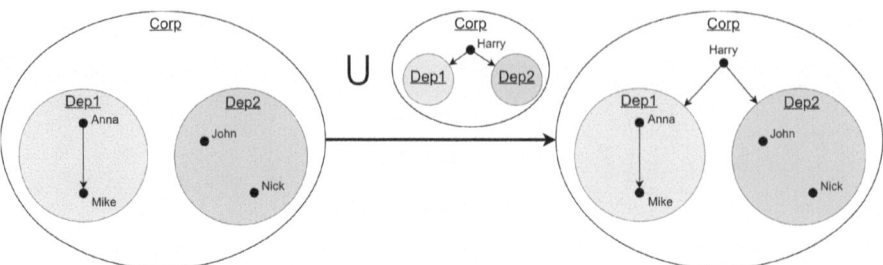

Fig. 3. Harry starts managing "Dep1" and "Dep2" departments

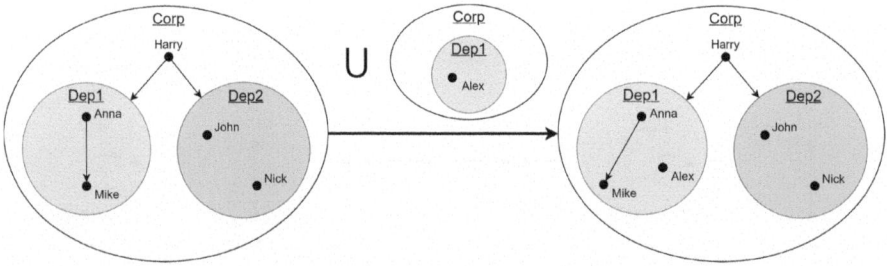

Fig. 4. Alex comes to "Dep1"

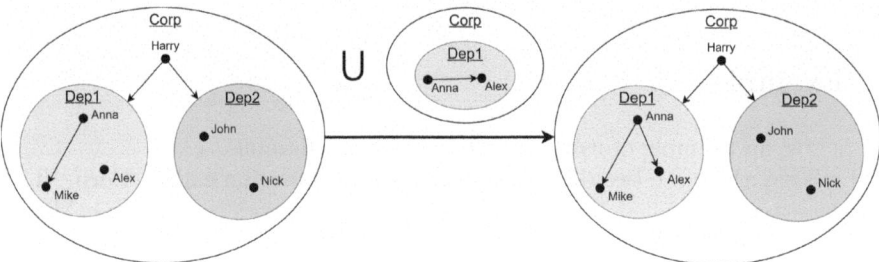

Fig. 5. Anna becomes Alex's supervisor

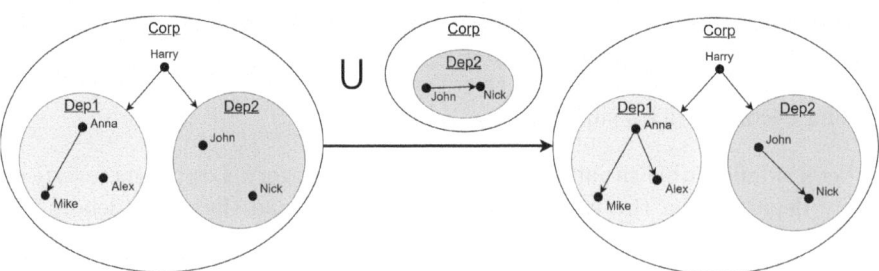

Fig. 6. John becomes Nick's supervisor

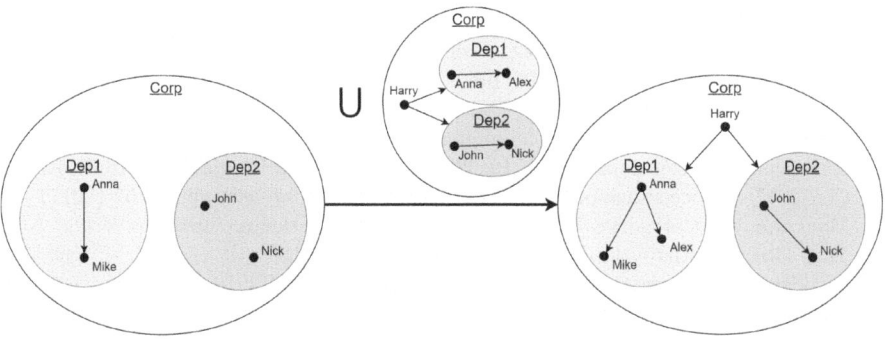

Fig. 7. Transition from the initial state to the final one through a single morphism

The right metagraph in Fig. 6 corresponds to the final state of the "Corp" after the new year. Note that we could achieve the same state with just one morphism using the composition operation in $MetGr_\cup$, as shown in Fig. 7.

Thus, using a metagraph data model and the $MetGr_\cup$ category, we are able to simulate real-life situations.

5   Conclusion

The metagraph data model is a modern tool for modeling systems of varying degrees of complexity. In addition to various data about the characteristics of the described system, metagraphs store information about the hierarchy of this system, which advantageously distinguishes it from other modeling methods.

The proof of the properties of the simple recursive structure allowed us to prove that a metagraph can be viewed as a category.

The obtained research results are applicable for the further development of the rewriting system on metagraphs. The rewriting system can use either metagraph operations or a categorical approach or their hybridization. The effectiveness of each of these approaches is the subject of further research.

References

1. Basu, A., Blanning, R.W.: Metagraphs and Their Applications. Springer (2007). https://doi.org/10.1007/978-0-387-37234-1
2. Gapanyuk, Y.: The development of the metagraph data and knowledge model. In: Russian Advances in Fuzzy Systems and Soft Computing: Selected Contributions to the 10th International Conference on "Integrated Models and Soft Computing in Artificial Intelligence (IMSC-2021)", pp. 1–7. Kolomna, Russia (2021)
3. Ehrig, H., et al.: Graph and Model Transformation. Springer (2015). https://doi.org/10.1007/978-3-662-47980-3
4. Allemang, D., Gandon, F., Hendler, J.A.: Semantic Web for the Working Ontologist: Effective Modeling for Linked Data, RDFS, and OWL. ACM Books (2020)

5. Baader, F., Horrocks, I., Sattler, U.: Description logics as ontology languages for the semantic web. In: Hutter, D., Stephan, W. (eds.) Mechanizing Mathematical Reasoning 2005. LNCS, vol. 2605, pp. 228–248. Springer, Berlin, Heidelberg (2005). https://doi.org/10.1007/978-3-540-32254-2_14

6. Chernenkiy, V., et al.: Using the metagraph approach for addressing RDF knowledge representation limitations. In: 2017 Internet Technologies and Applications, ITA 2017 - Proceedings of the 7th International Conference, pp. 47–52 (2017)

7. MacLane, S.: Categories for the Working Mathematician. Springer-Verlag, New York (1971). https://doi.org/10.1007/978-1-4757-4721-8

8. Awodey, S.: Category Theory. Oxford University Press (2010)

9. Leinster, T.: Basic Category Theory. Cambridge University Press (2014)

10. Yanofsky, N.S.: Theoretical Computer Science for the Working Category Theorist. Cambridge University Press (2022)

11. Grandis, M.: Category Theory and Applications: A Textbook for Beginners. World Scientific (2018)

12. Grami, A.: Discrete Mathematics: Essentials and Applications. Academic Press (2022)

13. Mendelson, E.: Introduction to Mathematical Logic. CRC Press/Taylor & Francis Group (2015)

Conceptual Data Model : Concept, Formal Bases, and Implementation Issues

Manuk G. Manukyan$^{(\boxtimes)}$ [iD]

Yerevan State University, 0025 Yerevan, Armenia
mgm@ysu.am

Abstract. The investigation subjects of this paper are the implementation issues of a conceptual data model. An approach to implement the concept of the considered conceptual data model is proposed. The discussed conceptual data model concept is based on the behavioral and data definition symbols to model conceptual entities. The result of formalization of these symbols is content dictionaries. An important feature of the considered conceptual data model is its extensibility property. The extension is achieved by introducing new symbols into the conceptual data model. Thus, the result of extension of the considered conceptual data model is reduced to create new content dictionaries to support new concepts. The conceptual data model is defined as the union of all content dictionaries.

Keywords: Conceptual Data Model · Hierarchical Relation · Hierarchical Relations Algebra · Data Integration · OPENMath

1 Introduction

The emergence of a new paradigm in science and various applications of information technology is related to issues of big data handling. Big data is a field that treats of ways to analyze, systematically extract information from, or otherwise deal with data sets that are too large or complex to be dealt with by traditional data-processing application software. The concept of big data is relatively new and is based on the following widely spread notions: *data volume, velocity, variety, veracity* and *value* [1]. This concept involves the growing role of data in all areas of human activity beginning with research and ending with innovative developments in business. In this connection, the creation of new information technology is expected in which data becomes dominant for new approaches to conceptualization, organization, and implementation of systems to solve problems that were previously considered extremely hard or, in some cases, impossible to solve. In this context, the issues of conceptual modeling of big data become relevant.

This work was supported by the RA MES State Committee of Science, in the frames of the research project No. 21T-1B326.

J. Baixeries et al. (Eds.): DAMDID/RCDL 2023, CCIS 2086, pp. 51–64, 2024.
https://doi.org/10.1007/978-3-031-67826-4_4

Conceptual data modeling has been the subject of intense research since the late 1970s. Prerequisites for such research is that database systems usually have limited knowledge about the *meaning* of the data stored in them [2]. In fact, they allow to manipulate data of certain simple types. Any more complex interpretation is left to the user. In this context, the issues of formal knowledge representation in the form of a set of concepts of some subject domain and relations between them are topical. Such representations are used for reasoning about entities of the subject domains, as well as for the domains description. Thus, conceptualization of subject domain assumes accessing and managing the data in terms of the conceptual entities. Some arguments in favor of creating a conceptual data model with extensibilty property are discussed below.

The new data management paradigms, as well as the many directions in which modern database systems are evolving, leads to the idea of developing a conceptual data model with extensibility property. In particular, the data integration concept and temporal databases concept are examples of very important directions in which modern database systems are evolving. Analysis of existing approaches to data integration can be found in [3]. Basically, these works are devoted to the problems of integrating homogeneous data sources (in the case of big data the data sources basically are heterogenous). Typically, an extended relational or object data model was used as the target data model. Details of temporal databases and current trends in their development can be found in [4].

The polyglot persistence approach is an another argument to develop a conceptual data model with extensibility property. This is explained by the fact that different databases are designed to solve different problems. Using a single database engine for all of the requirements usually leads to non-performant solutions; storing transactional data, caching session information, traversing the graph of customers and the products their friends bought are essientially different problems. Even in the RDBMS space, the requirements of an OLAP and OLTP systems are very different nonetheless, they are often forced into the same schema (for more details see [5]).

In this paper the formal bases of implementation of a conceptual data model concept are considered. The discussed conceptual data model was proposed in [6]. An important feature of the considered conceptual data model is its extensibility property. The extensibility property provides the ability to conceptually model arbitrary data sources. An informal example of extending the conceptual data model to support the concept of data integration is introduced. Also, the formal bases of the considered conceptual data model are discussed.

The paper is organized as follows: A conceptual data model and its formal bases are considered in Sect. 2. An approach to implement the concept of the considered conceptual data model is proposed in Sect. 3. Related work is presented in Sect. 4. The conclusion is provided in Sect. 5.

2   Formal Bases

In this section we will briefly analyze formal bases of a conceptual data model. A more detailed analysis of these formalisms can be found in [6,7].

2.1   OPENMath Objects

OPENMath is an extensible formalism for representing mathematical objects so that software packages can exchange these objects without losing semantic content. The possibility of the extensibility of the considered formalism is explained by the fact that the mathematical notation is constantly evolving. Moreover, mathematics and its applications are a growing field of knowledge, new ideas and notations appear constantly. Formally, an OPENMath object is a labeled tree whose leaves are basic OOPENMath objects. Examples of basic OPENMath objects are: Integer, Symbol and Variable. The compound objects are defined in terms of *binding* and *application* of the λ-calculus [8]. The following recursive rules for constructing compound OPENMath objects are proposed:

- Basic OPENMath objects are OPENMath objects.
- If $A_1, A_2, ..., A_n$ $(n \geq 1)$ are OPENMath objects, then $application(A_1, A_2, ..., A_n)$ is an OPENMath *application object*.
- If $S_1, S_2, ..., S_n$ are OPENMath symbols, and $A, A_1, A_2, ..., A_n$ $(n \geq 1)$ are OPENMath objects, then $attribution(A, S_1\ A_1, S_2\ A_2, ..., S_n\ A_n)$ is an OPENMath *attribution object* and A is the object *stripped* of *attributions*.
- If B and C are OPENMath objects, and $v_1, v_2, ..., v_n$ $(n \geq 0)$ are OPENMath variables or attributed variables, then $binding(B, v_1, v_2, ..., v_n, C)$ is an OPENMath *binding object*.

OPENMath objects have the expressive power to cover all areas of computational mathematics. The OPENMath application and binding objects for $sin(x)$ and $\lambda x.x + 1$ in tree-like notation are presented below:

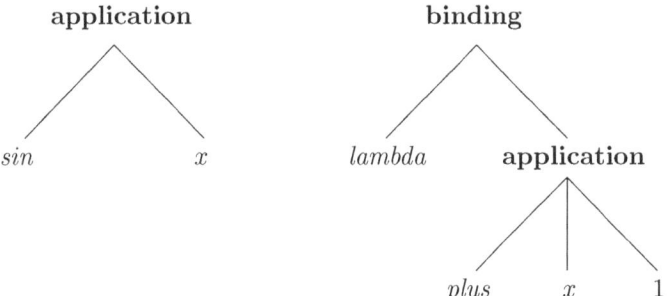

2.2   Conceptual Data Model

In the frame of the considered conceptual data model (as in the relational data model), a single concept is used to model subject domains, namely, hierarchical relation. Below we introduce the definitions of the hierarchical relation schema and the hierarchical relation. These definitions can be considered as strengthening the definitions of the relation schema and relation of the relational databases.

Definition 1. *A hierarchical relation schema X is an attribution object and is interpreted by a finite set of attribution objects $\{A_1, A_2, ... , A_n\}$. Corresponding* to each attribution object A_i is a set D_i (a finite, non-empty set), $1 \leq i \leq n$, called the domain of A_i.

Definition 2. Let $D = D_1 \cup D_2 \cup ... \cup D_n$. A hierarchical relation x on hierarchical relation schema X is a finite set of mappings $\{t_1, t_2,..., t_k\}$ from X to D with the restriction that for each mapping $t \in x$, $t[A_i]$ must be in D_i, $1 \leq i \leq n$. The mappings are called hierarchical tuples or simply tuples.

A hierarchical relation is an instance of a hierarchical relation schema. The considered concept of hierarchical relations allows to model as XML as well as JSON data.

Definition 3. A key of a hierarchical relation x on hierarchical relation schema X is a minimal subset K of X such that for any distinct tuples $t_1, t_2 \in x$, $t_1[K] \neq t_2[K]$.

Finally, a database schema S is a finite set of schemas of the hierarchical relations. A database d on database schema S is a collection of hierarchical relations $\{x_1, x_2,..., x_n\}$ such that for each schema of the hierarchical relation schema $s \in S$ there is a hierarchical relation $x \in d$ such that x is a hierarchical relation with schema s that satisfies every constraint defined in s.

3   Conceptual Data Model Concept Implementation

In this section we will consider the concept to model conceptual entities and its implementation issues.

3.1   Conceptual Schema

The conceptual schema is an instance of conceptual data model and is intended for formal knowledge representation in the form of a set of concepts of some subject domain and relations between them. Such representations are used for reasoning about entities of the subject domains, as well as for the domains description. A conceptual schema is defined as a set of OPENMath attribution objects. A distinguishing feature of the conceptual level is its stratification of the local and global levels to model the conceptual entities. On the local level, homogeneous representation of heterogeneous data sources is provided. The global level is intended to define derived conceptual entities. We consider the following formalisms to support conceptual entities:

- content dictionaries to define concepts (symbols) of conceptual data model (for example, type constructors, algebraic operations, etc.);
- signature files to formalize signatures of conceptual data model symbols to check the semantic validity of their representations.

A content dictionary (CD) which contains representation of symbols of the conceptual data model contains two types of information: one which is common to all CDs, and one which is restricted to a particular symbol definition. Definition of a new symbol includes name and description of the symbol, and also some optional information about this symbol. Below an example of a symbol definition is considered:

<CDDefinition>
      <Name> *add* < /Name>
      <Description> A n-ary commutative function addition < /Description>
      <CMP> $x + y = y + x$ < /CMP>
< /CDDefinition>

The above used XML elements have obvious interpretations. Only note, that the element "CMP" contains the commented mathematical property of the defined arithmetic function symbol. Specific information pertaining to the symbol like the signature is defined in additional files associated with CDs. CDs contain just one part of the information that can be associated with a symbol in order to stepwise define its meaning and its functionality. Signature files are used to formalize formats of conceptual data model symbols. The considering symbols can be divided into two groups: data definition symbols and behavioral symbols. Functional notation is used below to formalize the symbols of the conceptual data model. CDs of the conceptual data model are based on these formal definitions.

3.2   Behavioral Symbols

To define the behavior of entities of the conceptual level, an algebra of hierarchical relations and its formal semantics was developed [6]. The proposed symbols are analog to the relational algebra operations. Here the functional notation is used to define the semantics of these symbols. Detailed definitions of the semantics of the considered operations can be found in [6]. Let r be the set of all hierarchical relations expressible within conceptual data model.

To support n-ary associative operations union, we introduced the symbol *union*. The symbol *union* is used to denote the n-ary union of sets (hierarchical relations). It takes sets as arguments, and denotes the set that contains all the tuples that occur in any of them:

$$union : r^{*assoc} \to r$$

To support operations minus, we introduced the symbol *minus*. The symbol *minus* is used to denote the difference of sets (hierarchical relations). It takes two sets as arguments, and returns a set that is the difference between two sets:

$$minus : r \times r \to r$$

To support n-ary associative operation joining, we introduced the symbol *join*. The symbol *join* is used to denote the n-ary join of sets. It takes sets as arguments, denotes a set of tuples, and is interpreted analogously to the operation natural join of the relational algebra in general case (joins of many relations):

$$join : r^{*assoc} \rightarrow r$$

To support a filtering operation, we introduced the symbol σ. This symbol is used to denote a select operation on the set. It takes a set and a predicate as arguments, and denotes the set which contains all the tuples for which the predicate is satisfied:

$$\sigma : \{r \rightarrow \{p : \{tuple\} \rightarrow boolean\}\} \rightarrow r$$

Here p is a predicate which is applied to *tuple*.

To support a projection operation, we introduced the symbol π. This symbol is used to denote a unary operation on the set. It takes a set and a list of *attribution* object names as argument, denotes a set of tuples, and is interpreted analogously to the operation *project* of the relational algebra:

$$\pi : r[name^*] \rightarrow r$$

Here *name* denotes the name of an attribution object.

For processing data, aggregating functions play a significant role. We introduced the *min, max, count, sum* and *avg* symbols to support the corresponding aggregate functions of the relational algebra. Let $f \in \{min, max, avg, sum, count\}$, then

$$f : r[name] \rightarrow numericalvalue - string$$

Often, we need to consider the tuples of a hierarchical relation in groups. For this purpose, we introduced a grouping symbol γ. This symbol is used to denote a unary operation on the set. It takes a set, a list of *attribution* object names and aggregate functions as arguments, denotes a set of tuples, and is interpreted analogously to the operation *grouping* of the relational algebra:

$$\gamma : r[name^*(, f : (tuple[name^*])^* \rightarrow numericalvalue - string)^*] \rightarrow r$$

3.3   Data Definition Symbols

To model the hierarchical relation concept we introduce the symbols *sequence* and *choice* which have analogous semantics as *sequence* and *choice* elements in the XML Schema language. The arguments of these functions are typed attribution objects, and the return value is a typed attribution object[1]. Let R be the set of all hierarchical relations schemas in the frame of conceptual data model and $f \in \{sequence, choice\}$, then

$$f : R^* \rightarrow R$$

We introduced the symbol *card* for modeling a cardinality number concept:

---

[1] The attribution construct is used to define conceptual entities.

$$card \in \{?, *, +\}$$

To model constraint concept of databases a *constraints* symbol is introduced, which value is an attribution object.

We introduced the symbol *key* to model the hierarchical relation key's concept, which value is a set of attribution objects names.

To support the built-in data types concept of the XML Schema language the corresponding symbols were introduced (for instance, *integer*, *string*, etc.).

The above considered symbols form the CD for the conceptual data model which is defined in the previous section. Our concept to creating conceptual data model assumes that this model must be extensible. The extensibility concept of the considered conceptual data model coincides with the analogous concept of OPEN-Math. In other words, extension of the conceptual data model is reduced to defining new symbols and formalyzing these symbols using the CD mechanism.

3.4   An Informal Example of Extending the Conceptual Data Model to Support the Data Integration Concept

As mentioned above, the conceptual schema is stratified into two levels: local and global. To support this concept, the conceptual data model has been extended with *local* and *global* symbols. On the local level homogeneous representation of heterogeneous data sources is provided. In other words, data sources are represented by a set of hierarchical relations. Hierarchical relations of the local level are result data extraction from data sources. To support the data extraction concept (extract, transform, and load) the following symbols *etl*, *source* and *location* are introduced. These symbols are applying at the local conceptual level and have obvious semantics. Below, a formal example of local level attribution object is produced:

attribution(*local*, *source attribution*(Name, *type* applicationObject),
etl applicationObject, *location* LValue)

We have expanded the conceptual data model to support the data integration concept with the following data definition symbols: *med*, *whse*, *cube* and *rule* (they are nullary functions). These symbols applying on the global conceptual level. The values of the *whse*, *med* and *cube* symbols are attribution objects (the data warehouse, mediator, and data cube schemas, respectively) and the value of the *rule* symbol is an application object (an algebraic program) by means of which a mapping from data sources into data warehouse, mediator and data cube are defined. To define the entities of the global level the following construction is proposed:

attribution(*global*, *mwc attribution*(Name, *type* applicationObject),
rule algebraicProgram), where $mwc \in \{med, whse, cube\}$

The result of such extension of the conceptual data model is a new CD to model the data integration concept[2]. Below, an example of a data warehouse

---

[2] Here we did not consider all the symbols that are used to support the data integration concept.

in tree-like notation for an automobile company database is adduced [10] which is an instance of the conceptual data model with orientation to data integration. Suppose for simplicity that there are only two dealers in the Aardvark system and they respectively use the schemas:

Cars = {serialNo, Model, Color}, and

Autos = {serialNo, Model}

Colors = {serialNo, Color}

It is assumed that a data warehouse is created with the schema

AutosWhse = {serialNo, Model, Color}

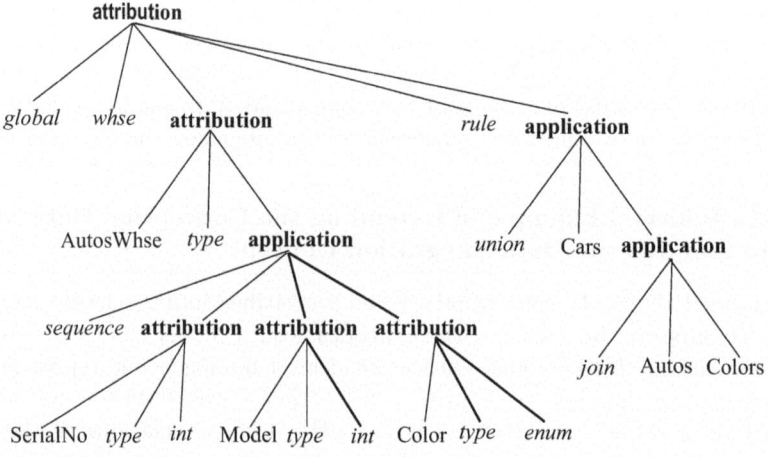

Examples of formalization of symbols by CD of the conceptual data model is given in Appendix A.

3.5   Signature Files for Conceptual Data Model

As is mentioned above, to check semantic validity of symbols representations we associate extra information with CDs, namely signature files. A signature file contains the definitions of all symbol signatures of the considered CD. We use Small Type System [9] to formalize the concept of signatures. Below the definition of the signature of the above considered symbol *add* is provided (see examples of the conceptual data model concept signature in Appendix B):

```
<Signature name = "add">
   <OMOB>
    <OMA>
     <OMS name = "mapsto" cd = "sts"/ >
     <OMA>
      <OMS name = "nassoc" cd = "sts"/ >
      <OMV name = "AbelianSemiGroup" cd = "sts"/ >
     < /OMA>
     <OMV name = "AbelianSemiGroup" cd = "sts"/ >
    < /OMA>
   < /OMOB>
```

< /Signature>
The above considered symbols *mapsto* and *nassoc* were defined in the OPEN-Math. The symbol *mapsto* represents the construction of a function type. The first n-1 children denote the types of the arguments, the last denotes the return type. The symbol *nassoc* constructs a child of *mapsto* which denotes an arbitrary number of copies of the argument of *nassoc*. The operator is associative on these arguments which means that repeated uses may be flattened/unflattened.

4 Related Work

In this Section we will consider popular notations for describing database designs, namely the E/R, UML and XML models [10]. Furthermore, a brief discussion of the data integration approach based on the conceptual data model is provided, along with known data integration approaches.

The most common conceptual data model is the E/R model in which two concepts (*entity set* and *relationship*) are used to model the subject domain. By means of this model, it is impossible to define the behavior of the conceptual entities. In addition, this model does not support means of extension.

UML offers much the same capabilities as the E/R model, with the exception of multiway relationships. Basic construction for modeling the subject domain is *class*. In contrast to the E/R model, it provides the ability to define the behavior of conceptual entities. There are three common extensibility mechanisms that are defined by the UML: stereotypes, tagged values, and constraints. The extensibility mechanisms allows to extend the UML by adding new building blocks, creating new properties, and specifying new semantics in order to make the language suitable for specific problem domain.

In the XML data model, basic construction to model the subject domain is *element*. The XML data model, like the E/R model, does not provide means to define the behavior of the conceptual entities. In contrast to the above considered data models, the XML data model supports means of extension. Extension is reduced to the creation of a new DTD. Due to this property, the use of the XML data model as the conceptual data model is preferred.

In [11], the methods and tools for equivalent data model mapping construction is proposed. The canonical data model is expanded axiomatically. In works [12–16], relational data sources are considered in the frame of the traditional approach of the data integration as well as in the frame of the paradigm of ontology-based data access and integration. In [17] an analysis to use machine learning techniques to solve different problems to integrate unstructured and semistructured data is provided.

In our case, a single concept (*attribution*) is used to model the conceptual entities. Like the XML data model, the considered conceptual data model is extensible. As UML, it provides the ability to define the behavior of the conceptual entities. The extensibility property is an argument to use this conceptual data model as a canonical data model for integrating arbitrary data sources. We used the considered conceptual data model as a canonical data model for supporting the data integration concept [6]. The conceptual level means are sufficient to model the data integration concepts proposed in the above considered works. Modeling means of the conceptual level are insensitive to the extension of the considered conceptual data model.

5   Conclusions

The investigation subjects of this paper are the implementation issues of a conceptual data model. The result of our research is an approach to implement the concept of the considered conceptual data model. A distinguishing feature of the discussed conceptual data model is its stratification into local and global levels to model the conceptual entities. On the local level, homogeneous representation of heterogeneous data sources as basic conceptual entities is provided. The global level is intended to define derived conceptual entities. A single concept is used to model subject domains, namely, hierarchical relation. The conceptual data model concept is based on the behavioral and data definition symbols to model conceptual entities. The considered behavioral symbols are analogues to relational algebra operations. The data definition symbols are used to support hierarchical relations and as a rule are nullary functions. The result of formalization of these symbols is CDs. An important feature of the considered conceptual data model is its extensibility property. The extension is achieved by introducing new symbols into the conceptual data model. Thus, the result of extension of the considered conceptual data model is reduced to create new CDs to support new concepts. The extensibility property provides the ability to conceptually model arbitrary data sources. A non-formal example of extending the conceptual data model to support the data integration concept is considered. The outcome of such extension is a new CD to support the data integration concept. A computionally complete language is used for data modeling in the frame of the considered conceptual data model. The conceptual data model is defined as the union of all CDs. As a development of this work, it is planned to construct an extension of the considered conceptual data model to support big data integration concept.

A   The *cdm* Content Dictionary File

<CD>
<CDName> *cdm* < /CDName>
<Description>
       This CD defines the symbols of conceptual data model.
< /Description>

<CDDefinition>
     <Name> *R* < /Name>
     <Description>
     This symbol represents the set of hierarchical relations schemas.
     < /Description>

<CMP>$x \in R \Leftrightarrow attribution(x,\ type\ A,\ S_1\ A_1,\ S_2\ A_2,...,\ S_k\ A_k), k \geq 0$<
/CMP>
< /CDDefinition>

<CDDefinition>
    <Name> $card$ < /Name>
    <Description>
    This symbol represents the cardinality number.
    < /Description>
    <CMP> $card \in \{?, *, +\}$ < /CMP>
< /CDDefinition>

<CDDefinition>
    <Name> $sequence$< /Name>
    <Description>
    This symbol represents the sequence of the attribution objects.
    < /Description>
    <CMP> $R^* \rightarrow R$ < /CMP>
< /CDDefinition>

<CDDefinition>
    <Name> r < /Name>
    <Description>
    This symbol represents the set of hierarchical relations.
    < /Description>
    <CMP> $R \rightarrow r$ < /CMP>
< /CDDefinition>

<CDDefinition>
    <Name> $union$ < /Name>
    <Description>
    An n-ary associative union operation.
    < /Description>
    <CMP> $r^{*assoc} \rightarrow r$ < /CMP>
< /CDDefinition>

<CDDefinition>
    <Name> $minus$ < /Name>
    <Description>
    A difference operation.
    < /Description>
    <CMP> $r \times r \rightarrow r$ < /CMP>
< /CDDefinition>

   ...

< /CD>

B   The *cdm* Signature File

\<CDSignatures type = "sts" cd = "*cdm*">

\<!−− Definition of signature of the R symbol. This symbol defines a finite set of all hierarchical relations schemas in the frame of conceptual data model.−− >
\<Signature name = "R">
   \<OMOB>
     \<OMV name = "Set"/ >
   \< /OMOB>
\< /Signature>

\<! − − Definition of signature of the *card* symbol. The value of this symbol is the cardinality number.−− >[3]
\<Signature name = "card">
   \<OMOB>
     \<OMS name = "*attribution*" cd = "sts"/ >
   \< /OMOB>
\< /Signature>

\<! − − Definition of signature of the *sequence* symbol. This symbol defines the sequence of the attribution objects. −− >[4]
\<Signature name = "sequence">
   \<OMOB>
    \<OMA>
     \<OMS name = "mapsto" cd = "sts"/ >
    \<OMA>
     \<OMS name = "nary" cd = "sts"/ >
     \<OMS name = "R" cd = "*cdm*"/ >
    \< /OMA>
     \<OMV name = "List" / >
    \< /OMA>
   \< /OMOB>
\< /Signature>

\<!−− Definition of signature of the r symbol. This symbol defines a finite set of all hierarchical relations expressible in the frame of conceptual data model.−− >
\<Signature name = "r">
   \<OMOB>
     \<OMV name = "*Set*"/ >
   \< /OMOB>
\< /Signature>

---

[3] An *attribution* object consists of pairs of keys and values. The use of the symbol *attribution* in a signature indicates that the symbol is to be used as a key.

[4] The symbol *nary* constructs a child of *mapsto* which denotes an arbitrary number of copies of the argument of *nary*.

```
<! - - Definition of signature of the union function. An n-ary commutative
function union.-- >
<Signature name = "union">
    <OMOB>
     <OMA>
      <OMS name = "mapsto" cd = "sts"/ >
      <OMA>
       <OMS name = "nassoc" cd = "sts"/ >
       <OMS name = "r" cd = "cdm"/ >
      < /OMA>
      <OMS name = "r" cd = "cdm"/ >
     < /OMA>
    < /OMOB>
< /Signature>
```

```
<! - - Definition of signature of the minus function. The value of this function
is the difference between two sets.-- >
<Signature name = "minus">
    <OMOB>
     <OMA>
      <OMS name = "mapsto" cd = "sts"/ >
      <OMS name = "r" cd = "cdm"/ >
      <OMS name = "r" cd = "cdm"/ >
      <OMS name = "r" cd = "cdm"/ >
     < /OMA>
    < /OMOB>
< /Signature>

...

< /CDSignatures>
```

References

1. Sharma, S., Tim, U.S., Wong, J., Gadia, S., Sharma, S.: A brief review on leading big data models. Data Sci. J. **13**, 138–157 (2014). https://doi.org/10.2481/dsj.14-041
2. Date, C.J.: An Introduction to Database Systems, 8th edn. Addison-Wesley, USA (2004)
3. Golshan, B., Halevy, A., Mihaila, G., Tan, W.: Data integration: after the teenage years. In: Proceedings of the 36th ACM SIGMOD-SIGACT-SIGAI Symposium on Principles of Database Systems, pp. 101–106, (2017). https://doi.org/10.1145/3034786.3056124
4. Gamper, J., Ceccarello, M., Dignös, A.: What's new in temporal databases?. In: Chiusano, S., Cerquitelli, T., Wrembel, R. (eds.) Advances in Databases and Information Systems. LNCS, Springer, vol. 13389, pp. 45–58 (2022). https://doi.org/10.1007/978-3-031-15740-0_5
5. Sadalage, P.J., Fowler, M.: NoSQL Distilled, 2nd edn. Addison-Wesley, USA (2013)

6. Manukyan, M.G.: On an extensible conceptual data model by a non-formal example. In: Chiusano, S., et al. New Trends in Database and Information Systems. Communications in Computer and Information Science, Springer, vol. 1652, pp. 3–13 (2022). https://doi.org/10.1007/978-3-031-15743-1_1
7. Dewar, M.: OpenMath: an overview. ACM SIGSAM Bull. **34**(2), 2–5 (2000)
8. Hindley, J.R., Seldin, J.P.: Introduction to Combinators and λ-Calculus. Cambridge University Press, Great Britain (1986)
9. Davenport, J.H.: A small OpenMath type system. ACM SIGSAM Bull. **34**(2), 16–21 (2000)
10. Garcia-Molina, H., Ullman, J., Widom, J.: Database Systems: The Complete Book, 2nd edn. Prentice Hall, USA (2009)
11. Kalinichenko, L.A.: Methods and tools for equivalent data model mapping construction. In: F. Bancilhon, et al. (eds.) Advances in Database Technology-EDBT 1990, pp. 92–119. Springer, Italy (1990). https://doi.org/10.1007/BFb0022166
12. Calvanese, D., De Giacomo, G., Lembo, D., Lenzerini, M., Rosati, R.: Tractable reasoning and efficient query answering in description logics: the $DL - Lite$ family. JAR **39**(3), 385–429 (2007). https://doi.org/10.1007/s10817-007-9078-x
13. Calvanese, D., De Giacomo, G., Lembo, D., Lenzerini, M., Rosati, R.: Ontology-based data access and integration. In: Liu, L., Özsu, M.T. (eds.) Encyclopedia of Database Systems, pp. 1–7. Springer, New York (2017). https://doi.org/10.1007/978-1-4614-8265-9_80667
14. Calvanese, D., De Giacomo, G., Lenzerini, M., Vardi, M.Y.: Query processing under GLAV mappings for relational and graph databases. Proc. VLDB Endowment **6**(2), 61–72 (2012)
15. Calvanese, D., et al.: The mastro system for ontology-based data access. Semant. Web **2**(1), 43–53 (2011)
16. Console, M., Lenzerini, M.: Data quality in ontology-based data access: the case of consistency. In: Proceedings of the Twenty-Eighth AAAI Conference on Artificial Intelligence, AAAI 2014, vol. 28, pp. 1020–1026 (2014)
17. Dong, X.L., Rekatsinas, T.: Data integration and machine learning: a natural synergy. Proc. VLDB Endowment **11**(12), 2094–2097 (2018). https://doi.org/10.14778/3229863.3229876 https://doi.org/10.14778/3229863.3229876 https://doi.org/10.14778/3229863.3229876

Construction of a Personal Knowledge Graph in a Digital Semantic Library LibMeta

Olga Ataeva[✉] ⓘ, Vladimir Serebryakov ⓘ, and Natalia Tuchkova ⓘ

FRS «Computer Sciences and Control», Russian Academy of Sciences, Vavilov Street, 40,
Moscow 119333, Russia
oli.ataeva@gmail.com

Abstract. The paper discusses an approach to constructing a personal knowledge graph based on an ontological description of a scientific subject area and its data, presented in the form of a knowledge graph. The presentation is based on concepts related to information and data mining, such as subject domain ontology, scientific subject area, thesaurus, semantic digital library, personal knowledge graph.

The procedure for constructing a personal knowledge graph is presented using the example of the mathematics subject area, the ontology of which is based on extensive material from the mathematical encyclopedia edited by academician I.M. Vinogradov within the framework of the semantic library LibMeta. The proposed approach will make it possible to use the content of the Mathematics semantic library for scientific research, minimizing the process of searching for information in the local subject area, without losing more general results contained outside this area.

This work refers to the experience of building semantic libraries based on thesauruses and ontological design. Construction of ontologies based on the thesaurus of the subject area is a necessary condition for constructing a personal knowledge graph of the scientific subject area.

Keywords: Knowledge graph · Subject domain ontology · Semantic digital library LibMeta · Personal knowledge graph

1 Introduction

Modern research in science is unthinkable without digitized data, publications, audio and video materials. This entire arsenal helps you quickly obtain information, follow the development of ideas from the scientific community, and move forward in your own research. However, the problem of obtaining and maintaining reliable information remains, which is especially important in scientific research. Anyone who works in science knows the situation when you have to spend time searching for previously used sources and even information about your own publications. In the meantime, more and more new works are appearing, which you also need to familiarize yourself with so as not to lose the relevance of the research. Therefore, the idea of organizing a researcher's personal information space was repeatedly discussed and began to be implemented in

J. Baixeries et al. (Eds.): DAMDID/RCDL 2023, CCIS 2086, pp. 65–76, 2024.
https://doi.org/10.1007/978-3-031-67826-4_5

the "pre-ontological" era in the form of a desktop metaphor defined by the graphical system of the Windows OS (https://www.microsoft.com/) in the 90s of the last century. Organizing documents into folders on a digital desk is an idea that helps all personal computer users find documents better than on their real desk. Now the capacity of a personal computer is not enough, and technologies have appeared that make it possible to build entire chains for searching for what you need and store them in digital libraries, thereby increasing the speed and reliability of finding.

The knowledge graph (KG) [1, 2] and its version in the form of the user's personal knowledge graph represents such a modern technology, which is implemented within the framework of this work in the semantic library LibMeta [3] based on the ontological representation of scientific subject areas [4–7].

In this work, the scientific field "Mathematics" is considered as a subject area. The LibMeta library contains extensive material on classical mathematics, in particular mathematical analysis, including a mathematical encyclopedia edited by academician Ivan Matveevich Vinogradov [8], an encyclopedia of mathematical physics [9] edited by academician Ludwig Dmitrievich Faddeev, the author's thesaurus on the subject area of ordinary differential equations (ODE))[10] and the dictionary of special functions [11], MSC (https://msc2020.org) and UDC (https://teacode.com/online/udc) classifiers. LibMeta is filled by adding thesauruses and ontologies of subject areas based on modern publications and classical primary sources.

The formal representation of data and their connections (data model) of the subject area in the LibMeta semantic library is based on an ontological approach. Today, a huge number of ontologies have been developed for subject areas that cover different aspects of human activity. By definition [12–14], the mandatory properties of describing knowledge in the form of an ontology are:

(1) a definitive and controlled vocabulary of concepts and terms that excludes their ambiguous interpretation;
(2) a strict hierarchy of relations between subclasses of concepts and terms that describe knowledge of the subject area.

The construction of a domain ontology makes it possible to identify *metadata for the design of specific data structures* of scientific subject areas and options for managing this data. To do this, it is necessary to structure and link various *resources*, extract from them and contextualize (define in context) *data*, giving them the *properties of knowledge*. To define the subject area, an ontology is built within which various data sources [15, 16] can be integrated and various taxonomies [17] of concepts and terms verified by recognized experts in the scientific field can be used. The ontology of the Mathematics library [18] was based on the reference materials listed above.

2 Definitions

A knowledge graph is a graph whose purpose is to represent knowledge defined by entities and the relationships between them, extracted from various data sources [19, 20]. The properties of knowledge graph vertices, arcs (relationships between vertices) and methods of working with them are determined by applications and can change from application to application.

A personal knowledge graph is a knowledge graph built on the basis of a personal library and a (personal) thesaurus of this library, generated during the user's work and at the user's request. The personal knowledge graph is built based on the content of the user of the personal semantic library, reflecting the scientific interests of this user [21]. The user's content includes thesaurus concepts, articles, classifier codes that the user has defined as "his" and related resources.

A semantic digital library is a digital library whose content elements are connected by hierarchical and associative relationships in accordance with the domain ontology [10].

The ontology of a digital semantic library is a formal description of a set of data (types of data elements, relationships) of a subject area. The W3C Web Ontology language (OWL, https://www.w3.org/OWL/) is used to define ontologies. An ontology in OWL is an Resource Description Framework (RDF, https://www.w3.org/RDF/) graph. An OWL description is a description of the data structure, not the data itself. Each data instance is an instance of an ontology element. In the LibMeta library, a thesaurus is represented as an ontology. LibMeta editing tools are ontology editing tools.

Developing semantic libraries, special attention is paid to the data model of the library contents. At the same time, the content of digital libraries can be described in various formats and presented in various ways. A library, defined using the LibMeta system, is considered as a repository of structured diverse data with the ability to integrate it with other data sources and assumes the ability to specify its content by describing the subject area. The ontology of the semantic library content acts as a tool of formalization [22–24].

Solving the problem of ontology design, we need to use metadata at different levels:

- metadata as universal concepts of the data structure of a digital library;
- metadata as part of the description of objects of an application domain or a subset of an application domain;
- application domain metadata.

In such an ontology, at the top level, concepts are used that are essentially related to high-level ontologies and are not related to the specifics of any particular subject area. At the second level, concepts are used that describe the subject area, while being instances of classes defined at the first level, but at the same time used as class definitions to describe data of the third level already in a specific subject area. The definition of a subject area is given by a thesaurus, which contains the main terms of this subject area, connected by hierarchical and horizontal connections. The contents of the library are specified by resource types, the description of which specifies a set of valid objects, possibly combined into various collections that, together with the thesaurus, make up its content.

In the ontology of a digital library, one can distinguish a "system" part, which in one way or another describes the structure of the library itself, and a "user" part, which describes the structure of the subject area data loaded into the library. The thesaurus is, in a sense, the top part of the user ontology, fully and explicitly visible to the user.

The digital library ontology defines the data structure of the library's content. Each data element loaded into the library can be associated with an ontology vertex, which determines the position of the data element in the ontology (the "data type" of the element). Based on the ontology connections and connections determined at the design stage, it is possible to construct a data graph, the structure of which is determined by the ontology: vertices (articles) are instances of ontology elements, connections are thesaurus connections. This is the knowledge graph of a digital library.

The user interface is also built on the basis of the personal library thesaurus. When constructing, displaying and using (navigating) a knowledge graph, the connections of each node of the graph are the connections defined in the thesaurus. They appear as links to related nodes for the current node (Fig. 1).

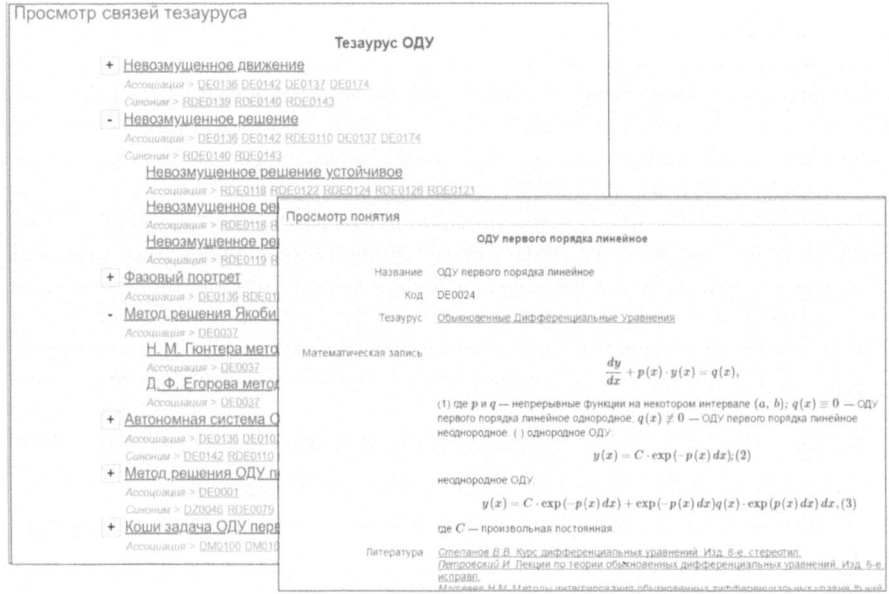

Fig. 1. The connections of each node of the graph are the connections defined in the thesaurus. They appear as links to related nodes for the current node

Thus, the role of ontology in the process of designing and operating a digital library can be summarized as follows:

- Based on the ontology, a library knowledge graph database is built;
- Based on the ontology, namely, the objects and relationships defined in the thesaurus, the user interface of the library is built.

The language for defining the ontology of the LibMeta digital semantic library is OWL [25].

3   Personal Knowledge Graph in LibMeta

The basis of the knowledge graph in LibMeta is the conceptual knowledge graph of a mathematical encyclopedia. Scaling the ontology (knowledge graph) determines the procedures for completing the graph and extracting the sub-graph. Completing the graph, data processing methods can be used for adding nodes. For example, the procedure for including a publication as a node in the knowledge graph in Libmeta involves adding several types of nodes, such as person, publication, affiliation, new thesaurus concept, etc. The extracting a subgraph procedure involves determining the central node type of the subgraph and then the specific node, the depth of the extracted subgraph is determined. For example, the choice of node type is determined during the search by specifying the "search only publications" restriction; the depth of the subgraph determines possible recommended nodes that can be displayed as additional/refining search results.

Creating a personal information environment for scientific work imposes certain conditions on the functionality of the semantic library. Let us dwell on the features of such an information resource in the context of the LibMeta semantic library.

A fundamental role in the creation and functioning of a library is played by its ontology, the top level of which is represented by a thesaurus. A thesaurus can be created in two ways:

1) loading an existing thesaurus, and
2) through the user interface.

In any case, the thesaurus can be edited by the user.

The semantic library LibMeta is positioned by the authors as personal. The "personality" of LibMeta is manifested in the fact that the user has the opportunity to create his own copy of the library with his own thesaurus and content.

A personal knowledge graph is a knowledge graph built on the basis of a personal library and the (personal) thesaurus of this library, formed during the user's work and at the user's request.

Let's consider when and how the need for a personal knowledge graph arises and how it differs from a search query.

3.1 When the Need for a Personal Graph Arises

A personal knowledge graph is a limited selection of data depending on the user's interests. Despite the fact that there is an association with search queries and its results in the classical sense, the significant difference is that the result of a classical search is the resources that most closely match the search query at the level of matching phrases and words. In the personal knowledge graph, thanks to semantic connections (which are specified at the library ontology level) between resources, we obtain the most complete slice of data related thematically at different levels, which allows us to achieve a wider scope of knowledge.

It is one thing to find among the publications those that are devoted to boundary value problems using the query "boundary value problems" in the search results; it is another thing to "see" connections with publications that discuss applications and solutions of such problems, for example, in the field of elasticity theory or connections with publications in the field of composite materials. These are different tasks, where in the second case, the data complements and enriches the results of search and research.

In order to be able to use these connections and explore them in the process of studying the subject area, there is a need for a personal knowledge graph.

Thus, when forming a personal knowledge graph, the user specifies the characteristics (objects) by which the library should compose information upon request.

For example, when searching, such objects can be various resources specific to the subject area, for example, information about persons, publications, terms, tasks, applications, solutions, etc.; they will form the vertices of the graph along which navigation will be carried out. When the graph is replenished, a search object can be added, which can become a new node for navigation.

3.2   How a Personal Knowledge Graph Is Created

Initially, when a search request is made in the library, it is processed and reformulated using synonyms [26] and the thesaurus terms are searched in it. After this, the user is given a list of results found and, thanks to the connections of the knowledge graph, is asked to determine/clarify the further direction of the search. Based on the connections of the subject area, with the search string "boundary value problems" a clarifying query "integral boundary value problems" can be proposed, but a clarifying query of the form "integral boundary value problem" will never arise, although the words "integral" and "boundary value problem" can be defined as synonyms by definition from [26], as co-occurring in the same context. That is, at this stage (query) the terms of the subject area are actually defined, which define the description of the personal knowledge graph and the connections between them based on the search query.

When replenishing the knowledge graph by adding a new node, new terms and concepts are added that are included in the thesaurus of the subject area and/or existing ones are used to establish connections and integrate the new node into the knowledge graph. That is, just as in the previous case (search query), subject area terms are defined that define the description of the personal knowledge graph.

As a result, the user in both cases can capture a set of terms corresponding to a new node or search query, combining them into a group and specifying their description. Or he can initially describe such a variety of concepts and thus begin to form his personal knowledge graph.

Figure 2 schematically shows a small fragment of a personal knowledge graph based on a personal thesaurus, the vertex of which is labeled "Text description/query". This text description (as a result of the user's search query) can be associated with thesaurus concepts and publications. As an example, we can consider the text description of the ODE thesaurus article. Such a thesaurus entry may include "keywords," "classifier code," "thesaurus concept," definitions, and navigable links. Explicit connections between these components are marked with solid lines in Fig. 2. In this case, a concept can be included in different publications (marked as "P1"), and also be related to other concepts (marked as "C1") by thesaurus relationships and have many keywords, (marked as "K1"). The "classifier code" and "keywords" may have the same connections. In this case, the sets P1, P2, P3 can intersect, which means that there is a connection between the ancestor nodes. The same is true for the sets K1, K2, K3 and C1, C2, C3. For example, a relationship may arise between a "classifier code" and a "concept" if the sets P1 and P2 contain a publication that includes this concept and is tagged with some classifier code. Each instance of the sets Pi, Ci, Ki can be the top of a subtree, which allows you to extract/establish new connections between graph elements within the thesaurus subject area at different levels.

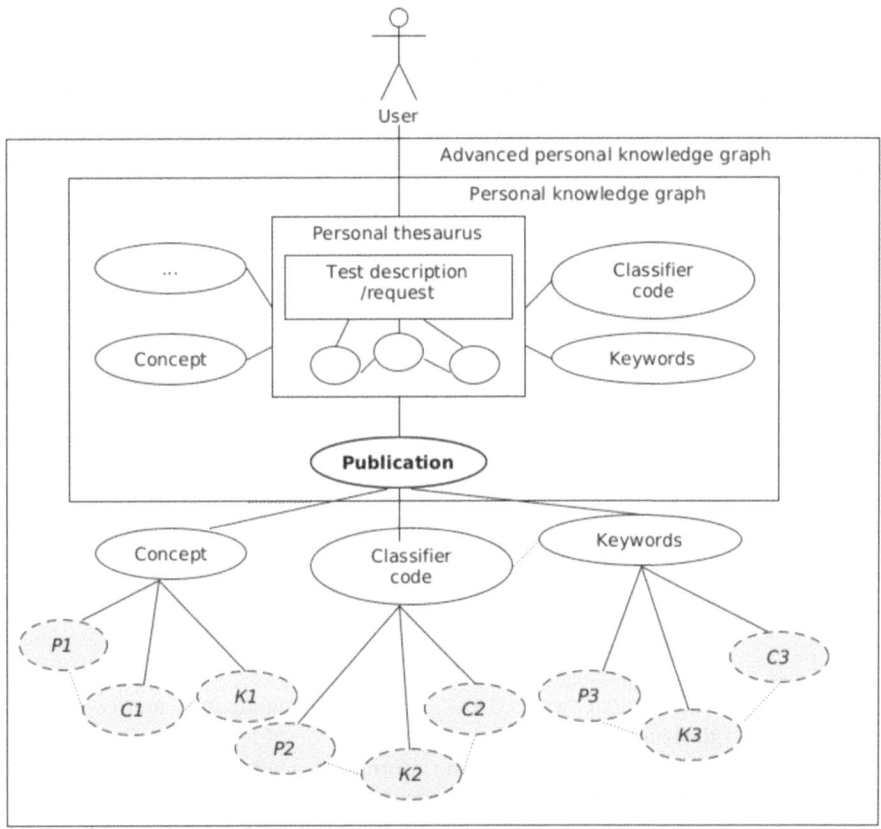

Fig. 2. Personal knowledge graph and its connections.

As an example, consider the work of a journal editor in a personal library environment created on the basis of the LibMeta ontology.

It is important for the editor to see articles corresponding to a specific subject area in the array of journal publications, how they are related to the main definitions of the mathematical encyclopedia and classifiers. The task is to find out whether there are contradictions there, since the inconsistency of terms with domain classifiers leads to difficulties for search queries and indexing in databases. Based on the connections in Fig. 2, all inconsistencies are easily revealed, and "essential" explanations, if desired, can be obtained by turning from the "concept" to encyclopedias. The editor sees the articles and analyzes the correctness of the use of classifier codes.

In Fig. 3 shows a screenshot of a personal library based on the content of the journal "Mekhanika Kompozitsionnykh Materialov i Konstruktsii" (MKMK, http://journal.iam ras.ru/). Keywords are associated with many publications and by moving through them the editor navigates through the "personal knowledge graph of the MKMK editor."

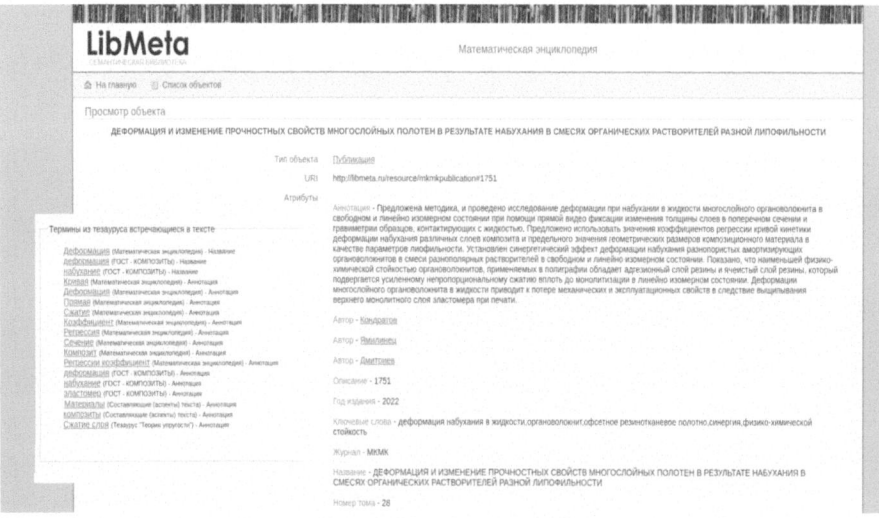

Fig. 3. Example of a publication page from MKMK with navigation

Figures 4–5 demonstrated two variants subgraphs of *publication* and *concept* as central nodes in LibMeta library.

The choice of central node depends on the information needs of the user. This choice is made at the request stage on the interface page. If the user is interested in the *publication/concept* and the connections of this object, then the *publication/concept* becomes the main nodes. Next, generally speaking there is a possibility to set the depth of connections, in accordance with the structure of the thesaurus of the subject area in the LibMeta library.

Figure 4 shows a visualization of a subgraph whose central node is a *publication* (labeled ***pub***). This publication is related to thesaurus *concepts* (*concepts* are labeled ***con*** and thesauruses containing them are labeled ***thes***). Which in turn are linked to other publications and classifier codes (marked as ***tax***).

Figure 5 shows a visualization of a subgraph whose central node is a concept of thesaurus (labeled ***con***). This concept is related to other thesaurus concepts, keywords (labeled ***key***), classifier code (labeled ***tax***). Which in turn are linked to other publications and classifier codes. In the figure, the further the nodes are located from the center of the subgraph, the lighter their background.

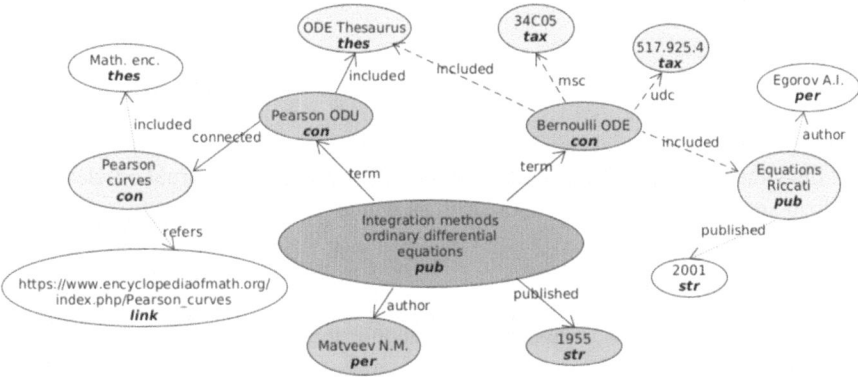

Fig. 4. The publication with connections

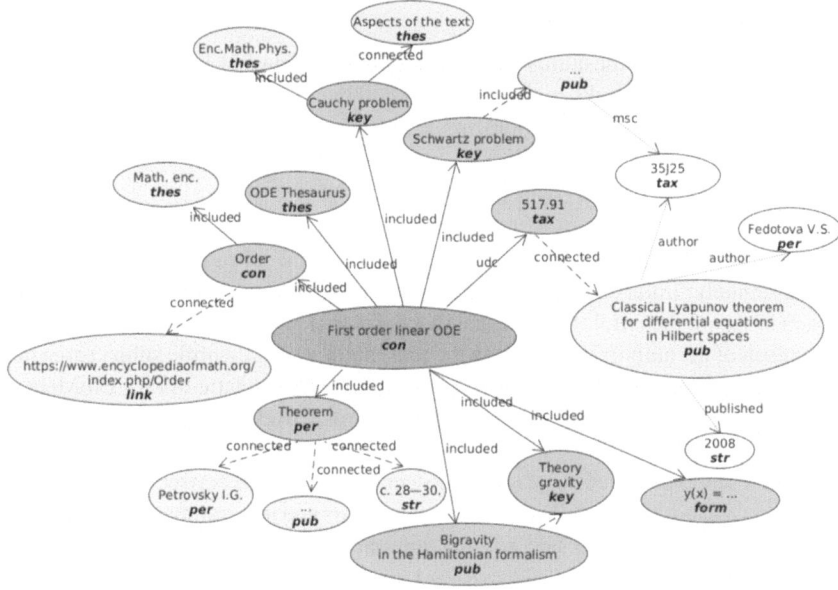

Fig. 5. The concept with connections

4 Statistics on the Knowledge Graph "Mathematics" in LibMeta Library

With the initial data set in the form of a list of encyclopedia articles [8] (about 6000) and a set of publications (about 5000), using the MSC and UDC classifiers, a knowledge graph was constructed in which the number of nodes in the data graph is 30628, of which:

- library data nodes – 13860:

- Journal articles – 4462,
- Persons – 4285,
- Formula – 4884.

The total number of node attributes is 83,616, of which 9,282 are keywords. At the same time, there are 13,758 connections between keywords and journal articles.

There are 16768 nodes in the thesaurus, of which

- encyclopedic nodes 8462,
- classifier nodes 8306.

Attributes of encyclopedic nodes 25166, connections in classifiers 5922. Explicit connections:

- encyclopedic nodes with data nodes 3873,
- classifier nodes with data nodes 6307.

At the same time it was built:

- connections between thesaurus concepts and library data 53139,
- number of horizontal links in the thesaurus 88005,
- number of hierarchical connections in the thesaurus 1957.

5   Conclusion and Further Research

The work meets the challenges facing modern scientific researchers. It offers to the user of a semantic library to organize a personal information scientific space using an approach based on the ontological design of a personal knowledge graph. This approach was the result of the authors' research on the presentation of a scientific subject area in the information environment. A pilot version of the personal semantic library LibMeta was implemented with downloaded content from the journal "Mekhanika Kompozitsionnykh Materialov i Konstruktsii".

Further research is aimed at establishing deeper connections between the nodes of the knowledge graph based on intellectual content analysis and integrating these connections into the library thesaurus and, accordingly, into the user interface.

The combination of expert (academic) knowledge and data analysis mechanisms using artificial intelligence algorithms constitutes the further direction of development of the LibMeta semantic library.

Current problems in the development of representation of scientific subject areas based on personal knowledge graphs include the following:

- deepening the classification of subject areas;
- creation of reference text corpora for subject areas;
- developing assessments to identify trends in subject areas over time;
- use of knowledge graphs to improve machine "understanding" of scientific texts;
- data collection into a knowledge graph in real time;
- obtaining "quality" knowledge from peer-reviewed primary sources;
- providing scientific communications based on personal knowledge graphs.

Acknowledgments. The work was presented within the framework of the Russian Ministry of Education and Science theme "Mathematical methods of data analysis and forecasting" of the Federal Research Center Institute of Management of the Russian Academy of Sciences.

The authors have no competing interests to declare that are relevant to the content of this article.

References

1. Paulheim, H.: Knowledge graph refinement: a survey of approaches and evaluation methods. Semantic Web J. (Preprint). 1–20, **8**(3), 489–508 (2016). https://doi.org/10.3233/SW-160218

2. Ataeva, O., Serebryakov, V., Tuchkova, N.: Ontological approach to a knowledge graph construction in a semantic library. Lobachevskii J. Math. **44**(6), 2229–2239 (2023). https://doi.org/10.1134/S1995080223060471

3. Ataeva, O., Serebryakov, V.A., Sinelnikova, E.: Thesaurus and ontology building for semantic library based on mathematical encyclopedia, DAMDID/RCDL 148–157 (2019)

4. Ataeva, O., Serebryakov, V.A.: Ontologiya cifrovoj semanticheskoj biblioteki Lib-Meta. Informatika i ee primeneniya **12**(1), 2–10 (2018)

5. Serebryakov, V.A., Ataeva, O.M.: Ontology based approach to modeling of the subject domain '"Mathematics"' in the digital library. Lobachevskii J. Math. **42**(8), 1920–1934 (2021). https://doi.org/10.1134/S199508022108028X

6. Allemang, D., Hendler, J., Gandon, F.: Semantic Web for the Working Ontologist. ACM Books (2020)

7. Elizarov, A.M., Kirillovich, A.V., Lipachyov, E.K., Nevzorova, O.A.: ONTOMATHPRO – ontology of mathematical knowledge. Doklady Math. **507**(1) 29–35 (2022). https://doi.org/10.31857/S2686954322700011

8. Vinogradov, I.M. (red.), Matematicheskaya enciklopediya (v 5 tomah) M.: Sovetskaya enciklopediya (1977—1985)

9. Matematicheskaya fizika. Enciklopediya / Gl. red. M34 L. D. Faddeev. M.: Bol'shaya Rossijskaya enciklopediya (1998)

10. Moiseev, E.I., Muromskij, A.A., Tuchkov, N.P.: Tezaurus informacionno-poiskovyj po predmetnoj oblasti: obyknovennye differencial'nye uravneniya. MAKS Press, Moscow (2005)

11. Brychkov, Y.: Handbook of special functions: derivatives, integrals, series and other formulas. CRC Press (2008). https://doi.org/10.1201/9781584889571

12. Hlav,a M.M.K.: The taxobook: history, theories, and concepts of knowledge organization, part 1 of a 3-part series. Synth. Lect. Inf. Concepts Retr. Serv. **6**(3), 1–80 (2014)

13. Hlava, M.M.K.: The taxobook: principles and practices of building taxonomies, part 2 of a 3-part series. Synth. Lect. Inf. Concepts Retr. Serv. **6**(4), 1–164 (2014)

14. Hlava, M.M.K.: The taxobook: applications, implementation, and integration in search: part 3 of a 3-part series. Synth. Lect. Inf. Concepts Retr. Serv. **6**(4), 1–156 (2014)

15. Gavrilova, T.A., Horoshevskii, V.F.: Bazy znanij intellektual'nyh sistem. Piter, St Petersburg (2000)

16. Rabunal, J.R., Dorado, J., Sierra, A.P.: Encyclopedia of Artificial Intelligence (IGI Global) (2009). https://doi.org/10.4018/978-1-59904-849-9

17. Dextre Clarke, S.G., Zeng, M.L.: Standard Spotlight: From ISO 2788 to ISO 25964: the evolution of thesaurus standards towards interoperability and data modeling. Inf. Standards Q. (ISQ) **24**(1), 20–26 (2012). https://doi.org/10.3789/isqv24n1.2012.04

18. Ataeva, O., Serebryakov, V., Tuchkova, N.: Development of the semantic space 'Mathematics' by integrating a subspace of its applied area. Lobachevskii J. Math. **43**(12), 29–40 (2022)

19. Kroetsch, M., Weikum, G.: Journal of Web Semantics: Special Issue on Knowledge Graphs (2016). http://www.websemanticsjournal.org/index.php/ps/announcement/view/19, last access 2023/12/26

20. Blumauer, A.: From Taxonomies over Ontologies to Knowledge Graphs (2014). https://blog.semanticweb.at/2014/07/15/from-taxonomies-over-ontologiesto-knowledge-graphs, last access 2023/12/26

21. Pinker, S.: The Sense of Style: The Thinking Person's Guide to Writing in the 21st Century. Penguin Books, London (2015)

22. Gruber, T.R.: The role of common ontology in achieving sharable, reusable knowledge bases. In: Allen, J.A., Fikes, R., Sandewell, E. (eds.) Proceedings of the Second International Conference on Principles of Knowledge Representation and Reasoning (KR'91), pp. 601–602 (1991). https://doi.org/10.5555/3087158.3087222

23. Vrandecic, D.: Ontology evaluation. In: Staab, S., Studer, R. (eds.) Handbook on Ontologies, International Handbooks on Information Systems, pp. 293–313 (2009). https://doi.org/10.1007/978-3-540-92673-3_13

24. Semantic Web. https://www.w3.org/standards/semanticweb, last access 2023/12/26

25. https://www.w3.org/2001/sw/wiki/OWL, last access 2023/12/26

26. Ataeva, O.M., Serebryakov, V.A., Tuchkova, N.P.: On Synonyms Search Model. In: CEUR Workshop Proceedings, M. Jeusfeld c/o Redaktion Sun SITE, Informatik V, RWTH Aachen (Aachen, Germany), 3066, pp. 13–22 (2021)

Global Cognitive Graph Properties Dynamics of Hippocampal Formation

Konstantin Sorokin[1]($\boxtimes$), Andrey Zaitsew[2], Aleksandr Levin[1], German Magai[1], Maxim Beketov[1], and Vladimir Sotskov[3]

[1] Higher School of Economics, International Laboratory of Algebraic Topology and Its Applications, Moscow, Russia
`ksorokin@hse.ru`
[2] Higher School of Economics, Faculty of Computer Sciences, Moscow, Russia
[3] Laboratory of Neuronal Intelligence Institute for Advanced Brain Studies, Lomonosov Moscow State University, Moscow, Russia

Abstract. In the present study we have used a set of methods and metrics to build a graph of relative neural connections in a hippocampus of a rodent. A set of graphs was built on top of time-sequenced data and analyzed in terms of dynamics of a connection genesis. The analysis has shown that during the process of a rodent exploring a novel environment, the relations between neurons constantly change which indicates that globally memory is constantly updated even for known areas of space. Even if some neurons gain cognitive specialization, the global network though remains relatively stable. Additionally we suggest a set of methods for building a graph of cognitive neural network.

Keywords: hippocampal formation · cognitome · spatial navigation · network analysis

1 Introduction

The idea of building a graph of connections in terms of contextual connectedness is inspired by ideas of functional connectome and cognitome of K.V. Anokhin [1]. The initial dataset we are working with are neuronal activities of CA1 region of hippocampal formation registered using calcium imaging. Then using different correlation dynamics of the neuronal dynamics time series the graph is constructed and natural graph properties are studied. We intentionally didn't focus on neurons' specializations and checked if the properties of global networks would change while an animal is performing a learning of new environment. We have already shown in the previous study that neurons (place cells to be particular) have quite rapid tuning dynamics [2] in terms of specialization. Current study is aimed at answering the question: does it affect the global properties of a neural network, or, say, cognitive map? Thus question comes naturally from the referenced study as it is shown there that some neurons, which are attached to certain places instantly gain their specialization, whereas some need several

J. Baixeries et al. (Eds.): DAMDID/RCDL 2023, CCIS 2086, pp. 77–87, 2024.
https://doi.org/10.1007/978-3-031-67826-4_6

more visits of a certain place on an arena to stabilize. We have also previously showed that cognitive map is very useful in tasks of topology reconstruction of the ambient environment using place cells activities only [3]. In this study we also used a time series of preselected place cells as one of methods of constructing the covering of a arena, where an animal was freely navigating. This method showed to be meaningful but global network forming dynamics is still yet to be learned. Studying the graph of neural network contextual coactivation connections would help understanding the process of learning (in general) and exploring novel environment of arena by a rodent. Looking at such graphs in dynamics may be extremely useful for describing the way how rodent understands and learns the space around as well as how the space memory is translated into the synapses and synaptic strengths.

Having in mind all these results, we can discuss the way the network would perform, in general, on a cells level. It is clear that it has to retain some stability of it's properties, by design, and yet it should be efficient from the signal translation point of view. But how does it change when the network performs active reconstruction while animal is learning something new (we have shown [3] that it in fact does)? The efficiency of a network can be shown with the small-worldness [4], for example which is characterized by certain properties, like the clusterization coefficient and average path length.

The main principle of determining the connections properties between the active neurons and, hence, reconstructing the neural network graph, is STDP (Spike-timing dependent plasticity) [5,6]. The STDP suggests that if there are two neurons connected with a synapse and the presynaptic is fired after post-synaptic one, the synapse strength reduced and if vice-a-versa the synapse-strength is increased. In computer science words, the brain tries to greedily reduce path length.

Another important metric showing the effectiveness of the network is presence of, so called, rich-clubs [7]. Rich-clubs are observed in many biological systems, but for cognitome they are especially reasonable as a system of distributing information between different hierarchical subsystems. Such rich clubs may also migrate, i.e. change the neural connections, though we would expect our graph to have at least one subgraph with very rich collection of connections.

Additionally we expect that most of neurons will have synapses as the neural network should be connected. In case if the connections amount is too small, this means that actually neural network is unable to store and transmit information. Moreover, these networks wouldn't perform robustly in case of reconstruction of unexpected damage of some parts. Such networks would not give a significant results which would be useful in further research.

2   The Experiment and Data Preparation

2.1   Experimental Setup

During the experiment, four different mice (C57Bl/6J) were each placed in a cylindrical arena with obstacles. A single-photon detector was installed on each mouse's skull to record neuronal activity in the CA1 region of the hippocampus. An ion-dependent dye was used for detection, which fluoresces when interacting with calcium ions emitted in synapses. More detailed information on the surgical interruptions performed on animals can be found in our previous paper [3].

The experiment was conducted in three steps with the same four mice, and each day a different number of obstacles were placed in the arena. So, for each mouse three experiments were performed. While the mice explored the arena, their activities were recorded on a vertically oriented camera. The video with the movements of the animals was processed, adjusting the brightness and contrast to distinguish the mouse from other objects falling into the frame. Two colored LEDs were attached to each mouse's head to restore the direction of the animal's head.

The wire coming from the camera miniscope attached to the mouse's head was deliberately made white. Animal's movements were synchronized with the readings of the neural activity sensor and analyzed.

2.2   Highlighting of Neurons

Data from the camera attached to the miniscope is a black and white video recorded in 20 frames per second (see Fig. 1). There are multiple problems related to this: firstly, due to mechanical influences on the miniscope, the use of these recordings is impossible without post-processing like stabilization and the overall motion correction; secondly, this is not really a computer-friendly data format. In this work the CaImAn package [8] is used to solve these problems.

Fig. 1. Mouse and data from the miniscope attached to it

For the motion correction the NoRMCore is used. This algorithm divides the area into intersecting sections, which are processed separately, and are then combined using smooth interpolation.

After the stabilization the neurons are identified with CNMF-E, a one-photon extension of the CNMF framework, which stands for Constrained Nonnegative Matrix Factorization.

After that we obtain a smooth signal of each candidate expected to be a real neuron, which needs even more processing. We used several criteria to filter out the "fake" neurons which do not glow uniformly, have inconsistent activation patterns, etc. At the following step the chosen neurons are normalized and uniformed. Finally, we have a set of time series for each neuron represented as $a = \{a_1, a_2, \ldots, a_t | 0 \le a_i \ge 1\}$ and $i \in I$ which is set of all frames of video.

3   Graph Constructions and Similarity Metrics

The idea behind graph $(G = (V, E))$ construction is very simple and it is mostly about the matter of measuring similarity distance between vertices $V = \{0, 1, \ldots, n\}$ of a graph, which are exactly the neurons with their activity. The edges E then are directed and weighted and should be represented by a triplet $(a, b, \hat{\rho}) \in E$, where $a, b \in V$ and $\hat{\rho}$ is a weight of an edge. From neurobiological perspective it contains presynaptic neuron, post-synaptic neuron and the estimate of synaptic strength. In current definition graph is full, i.e. contains all possible edges. Such graph is unlikely to be useful in a research so we limit it's connections by the lower bound threshold ρ for the edge existence. then the set of edges is defined as $E = \{(a, b, \rho) | \rho(a, b) > \rho\}$.

However, we are interested in network dynamics, that's why we would slice the time series of neural activities. As previously mentioned we would like to design a metric which can assess the influence of one's neuron activity on another one. In order to do this we would need to compare the "previous" value of activity of presynaptic neuron candidate with activity on a post-synaptic candidate neuron. So we would need to shift our vectors by k entries (k should be adjusted with according to the discretization rate, in our case $k = 1$ unless stated otherwise). The shift operation is defined as follows: $s(\overrightarrow{U}, k) = (U_1, U_2, \ldots, U_{T-k})$, where $k \in \mathbb{Z}_{>0}$, $T = \dim(\overrightarrow{U})$. For $k \in \mathbb{Z}_{<0}$: $s(\overrightarrow{U}, k) = (U_{k+1}, U_{k+2}, \ldots, U_T)$. Intuitively saying the value i if $s(\overrightarrow{U}, k)$ would correspond to value $i + k$ in $s(\overrightarrow{u}, -k)$ for any $k \in \mathbb{Z}, k > 0$.

So, combining it into a value to measure the synaptic strength we define a function which maps two vectors of neural activity to some real number preserving synaptic strength. Also we should remember that our vectors needs to be shifted as the effect of presynaptic neuron on the post-synaptic neuron is slightly delayed. The function is defined as follows: $\rho \colon \mathbb{R}^{T-k} \times \mathbb{R}^{T-k} \to \mathbb{R}$, where T is the duration of the recording, and k is fixed number by which the time series is shifted (Intuitively saying this is our expectation on delay between the effect of presynaptic neuron on post-synaptic one).

In order to build the graph we have used the following algorithm for given ρ, ρ, k, N, where N is a matrix of all neural activities

For the function it's only a matter of defining the ρ, which we have several candidates for.

Algorithm 1. Graph construction

---

function BUILD GRAPH($\rho, \underline{\rho}, k, N$)
    $h \leftarrow \dim(N^1)$ ▷ Obtain a height of matrix
    $w \leftarrow \dim(N_1)$ ▷ Obtain a width of matrix
    $V \leftarrow \{1, \ldots, w\}$
    $E \leftarrow \{\}$
    for $i \in \{1, \ldots, h\}$ **do**
        for $j \in \{1, \ldots, h\}, i \neq j$ **do**
            $\tilde{\rho} \leftarrow \rho((s(N^{(i)}, k)), s(N^{(j)}, -k))$
            if $\tilde{\rho} \geq \underline{\rho}$ **then**
                $E \leftarrow E \bigcup \{(i, j, \tilde{\rho})\}.$
                $i + 1 \leftarrow i.$
            end if
        end for
    end for
    $Result \leftarrow (V, E)$
end function

3.1 Correlation

The correlation is the most common, obvious and simple way to measure effect of one variable to another. The correlation coefficient is a statistical measure that describes the degree to which two variables are linearly related. In context of our research, the correlation between the firing rates of two neurons over time can be an efficient as a proxy variable for synaptic strength. For two vectors of neural activity $\vec{a}$, $\vec{b}$ the Pearson correlation measure would be as follows:

$$\rho(\vec{a}, \vec{b}) = \frac{\sum_{i=1}^{n}(\vec{a}_i - \overline{a})(\vec{b}_i - \overline{b})}{\sqrt{\sum_{i=1}^{n}(\vec{a}_i - \overline{b})^2 \sum_{i=1}^{n}(\vec{b}_i - \overline{b})^2}},$$

where $\overline{a} = \dfrac{\sum_{i=1}^{T-k} a_i}{T - k}$ and $\overline{b} = \dfrac{\sum_{i=1}^{T-k} b_i}{T - k}$

3.2 Granger Causality

The Granger causation [9] is a way to measure the effect of one time series on another by building a predictive models for each of time series. It was initially developed for econometric applications, where the signals may not have obvious and immediate interactions between each other and as the casualty is very important in neurons interaction we decided to test it. Granger causality may be useful as it already implies directionality of the interaction, i.e. we do not need to shift time series. According to the Granger causality test, if a signal X "Granger-causes" (or "G-causes") a signal Y, then past values of X should contain information, which helps predict Y above and beyond the information contained in past values of Y alone.

The idea of Granger causation lies in fitting two autoregressive models and if $y(t)$ prediction is significantly improved by adding $x(t)$ to the model then $x(t)$ G-causes $y(t)$. The models are as follows: $y(t) = \sum_{i=1}^{n} \alpha_i y(t-i) + \epsilon(t)$, $y(t) = \sum_{i=1}^{n} \alpha_i y(t-i) + \sum_{i=1}^{n} \beta_i x(t-i) + \epsilon'(t)$.

Here, α_i and β_i are the model parameters, n is the model order, and $\epsilon(t)$ and $\epsilon'(t)$ are error terms. If the second model significantly improves the prediction of $y(t)$ over the first model, then we conclude that $x(t)$ Granger-causes $y(t)$. Obviously, we can consider $x(t) = \vec{a}$ and $y(t) = \vec{b}$

Though the significance as is can not be used to measure and compare synaptic strength between neurons, as we actually would like to see only edges which are results of actual presence of synapses we may use a p-value of a test as a measure of synaptic strength along with setting the p to 0.95 which corresponds to the 95% level of statistical confidence. Additionally we set shift to $k = 0$ in this approach due to the autoregressive nature of models used in Granger casualty it already takes in account all required shifts of data.

3.3   Coherence

Coherence is another measure of the linear correlation between two signals. The coherence $C_{xy}(f)$ between two signals $x(t)$ and $y(t)$ is defined as follows:

$$C_{xy}(f) = \frac{|G_{xy}(f)|^2}{G_{xx}(f)G_{yy}(f)},$$

where $G_{xy}(f)$ is cross-spectral density of $x(t)$ and $y(t)$, $G_{xx}(f)$ and $G_{yy}(f)$ are the power spectral densities of $x(t)$ and $y(t)$, respectively. The cross-spectral density is the Fourier transform of the cross-correlation function, and the power spectral density is the Fourier transform of the autocorrelation function, so f here denotes frequency. The cross-spectral density is defined as follows: $G_{xy}(f) = F(R_{xy}(\tau))$, where F denotes the Fourier transform, and $R_{xy}(\tau)$ is the cross-correlation function of $x(t)$ and $y(t)$. The coherence also ranges from 0 (no coherence) to 1 (perfect coherence).

In the context of the neural activity analysis the coherence would show the consistency in activations of different neurons on different frequencies. As synapse is working on different rate in the same manner we expect to observe a high coherence of the connected neurons on all of the rhythms (i.e. frequencies) in which hippocampus has worked. The synaptic strength does not depend on the frequency and this means that the connections having the higher average coherence indicate the most powerful synaptic strength. In spite of this we decided to use the mean of coherences calculated on frequencies from 1 Hz to 100 Hz (the frequencies range choice is widest possible meaningful assuption within our dataset). The following formula was used to calculate coherence between all possible activities of neurons' pairs:

$$\rho(\vec{a}, \vec{b}) = \frac{1}{\overline{f} - \underline{f}} \sum_{i=\underline{f}}^{\overline{f}} C^D(\vec{a}, \vec{b}),$$

where C^D is discretized version of C_{xy}, $\underline{f} = 1, \overline{f} = 100$.

3.4   Performance Comparison

We have used dataset recorded when rodent was on 3-obstacles (which is a maximal number we used across experiments) arena for evaluation of the methods suitable for constructing graph of neural activities. Indeed, conjecturally, it's the most complex environment for an animal to explore and memorize.

Even on preliminary part of testing classical Pearson correlation showed very high level of connectivity, but at the same time the network had an issue of multi-edges. We assumed that the reason for excessive multi-edges could be too low value of ρ. Moreover, as the graph is not simply-connected (which is perfectly fine) even the small graphs were with multi-edges. In order to verify that the reason for multi-edges is not low value of ρ we have also tried use higher correlation boundary. The graph had much less connectivity between edges, but multi-edges persisted. This indicates that correlation can not be used as metrics for building graphs with use of this approach as it can not be used for distinction of post-synaptic and presynaptic neuron creating multi-edges which leads to loss of the information on data directionality in neural network. This is due to the correlation's nature as, even on a smaller time scale, it still doesn't distinguish the temporal connectedness orientation - it relies only on the intersections of peaks - no matter which one is earlier, hence causality.

The results of Granger-casuality also gave poor results. In fact, it performed the worst results among the methods suggested. Taken the values of ρ for Granger causation at Fig. 2 (left) shows that after 0.3 the distribution becomes almost flat. This is likely means that the graph would have too many edges. This is confirmed when the graph is built Fig. 2 (on the right).

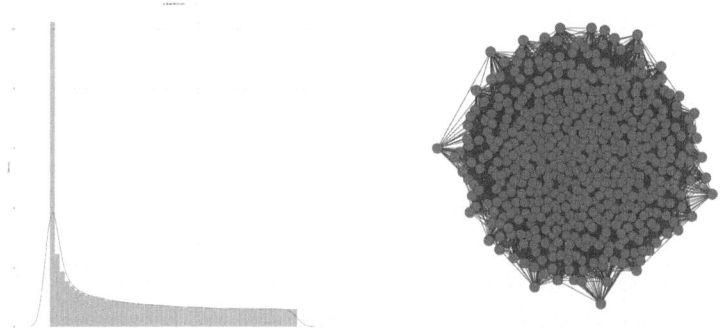

Fig. 2. Distribution of ρ for Granger causality (on the left) and the graph built with this method with $\rho = 0.95$ (on the right)

The likely reason for this is stability of a series and noise dependency of observed values of neural activity. This leads to that the autoregressive model predicts well just having one previous value of neural activity and may improve the prediction being supplied with other neural data. The improvement of such method may be in filtering out only spikes making activity completely binary. Though such would also lead to too good predictions of autoregressive models as most neurons do not express spike activity often enough.

The coherence being theoretically the most suiting solved the problems of the previous approaches. It does not give multi-edges and has very stable graph connections across slices Fig. 3. The computational check for multi-edges shows relatively small amount of them (about 0.5% from overall edges generated on all slices had complementary counter edge). Also, the significant part of neurons are included into the path-connected network and the clusters of the local "rich-club" are permanently formed. This all is in accordance to what was expected for the contextual network design.

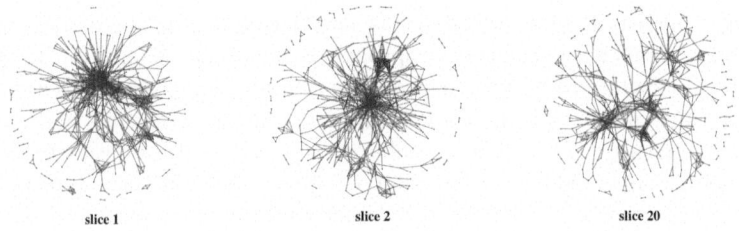

slice 1   slice 2   slice 20

Fig. 3. Coherence based graph on time-partitioned data

Even from the several slices we checked during the coherence evaluation we can see the stability of the network properties, at the same time of high plasticity of connections. That's why in the following part we will study the global properties more precisely.

4   Dynamics

On Fig. 4 the average coherence "metric" values is shown in dynamics. Each point on this plot is given for a graph computed for S_i with 'BuildGraph' algorithm we described above, namely $G_i = (V, E_i)$ and calculated as follows:

$$\bar{\rho}_i = \frac{1}{|V|^2 - V} \sum_{i \in V} \sum_{j \in V, j \neq i} \rho(i, j).$$

We may see from the graph Fig. 4 that $\bar{\rho}$ oscillates between 0.26 and 0.27 which indicates that on average synaptic strength has no trend and does not tend increase or decrease. Such behaviour along with the graphs presented earlier may indicate that rodent's hippocampus tend to distribute the neural network.

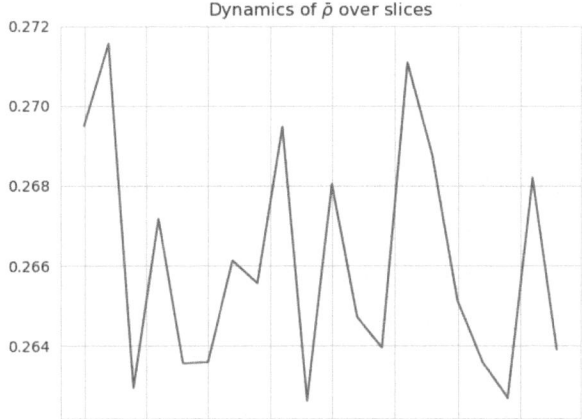

Fig. 4. Distributions of $\bar{\rho}_i$ of neural dynamics across slices of a single mouse

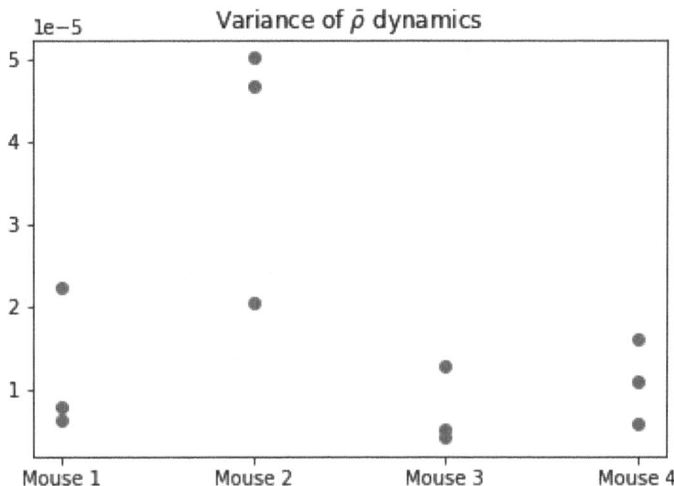

Fig. 5. Distributions of variance of neural dynamics across slices across all experiments with all mice

We applied this approach to all the experiments we had. So we had distributions for all four mice (see Fig. 5), by 3 experiments on arenas with 1, 2 and 3 "holes", so the environments mice explored were not the same. The distribution of variance of $\bar{\rho}_i$ for each mouse shows strong stability across experiments with the outliers still being very minor in value.

We also decided to check the network dynamics property by calculating the clustering coefficient for vertex u in G_i (a graph of certain slice of time series) as follows:

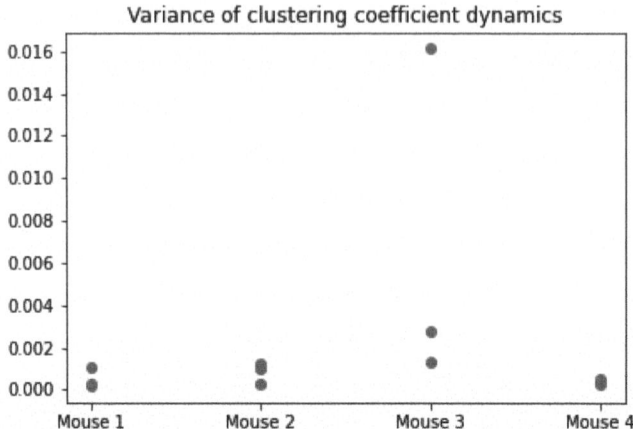

Fig. 6. Distributions of variance of graphs clustering dynamics across slices across all experiments with all mice

$$c_{u,i} = \frac{1}{\deg(u)(\deg(u) - 1)} \sum_{v,w \in V} (\widetilde{W}_{u,v} \widetilde{W}_{u,w} \widetilde{W}_{v,w})^{1/3}$$

where $\deg(u)$ is a vertex degree in corresponding graph, $\widetilde{W}_{u,v} = \dfrac{W_{u,v}}{\max(W_{u,v})}$ is normalized weight of edge between u, v and equals 0 if there's no edge between them. The clustering coefficient for all graph is the averaged clustering coefficient of all vertices: $c_i = \dfrac{1}{V} \sum_{u \in V} c_{u,i}$. The variance of clustering coefficients for all experiments is shown at Fig. 6.

It varies in a wider range, obviously, but still it is clearly very stable with only one outlier out of 12 experiments. In general clustering coefficient trends and values demonstrate that the network is constantly changing, which means that even when mouse learns the arena completely, the network in general recombines in other areas, not related to place cells. It might be related to other mice's activities. But the general network properties also on the clustering coefficient level remain within a very certain limits.

5   Conclusion

In this work we have shown that the coherence is very promising method for constructing the neural network in dynamics. It performs gradual reconstructions of a network and gives a set of graphs usable for studying. The network analysis showed that it's properties, even during reconstruction of local parts when learning the environment, are very stable and remain within tight limits. It shows the robustness of a hippocampal neural network and gives a new perspective to future modeling of such networks. As further research we may suggest the following: analyse graphs in more depth, develop models for activity (Bayesian inference networks to be specific) and develop model of synaptic strength change. The further analysis may be focused on developing of more quantitative methods of assessing resulting graphs. Additionally the analysis can include different parameters of a network like graph hyberbolicity, small-worldness and hierarchical properties. Finally, it may be of interest to compare graphs we obtained with graphs based on Bayesian networks and structural causal models.

Acknowledgements. The article was prepared within the framework of the project "Mirror Laboratories" HSE University, RF. This research was supported in part through computational resources of HPC facilities at HSE University.

Ethics Statement. The animal study was reviewed and approved by all methods for animal care and all experimental protocols were approved by the National Research Center "Kurchatov Institute" Committee on Animal Care (NG-1/109PR of 13 February 2020) and were in accordance with the Russian Federation Order Requirements N 267 M3. Three C57BL/6J mice were used in this study, ages 2–3 months. Mice were used without regard to gender.

References

1. Anokhin, K.V.: The cognitome: seeking the fundamental neuroscience of a theory of consciousness. Neurosci. Behav. Physiol. **51**(7), 915–937 (2021)
2. Sotskov, V.P., Plusnin, V.V., Sorokin, K.S., Anokhin, K.V.: Rapid tuning dynamics of CA1 place codes in one-and two- dimensional free exploration tasks. In: Fourth International Conference Neurotechnologies and Neurointerfaces (CNN). Kaliningrad, Russian Federation, vol. 2022, pp. 168–171 (2022). https://doi.org/10.1109/CNN56452.2022.9912511
3. Sorokin, K., et al.: Topology of cognitive maps/Cornell University. Series Computer Science "arxiv.org" (2022)
4. Watts, D., Strogatz, S.: Collective dynamics of 'small-world' networks. Nature **393**, 440–442 (1998). https://doi.org/10.1038/30918
5. Bush, D., Philippides, A., Husbands, P., O'Shea, M.: Spike-timing dependent plasticity and the cognitive map. Front. Comput. Neurosci. **4**, 1475 (2010)
6. Roberts, P.D., Bell, C.C.: Spike timing dependent synaptic plasticity in biological systems. Biol. Cybern. **87**(5), 392–403 (2002)
7. Griffa, A., Van den Heuvel, M.P.: Rich-club neurocircuitry: function, evolution, and vulnerability. Dial. Clin. Neurosci. **20**(2), 121–132 (2018). https://doi.org/10.31887/DCNS.2018.20.2/agriffa. PMID: 30250389; PMCID: PMC6136122
8. Giovannucci A., et al.: CaImAn: an open source tool for scalable calcium imaging data analysis bioarXiv preprint https://doi.org/10.1101/339564 (2018)
9. Granger, C.W.J.: Investigating causal relations by econometric models and cross-spectral methods. Econometrica **37**(3), 424–38 (1969). https://doi.org/10.2307/1912791

Databases in Data Intensive Domains

Flexible Materials Properties Management System as a Basis for Data-Centric Systems in Inorganic Materials Science

Victor Dudarev[1]([✉]) [iD], Nadezhda Kiselyova[2] [iD], and Alfred Ludwig[1] [iD]

[1] Ruhr-University Bochum, 44801 Bochum, Germany
`{victor.dudarev,alfred.ludwig}@rub.de`
[2] Baikov Institute of Metallurgy and Materials Science the Russian Academy of Sciences, Moscow 119334, Russia
`kis@imet.ac.ru`

Abstract. Efficient harvesting of research data with proper data management and data utilization is essential for scientific progress in materials science. On the one hand, the continuous production of huge amounts of data from measuring devices and the lack of industry standards for the data format for storage and processing necessitates the development of reliable information systems capable of organizing, analyzing, and using diverse materials data. This article focuses on the development of Internet-accessible data management systems for inorganic materials science, addressing typical domain requirements. These systems enable researchers to store diverse tabular data and different property types to support compatibility with wide a variety of research data. At the same time, typical data structures to address chemical systems, compositions, and their modifications are available out of the box together with a flexible graphical user interface inspired by the periodic table. The key emphasis is on data integrity, ensuring accurate tracking and preservation of materials data, and interoperability by providing data type templates and rich import/export capabilities. A case study with bandgap data highlights the system's effectiveness in materials properties storage, effective data modification, and fast user-friendly search to support data-driven decision-making. Considering the significance of a strong data infrastructure for flexible materials properties management we offer an open-source solution to facilitate and accelerate similar data-centric projects in inorganic materials science by encouraging you to deploy our solution in building your own customizable materials research infrastructure you could rely on.

Keywords: materials science data infrastructure · materials data management system · inorganic materials properties · materials information search

1 Introduction

In modern materials science we could observe intensive data grow which brings us to the problem of efficient management and utilization of research data. Solution of this task is crucial not only for scientific progress in this domain, but also in many

J. Baixeries et al. (Eds.): DAMDID/RCDL 2023, CCIS 2086, pp. 91–103, 2024.
https://doi.org/10.1007/978-3-031-67826-4_7

industrial applications for effective data sharing and information exchange for data-driven decision-making tasks related to modern materials [1].

With the continuous generation and accumulation of big data volumes in materials science, the need for robust information systems that can effectively organize, analyze, and leverage this valuable information has become paramount [2, 3]. Our focus lies specifically in the development of applicable in inorganic materials science data management systems, which are accessible via the Internet. Our extensive experience, accumulated over the years, has been demonstrated in the successful development and utilization of several information systems on inorganic materials properties [4]. With a **deep understanding** of the complexities of data in this field, especially regarding modern high-throughput techniques in materials research [5] we have been developing an open-source pioneering solution that encompasses the essential functionality required for modern materials science data management.

Since materials science core data must be strictly defined using strong types and all the problem domain **constraints** preserved, we rely on classical **relational** database structures, known for their sustainability and maturity (e.g., see ACID principles) [6]. Therefore, all general top-level chemical objects (systems, compositions) are described in terms of relations to ensure the utmost data integrity available at data level in computer science, which is a crucial aspect in accuracy of tracking and preserving materials science data. We hope that our commitment to development of transparent open-source solutions together with data integrity for core chemistry data structures will enable researchers to confidently rely on systems based on our solution for their critical data management needs, thereby enabling accurate tracking and preservation of materials data.

Despite that this article aims to highlight the significance of a strong-types data at the top level of materials infrastructure, we certainly need **flexibility** for defining materials properties data types as the foundation for flexible materials properties management in materials science. And here once again we are capitalizing on the power of relational databases to decompose arbitrary tabular research data into a set of predefined type-aware properties tables, defining algorithms for data transformations to preserve flexibility, required for diverse materials property types, including combined from tables with different column types: integers, floats, and strings. These three type primitives for table columns allow us to take care of data structures unknown at system design time and then reconstruct them on-line and facilitate search on different data types of columns comprising a property value.

Materials management system must facilitate not only storage but efficient search for persisted materials data. To this extend we rely on periodic table-based interface that empowers researchers to interact with a system for data search. Search form is automatically adjustable depending on stored data types and their corresponding properties, which makes system flexible out of the box. This innovative approach in materials science management systems simplifies the search process and enhances user experience, facilitating rapid access to relevant materials data, providing a robust solution for managing the ever-expanding array of materials properties.

To illustrate functional capabilities of developed system in real-world scenarios we provide case study by transferring materials properties data from existing bandgap data source and thus leveraging search facilities and delivering better user experience, that was available in initial proprietary solution.

2 Problem Domain and Key Requirements Definition

Materials science orientation brings to the system challenges concerned with representation of chemical entities from the problem domain. Considering vast variety of requirements gathered from diversity of problem subdomains it's clear that the general-purpose materials data management system should be capable of describing basic entities of the problem domain. Hence, we consider the very top (basic) level of materials description (chemical system, compositions, modifications) to be a part of functionality available out of the box. Any further details require developing corresponding data models and workflows to support them are subject for discussion and further implementation (system extension). Implemented three level hierarchy of chemical entities is described [7] as a model comprising:

- chemical system – a set of chemical elements that built up a material denoted by the symbol S. If each chemical element is represented by a unique symbol or identifier e_i, then mathematically, the chemical system S is a set $\{e_1, e_2,..., e_n\}$, where e represents a chemical element in the chemical system (qualitative composition).
- composition – a set of chemical elements with their concentration (if solid solution is considered) or particular content (in a case of single phase) that built up a material. Let n_i represents the quantity or coefficient associated with a particular element e_i, then composition, designated C, is a set of pairs: $C = \{(e_1, n_1), (e_2, n_2),..., (e_n, n_n)\}$ (quantitation composition).
- modification – a particular crystal or amorphous structure realization of a composition of other subtype / subdivision, expressed as a string (further details) or in a more formal way as lattice parameters, space groups and atomic coordinates (matrices, vectors, and equations).

If we omit typical functional requirements (user registration and authentication, API and user interface for CRUD operations, import/export functionality, data sharing and notifications) focusing on essential data-oriented functionality for material data management we conclude with necessity to provide facilities to support for chemical entities within above-defined hierarchy and arbitrary materials data types, e.g., properties values, tables.

3 Key Data Structures to Address Materials Data Management Tasks

Since data integrity is our primary concern we rely on relational databases, designed with intention to keep data in as strict a format as possible, but at the same time ensuring capabilities for flexible extension in terms of materials properties description.

3.1   Chemical Entities

Every object that is stored within materials management system holds its record with-ing **ObjectInfo** table, that encompasses all mandatory metadata for every object type such as creation and modification datetime stamps and corresponding user references (_* columns), reference to a particular object type (TypeId), defined in an external table, reference to a container for object (RubricId), that allows to structure all objects within a project tree, sort code to define a certain order in a list of objects (Sort-Code), object's accessibility: public/protected/protectedNDA/private (AccessControl); obligatory object name (ObjectName) (see Fig. 1).

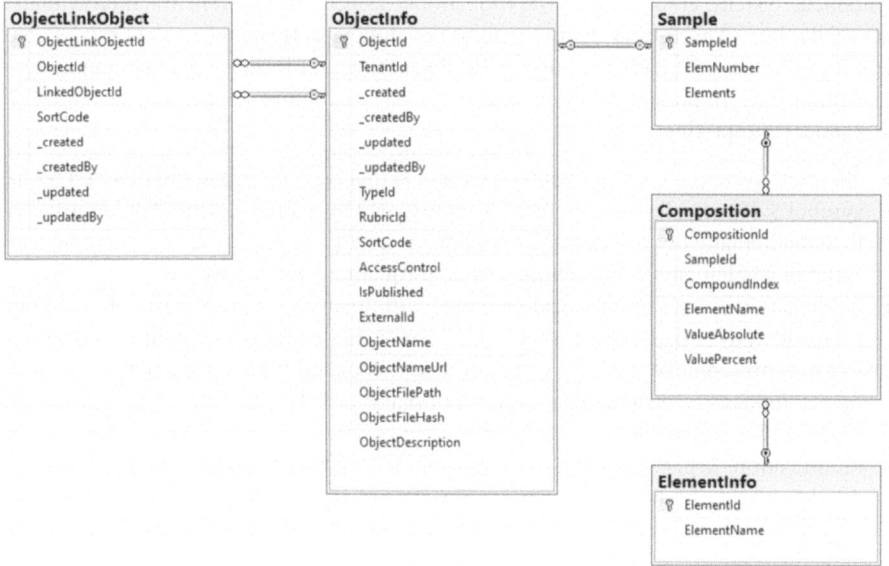

Fig. 1. Microsoft SQL Server diagram of objects storage structure within materials data management system.

In terms of data every chemical entity is an object and thus it must have a record in ObjectsInfo, but apart from that it is at least a chemical system, implying that we know chemical elements set, which is to be stored in **Elements** column of a **Sample** table as an alphabetically ordered sequence of elements separated with hyphen sign and starting and ending with hyphen also for a search simplicity, e.g., "-Li-Nb-O-" or "-As-Ga-" (additionally number of chemical elements in a system is stored (ElemNumber) to optimize queries for chemical systems). Denormalization for chemical system is used here to optimize search speed [8]. It's worth mentioning that a relationship between **ObjectInfo** and **Sample** table is one-to-one (ObjectId = SampleId) apart from other relationships, emphasizing that all complex types (in our example – chemical systems), derived from a base type, store data shared between several tables. Here **ObjectInfo** persists base type data that are common to all types and other tables – data related to particular derived types.

According to definition, compositions additionally contain quantity for every element either in terms of absolute (ValueAbsolute) or relative (ValuePercent) values. So, to describe chemical composition in addition to ObjectInfo and Sample tables one should feed several rows to **Composition** table, according to number of unique chemical elements. It's worth mentioning that ValueAbsolute is always calculated to enable that sum of elements within a composition is 100% and to make search more flexible. **ElementInfo** on the scheme is an auxiliary table to ensure element designations correctness (unknown ElementName is rejected).

ObjectLinkObject table is worth mentioning since it serves as linking opportunity between objects enabling to establish a dependency graph if one is required. By default, link is interpreted as a directed relation (arc, directed edge), implying that reverse relation should have opposite meaning, but in general link semantic could be defined on linked object types basis.

3.2 Object's Properties Database Structure

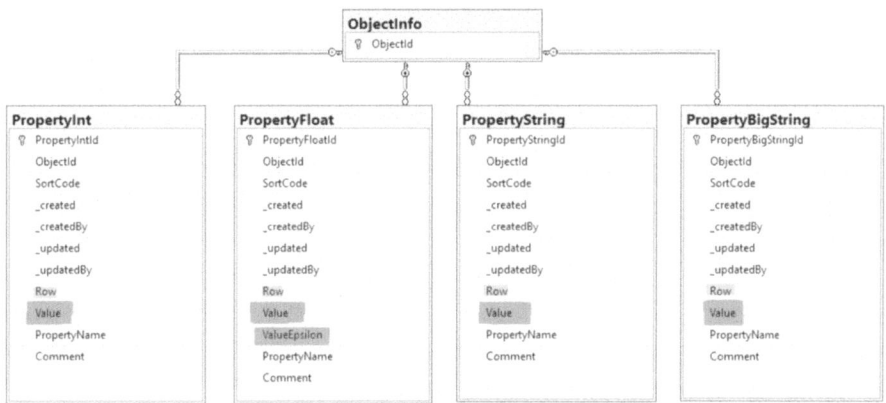

Fig. 2. Extended properties to describe objects.

As it was shown in previous section with chemical compositions or systems, additional information on object can be obtained from other tables, but it requires additional programming to support corresponding data type. In most of the cases it's enough to have an opportunity to add several additional values for object as a set of key-value pairs, respecting corresponding value's type (float, integer, string).

To provide a proper storage for different types of values and considering specifics of storage in SQL Server for string of different size we designed four tables **Property*** with obvious designation, see Fig. 2. Every table has similar structure, the only difference is in **Value** column data type and, in addition, **PropertyFloat** table contains optional **ValueEpsilon** field to specify measurement error, if possible.

Every property record (PropertyName, Value) references an appropriate object by specifying its ObjectId identifier and some common data regarding creation and modification timestamps (underscore columns) and provides, by manipulating SortCode value, showing properties in desired order by SortCode ascending. Comment field is an optional one and could be used to attach some remarks to the property value.

However, the way more interesting use case we have when it's required to represent not just a set of properties, but a table with rows and columns. To support this use case in Properties* tables it is required to use **Row** column, which is *null* by default, for non-tabular properties. If it's required to represent tabular data, then table is decomposed into a number of cells and every cell is represented as a property with name of the column and particular row value. This enables us to decompose tables of any structure (sparse tables are packed reasonably well, since empty cells could be omitted) and provide strong type check for every column according to selected property type (see Fig. 3).

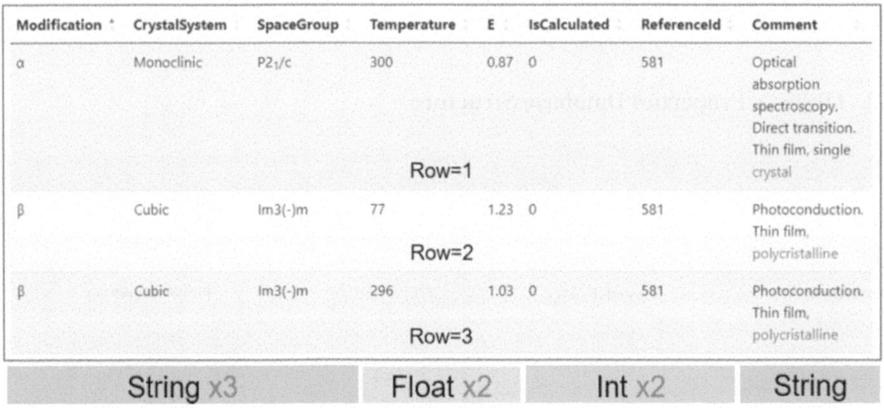

Modification *	CrystalSystem	SpaceGroup	Temperature	E	IsCalculated	ReferenceId	Comment
α	Monoclinic	P2₁/c	300	0.87	0	581	Optical absorption spectroscopy. Direct transition. Thin film, single crystal
			Row=1				
β	Cubic	Im3(-)m	77	1.23	0	581	Photoconduction. Thin film, polycristalline
			Row=2				
β	Cubic	Im3(-)m	296	1.03	0	581	Photoconduction. Thin film, polycristalline
			Row=3				
String ×3			Float ×2		Int ×2		String

Fig. 3. Extended properties for description of tabular data (Ag_2S bandgap).

3.3 Properties Templates and Import/Export Features

Although materials management system offers user interface to perform CRUD-operations on object's properties on individual property value basis and from formal perspective full functionality is covered, but if you must deal with a plenty of object properties, then this could be inefficient and time-consuming task and therefore could cause problems. To overcome this, we introduced properties templates and properties import/export from/to Excel.

Properties templates is a mechanism that enables to define a particular object type description in terms of an ordered set of properties. Having defined a template as a special object with a reserved "_Template" name of a required type we specify a set of properties bound to it (including their names and types) that characterizes objects of a type. When template is defined, any object of a given type "knows" what properties it should have.

As it was described earlier object properties could be considered as key-values pairs belonging to the object or representing a table data (if **Row** is greater than zero). In this regard within a template object, one could define two categories of properties:

- **table** properties (**Row** must be set to −1) – to specify property as a table attribute or a column).
- **scalar** properties (**Row** must be set to *null*) – to specify a scalar property that should be filled oud for objects of a given type.

Properties templates is a mechanism of flexible table definition that could be applied to a particular object data type. Using table properties template, user should define structure of a table as an ordered (SortCode is used) sequence of column names and types. Once property template is defined for a type, user could download Excel table template to facilitate work with properties. Thus, it's easier to fill desired table data in Excel document and import them to the system for selected object. Or if you need some data modifications in table properties you could export properties in Excel and upload adjusted version after modification. We extensively tested this option even for relatively large tables, e.g. properties from table with 16 columns and 468 rows (about 7.5K properties) were updated in a few seconds.

Initially, scalar properties look simpler than table properties, but from our experience scalar properties that characterize object could be numerous (approximately 500 key-value pairs are required to characterize sputter parameters in one of instances) and therefore complex to overview. This introduces a task to provide a structuring opportunity for a number of scalar properties enabling to hierarchically group them into categories, subcategories and so on. To solve this task, we rely on several considerations regarding properties attributes:

- **SortCode** should provide order for all properties participating in a template (unique values are recommended);
- to provide a section with a given name, that contains nested properties, user should create a special *integer* property with a **Name** corresponding to a section name, a fake **Value** (-1 is recommended) and a **Comment** equal to "SEPARATOR";
- a nested property should have a name prefixed by a name of section with " = >" separator from the current name. Aware that, its SortCode should be greater than a SortCode of a corresponding section property (for example, if section name is "Engine" than nested property should start with "Engine = >" and could be "Engine = > Model").

To illustrate a template for both table and scalar properties one could consider an easy-to-understand example of automobiles that have some scalar properties character-izing engine and safety along with associated table that defines prices for Model-Color combination (see Fig. 4 – section names are highlighted with blue background and all scalar properties names prefixed with section names; three table properties have table icon indicating effective table columns).

After a template definition any object of a given type could take advantage of a sectioned form for data modification (see Fig. 5 – *Engine* section is collapsed, while *Safety* section is wide open enabling user to input data). Color scheme used in properties backgrounds:

- blue – section header (could be used to collapse/expand the section);
- green – saved in database (property is in the template);
- white – no value in database for property defined in the template;

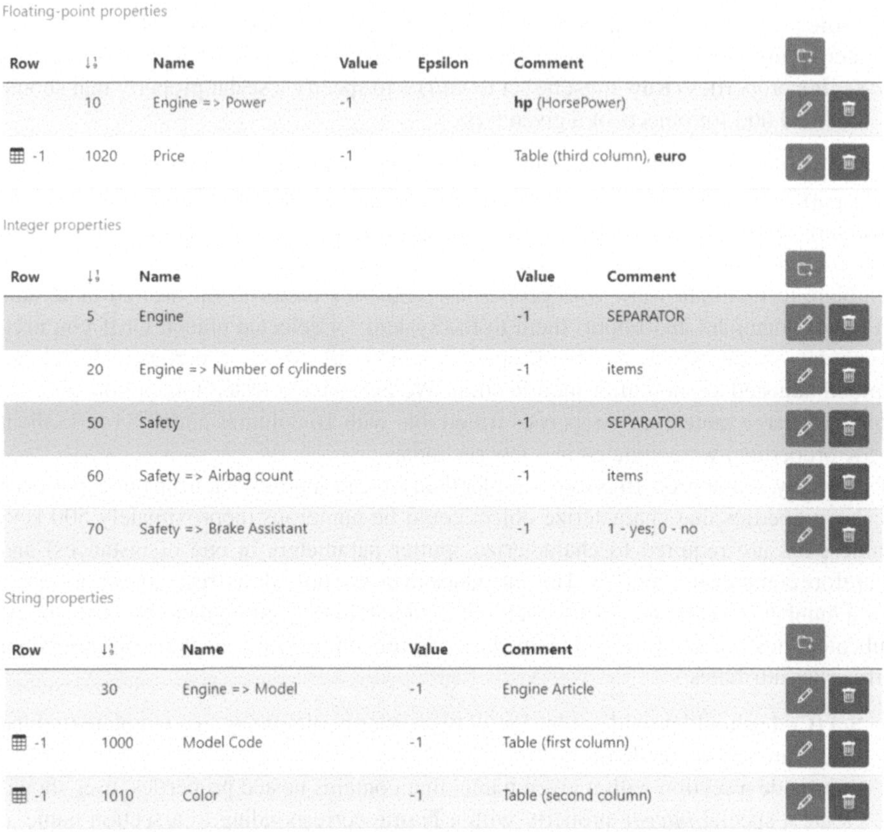

Floating-point properties

Row	↕	Name	Value	Epsilon	Comment	⊡
	10	Engine => Power	-1		**hp** (HorsePower)	✏ 🗑
▦ -1	1020	Price	-1		Table (third column), **euro**	✏ 🗑

Integer properties

Row	↕	Name	Value	Comment	⊡
	5	Engine	-1	SEPARATOR	✏ 🗑
	20	Engine => Number of cylinders	-1	items	✏ 🗑
	50	Safety	-1	SEPARATOR	✏ 🗑
	60	Safety => Airbag count	-1	items	✏ 🗑
	70	Safety => Brake Assistant	-1	1 - yes; 0 - no	✏ 🗑

String properties

Row	↕	Name	Value	Comment	⊡
	30	Engine => Model	-1	Engine Article	✏ 🗑
▦ -1	1000	Model Code	-1	Table (first column)	✏ 🗑
▦ -1	1010	Color	-1	Table (second column)	✏ 🗑

Fig. 4. A template for scalar and table properties in administrative interface.

- yellow – property value is saved in database, but the property name with respect to its type is not part of the type template).

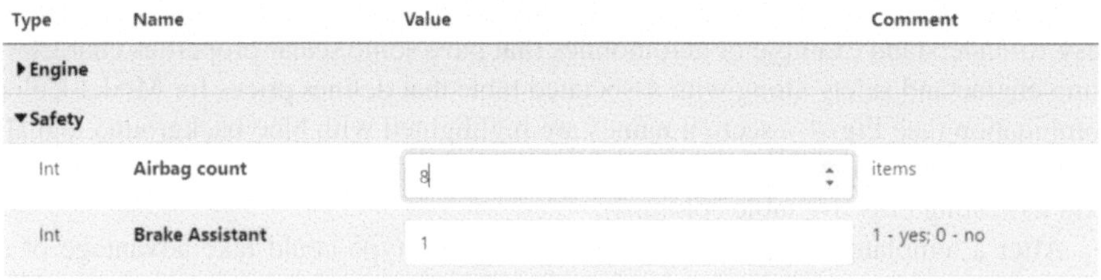

Type	Name	Value	Comment
▶ **Engine**			
▼ **Safety**			
Int	**Airbag count**	8 ⬍	items
Int	**Brake Assistant**	1	1 - yes; 0 - no

Fig. 5. Sectioned scalar properties input form after defining a template.

After filling all properties an object could expose this information as shown on Fig. 6: all scalar properties are displayed as a list with collapsible sections and table data is shown in a sortable table.

All properties (except table)

| | Edit by template | ☑ Include all template properties | | Choose File   No file chosen |
| | | ⬇ Download properties in Excel | | ⬆ Upload / Replace properties from Excel |

Type	Name	Value	Comment
▼ **Engine**			
Float	**Power**	140	**hp** (HorsePower)
Int	**Number of cylinders**	4	items
String	**Model**	SomeModel123	Engine Article
▼ **Safety**			
Int	**Airbag count**	8	items
Int	**Brake Assistant**	1	1 - yes; 0 - no

The table

| | ⬇ ⊞ Download properties in Excel | | Choose File   No file chosen |
| | | | ⊞ ⬆ Upload / Replace table from Excel |

Model Code	Color	Price
ABC	Red	10000
ABC	Blue	10050
XYZ	White	11500
XYZ	Black	11550

Showing 1 to 4 of 4 entries

Fig. 6. User interface showing an object with scalar and table properties.

Moreover, all the properties are exportable (blue buttons) to and importable (red buttons) from Excel documents of corresponding formats (defined by a template). That enables user-friendly properties modification for an end user.

4 Architecture: Scalability, Security, Flexibility

From an architecture perspective a system is a classical ASP.Net Core 3-tier web application, relying on a Microsoft SQL Server as database engine and Internet Information Services as an application server to provide its functionality. This platform is known for its excellent performance, support level, enterprise features and other benefits.

Security is a primary concern and therefore our management system relies on ASP.Net Core Identity (with persistence in SQL Server), which is a solid API that supports users, roles, profile data, claims, e-mail confirmation to make the system secure. We also implemented external authentication by means of OpenID Connect protocol via Google Identity, this enables Google users to use their accounts to authorize in the system and thus storing no passwords.

From a scalability perspective – apart from an ASP.Net Core platform which is highly scalable by itself – we implemented a multitenancy concept (database has **Tenant** table, and **ObjectInfo** has **TenantId** attribute). This gives an opportunity to flexibly spawn a new instance of a system in either of two ways:

- **sharing a database** instance (between several tenants) – users and data types are shared among all tenants in a database (rubrics and objects are not shared – providing an isolated data for a tenant);
- an **isolated database** instance (isolated tenant) – completely isolated environment, independent from others.

This helps to flexibly provide the management system as a software as a service (SaaS) for a particular user group and spawn a new instance in virtually on-demand mode.

Flexibility was partially discussed in a previous section, devoted to data structures and type-bound table and scalar properties. The flexibility taken from database schema opens opportunities to configure and introduce your own object type derived from knows ones and set it up including properties templates. Nevertheless, further customization is possible on a tenant basis due to HTML/CSS/JavaScript injection thus enabling to adjust and interact with Document Object Model.

Other type of flexibility is offered via external REST web services (satisfying the OpenAPI specification) to support object type data. Every object can have a data attached to it as a file. Due to extensibility of an object type system, new types can be introduced by a privileged user along with their configuration including external URL to support corresponding data file types. So, the managements system relies on three types of external services:

- **validation service** – to validate documents of a certain object type, the service provides binary decision (only valid files are allowed in the system);
- **data extraction service** – to extract data (chemical compositions and properties) in JSON format from a valid file (and to upload into the management system);
- **visualization service** – to visualize valid file and show the result in a client browser as an HTML page.

5 Use Case: Data Import from Bandgap Database

Described data structures and architecture are to be proved by real life data to make sure that they are flexible to deal with real materials data coming from existing data source in relatively significant volumes and provide enough expressiveness to tailor all data details from externals sources. To test developed materials data management system, we decided to import data from previously developed information system on bandgap of inorganic compounds https://bg.imet-db.ru [9]. Basically, we required to transfer data on inorganic compounds, corresponding bandgap values (which depend on compound crystal structure, temperature, measurements method, etc.) that were extracted from literature sources, that also good to have imported, since these are essential references to discover necessary details and keep track on data provenance.

Since literature reference is a common data type for data management system, we also introduced an additional data table, called **Reference**, that extends standard object by meaningful publication's data (title, journal, year, volume, pages, DOI, etc.). So, the task was to map bandgap database schema to the schema of materials management system, and it was solved with the following key points: compounds were mapped to Composition data type (ObjectInfo + Sample + Composition tables); literature references – to Reference data type (ObjectInfo + Reference tables); table data on bandgap values for a particular composition – to properties (Property* tables) and connections between literature reference and composition were established through ObjectLinkObject table. All data (about 5K compounds, 2K literature references and 10K bandgap values) were transferred by running a developed SQL-script. This is not a "documented way" of transferring the data, but a good one for quick data import from existing relational data source to demonstrate the overall system capabilities. We don't consider this script to be of much interest since it hardly relies on the Bandgap database structure, but for the sake of truth it's available "as is" [10].

6   Search Features in Material Data Management System

Data representation is important, but it makes no sense if data are not searchable and easily accessible. This means both that stored chemical entities are to be easily searchable by means of intuitive user interface, based on the periodic table and other common input forms, and all custom-defined properties should be also available for search with the help of dynamically adjustable search form (see Fig. 7 – search for compositions with bandgap in range [2; 2.1] eV in Ga-As system with 50% Ga contents).

In general, out of the box search works for: chemical system, composition, object type, phrase in object's name or description, properties values, person created record, creation date. For every object type user could input or select property name from a drop-down (property names are dynamically fetched for autocomplete list) and specify property value to search for. There is an option of search form adjustment by adding other criteria for property search.

Valuable feature on search is a persistent URL, that reflects search criteria, therefore one could easily provide a link to share with collaborators (e.g. link https://demo.mdi.ruhr-uni-bochum.de/search/?system=As-Ga&typeid=8&Gapctmin=50&Gapctmax=50&pr0name=E%3Csub%3Eg%3C%2Fsub%3E&pr0type=Float&pr0min=2&pr0max=2.1&prcnt=1 provides permanent link to search result from Fig. 7). To test user interface, you could visit https://demo.mdi.ruhr-uni-bochum.de without necessity to register. In general, current graphical interface offers sufficient functionality and user experience that satisfies requirements for data search offered by specialized solutions, for example original https://bg.imet-db.ru.

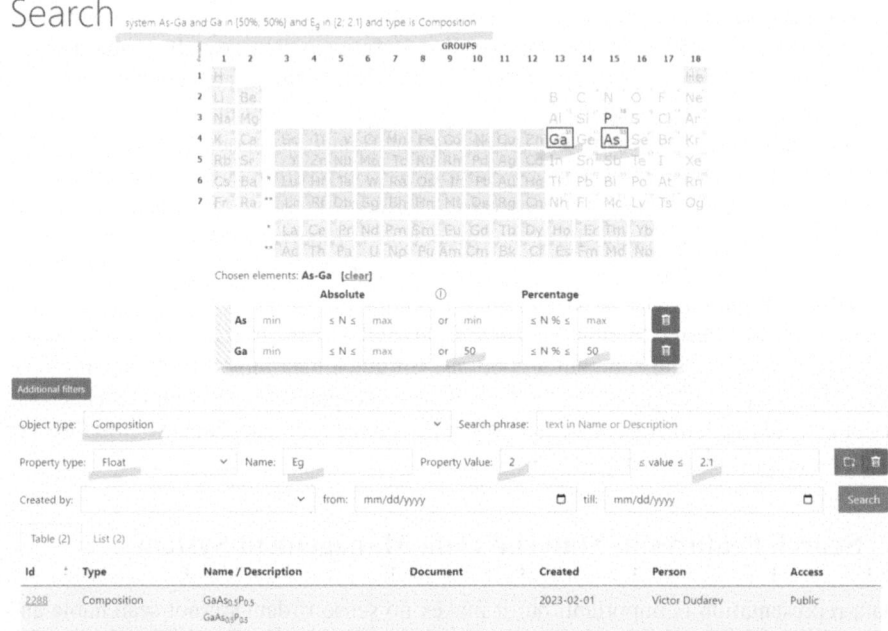

Fig. 7. User interface (search for bandgap in Ga-As in [2; 2.1] eV range).

7   Conclusion

In current paper we touched only the core parts of materials management system being developed as an open-source project [11] to facilitate deploying and usage of data infrastructure for inorganic materials science domain. Although the system is still under development, all core functions are already on their places. We struggle to grow project as extensible solution without plans for breaking changes in future. Through this article, we encourage researchers to establish their data-centric material infrastructure and unlock the full potential of materials science through efficient research data management. We believe that the system discussed, in its current state, is ready to support data-based materials research and can be easily adjusted to fulfill specific requirements. Developed architecture together with system extensibility provides a solid basis for a wide variety of materials science applications (currently we run seven tenants and methodically improving the system).

A lot of work is still to be done in future. Additional efforts to support **specific** data types and use cases are to be undertaken within current INF project, so we could expect more use cases and other useful scenarios to address materials research requirements, including API introduction to facilitate data exchange with external systems (materials repository, ML and data analysis tools) and measurement devices, establishing and monitoring materials research pipelines in collaborative research centers.

This research was financially supported by the Deutsche Forschungsgemeinschaft (DFG, German Research Foundation) Project-ID 388390466-TRR 247 (subproject INF).

The study was carried out as a part of the state assignment (project nos. 075-01176-23-00).

References

1. Klingner, C., Denker, M., et al.: Research Data Management and Data Sharing for Reproducible Research—Results of a Community Survey of the German National Research Data Infrastructure Initiative Neuroscience. eNeuro 7 February 2023, 10 .(2) ENEURO.0215-22.2023; https://doi.org/10.1523/ENEURO.0215-22.2023
2. Herres-Pawlis, S., Liermann, J.C., Koepler, O.: Research data in chemistry – results of the first NFDI4Chem community survey. Z. Anorg. Allg. Chem. **646**, 1748–1757 (2020). https://doi.org/10.1002/zaac.202000339
3. Junkes, H., Oppermann, P., Schlögl, R., Trunschke, A., Krieger, M., Weber, H.: FAIRmat-a Consortium of the German Research-Data Infrastructure (NFDI). ICALEPCS2021 (2021). https://doi.org/10.18429/JACoW-ICALEPCS2021-WEBL01
4. Kiselyova, N., Dudarev, V.: The distributed system of databases on properties of inorganic substances and materials. Int. J. "Inf. Theor. Appl." **12**(3), 219–224 (2005)
5. Banko, L., Ludwig, A.: Fast-track to research data management in experimental material science-setting the ground for research group level materials digitalization. ACS Comb. Sci. **22**(8), 401–409 (2020). https://doi.org/10.1021/acscombsci.0c00057
6. Nance, C., Losser, T., Iype, R., Harmon, G.: NoSQL vs RDBMS – Why there is room for both. In: SAIS 2013 Proceedings. 27 (2013). https://aisel.aisnet.org/sais2013/27
7. Kiselyova, N., Dudarev, V.: Creating inorganic chemistry data infrastructure for materials science specialists. Commun. Comput. Inf. Sci. **706**, 222–236 (2017). ISSN 1865-0929. https://doi.org/10.1007/978-3-319-57135-5
8. Sanders, G., Shin, S.: Denormalization effects on performance of RDBMS. In: Proceedings of the 34th Annual Hawaii International Conference on System Sciences, Maui, HI, USA, 2001, p. 9. https://doi.org/10.1109/HICSS.2001.926306
9. Kiselyova, N., Dudarev, V., Korzhuev, M.: Database on bandgaps of inorganic substances and materials. Inorganic Mater. Appl. Res. 7 №1, 34–39 (2016). https://doi.org/10.1134/S2075113316010093
10. SQL-script to transfer bandgap data to the materials properties management system demo tenant: https://gitlab.ruhr-uni-bochum.de/vic/infproject/-/blob/master/Adds/SQL/SQLQuery_BandGapToINF.sql, last accessed 2024/01/08
11. GitLab repository, https://gitlab.ruhr-uni-bochum.de/vic/infproject, last accessed 2024/01/08

Database for Properties of Nuclear Reactor Materials Based on the Ontology and NoSQL Data Format

Sergey A. Dyachkov[1,2], Adilbek O. Erkimbaev[1], Sergey Yu. Grigoryev[1,2],
Pavel Yu. Korotaev[2], Andrey V. Kosinov[1], Pavel R. Levashov[1],
Maxim A. Maltsev[1], Dmitry V. Minakov[1], Igor V. Morozov[1(✉)],
Mikhail A. Paramonov[1], Aleksey V. Yanilkin[2],
and Vladimir Yu. Zitserman[1]

[1] Joint Institute for High Temperatures of Russian Academy of Sciences, Moscow, Russia
morozov@jiht.ru
[2] Dukhov Automatics Research Institute, Moscow, Russia

Abstract. In the paper, the development of a new generation database for the properties of nuclear reactor materials is discussed. The goal of this project is to accumulate experimental and theoretical data on the mechanical, thermophysical, and service properties of different materials, their microstructure, and transformations due to radiation-induced swelling, corrosion, and other degradation processes for future use in data analysis and computer simulations that can predict unknown material properties and therefore accelerate the development of new materials for contemporary and planned nuclear reactors. Due to the large variety of data standards, types, and formats, as well as the various physical quantities and their dimensions used in different data sources, the use of typical relational databases requires a huge amount of handwork for restructuring and fitting the data into the common tables. Moreover, adding new data types may require a redesign of the database architecture. In this case, NoSQL databases where the information is stored as a set of JSON documents might greatly improve the flexibility of the data storage, although searching, extracting, and merging the data from different documents becomes more complicated. As a solution for this problem, we propose using an ontology for data annotation and integration that allows one to store the datasets in their original forms and convert them on-the-fly when processing a particular user query.

Keywords: database · NoSQL · ontology · semantic search · machine learning · nuclear reactor materials

1 Introduction

Materials informatics is a rapidly developing branch of the material science concerned with the issues of data storage, annotation, and management suit-

J. Baixeries et al. (Eds.): DAMDID/RCDL 2023, CCIS 2086, pp. 104–114, 2024.
https://doi.org/10.1007/978-3-031-67826-4_8

able for further processing using contemporary statistical and machine learning techniques [5, 8, 14, 15]. Along with experiments and computer simulations, the machine learning algorithms has become an indispensable part of scientific research related to the discovery, optimization, and development of new materials. Perhaps one of the most prominent projects of such kind is the Materials Genome Initiative [21] which introduces a new paradigm in the material sciences. In particular, the creation of modern nuclear reactors relies greatly on advances in the material science, including studies of engineering materials, fuels, and coolants [19, 20].

At present, there are a few nuclear reactor material databases available [9], that have been built using the traditional relational scheme. Although these databases are very fast and reliable, they lack flexibility when a broad range of materials, properties, and experimental conditions are to be considered. It is of particular importance when storing large amount of raw experimental and simulation results for fundamental or applied research rather than developing a ready-to-use engineering database with complete and verified datasets. In the first case, the so-called non-relational or NoSQL databases [11] may be used, as demonstrated by such projects as Citrination [18] or NOMAD [12].

In document-oriented NoSQL databases, the data is stored in collections of documents or records, usually represented in the JSON format. Their advantages are as follows:

- the source data structure can be retained provided that it is represented as a set of JSON documents (which is possible for most of the data structures);
- a database can be appended by new data types without the need to redesign the structure of tables and their relations;
- updates and exchanges of data between different databases are fast and straightforward;
- NoSQL databases can be scaled almost infinitely with the growth of the information volume, particularly with the use of distributed computer systems, in contrast to the relational databases, where the extension is limited and strongly affects the performance;
- there is native support for storing binary files along with structured JSON documents for many of NoSQL databases (see, e.g. MongoDB [3]).

At the same time, searching and analyzing data in NoSQL databases can be quite tricky. As soon as the data is saved in different formats using various physical quantities and dimensions, the database becomes disintegrated unless special metadata or annotations are provided. In this work, we propose to use an ontology as a supplementary metadata subsystem.

The ontologies being generalized semantic data models are used in many scientific areas, including material sciences [6, 7, 17]. If the particular data items (JSON key-value pairs) are linked to the ontology classes, the data from different collections can be related to each other, and moreover, the internal links between ontology entities enable one to use a wide range of instruments for data processing and discovery without rearranging the data in the original storage. This way of data integration is discussed, e.g., in [13, 16]. Our idea of ontology-based data linkage is somewhat similar to the one presented in [6,7], although it has many differences in its particular implementation.

In this paper, we briefly describe the idea and implementation of a new information system for nuclear reactor materials based on a NoSQL database, a custom-designed ontology, and a user-friendly web interface.

2   Database Structure and Ontology

The scheme of the information system is given in the Fig. 1. The users can interact with the database via a web interface, a Python API or command-line access (for administration and maintenance). The data is put into the storage subsystem, whose structure is presented in the Fig. 2.

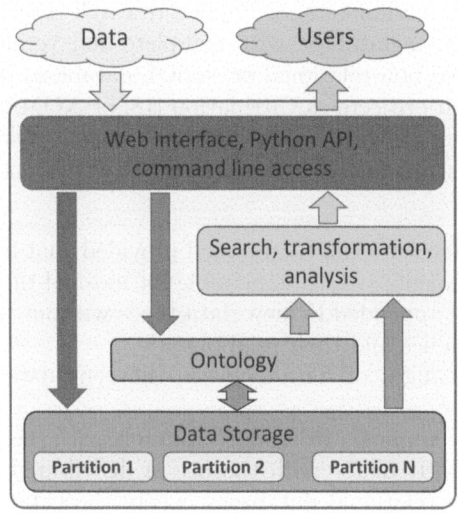

Fig. 1. The general scheme of the data flow in the information system for nuclear reactor materials.

The elementary data items in the database include key-value pairs, tables in the special format, and links to the binary files saved in the same storage. The hierarchical structure provided by the JSON format allows sets of data to be combined into objects that contain other objects, etc. The records represent the minimal amount of data that can be inserted into or extracted from the database. A set of records forms a collection that is somehow related to a table in a relational database (the records are related to the table rows in this case). There are no constraints on the internal structure of JSON documents, but typically, all records in the same collection have similar JSON objects. For instance, when storing the results of a series of experiments or simulations of the same type, each record usually contains the results of a single experiment or simulation.

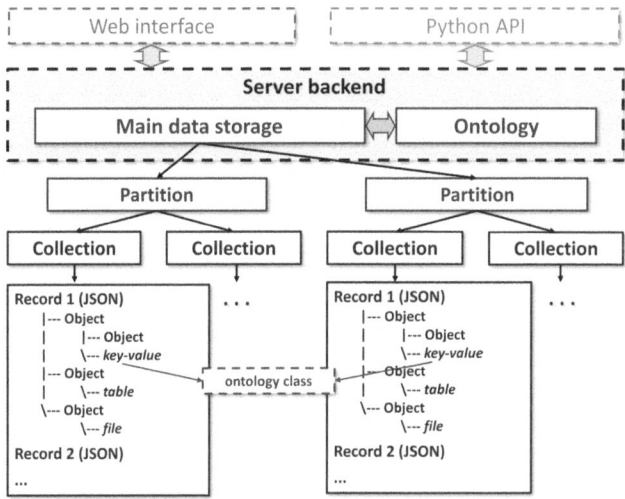

Fig. 2. The structure of the data storage subsystem. The red arrows show the linking of the data from different collections using the ontology class. (Color figure online)

The partition contains one or more data collections. It is the minimal data structure for which the access rights of users and groups are defined. Typically, the partition is assigned to a specific organization or research group that provides the data. It can also be related to specific research or a literature review project.

The key feature of the developed system is the method of linking data from different collections. In our case, it is done by changing the names of JSON keys and/or values to the names of the corresponding ontology classes. It can be done for the keys that contain numerical or literal data as well as for the dimensions. Therefore, the ontology is used to annotate all or a part of the data in the collections. The linkage process is described in Sect. 4 in more detail.

The fragment of the ontology is shown in the Fig. 3. It is usually represented as a set of classes and their subclasses that form an acyclic graph of linked entities. Each class can contain instances, e.g., the entity "Oxygen" is an instance of the classes "Atom", "Molecule" and "ChemicalSubstance". In course of this project, a special ontology for nuclear reactor materials was developed that contains more than 1000 classes and more than 3000 class instances describing various substances, engineering (structural) materials, devices, material properties (thermophysical, mechanical, microstructural, etc.), radiation-induced and general degradation processes, experiment and simulation types, as well as some general concepts. An early concept of such an ontology was presented in [10].

The upper level of the developed ontology contains the following classes: "Object", "Process", "Quality", and "Unit". The "Object" class describes both material and information objects, where the material objects include elementary particles, chemical substances, complex materials like alloys, fuels, etc., various samples, constructive objects, typical equipment of nuclear reactors and testing

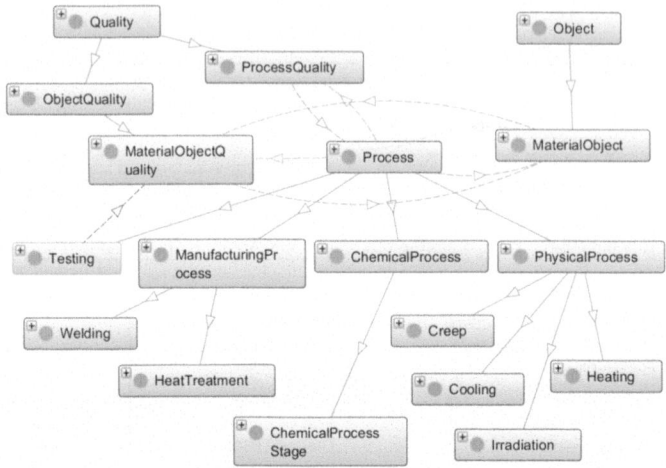

Fig. 3. A fragment of the ontology for nuclear reactor materials. The square boxes denote ontology classes; the solid lines correspond to the relations "class-subclass" while the dashed ones correspond to all other semantic relations.

laboratories. The information objects represent auxiliary classes for describing different data types, tables, files, and software. The next upper-level class "Process" is responsible for the description of general chemical and physical processes, manufacturing processes, and material treatment during their operation or testing. According to the main purpose of the database, the focus is on material degradation processes like radiation swelling, embrittlement, or corrosion. The class "Quality" contains subclasses for the properties of material and information objects, as well as the process characteristics; it also includes most of the general physical quantities. Finally, the "Unit" class has subclasses for all types of physical quantities used in the database, grouped by their dimensions. Each of these subclasses has a few instances representing individual units that can be converted to each other using the coefficients and offsets stored as numeral properties of the instance. In particular, the unit class instances are used by the web interface for conversion of units "on the fly" when retrieving information from the database.

The ontology classes can be linked to each other in different ways. Along with the links of the type "class-subclass" also known as "is a" relation, that form the main (vertical) structure of the ontology, there could be horizontal links between the classes that define their semantic relations. In this work, we used the following types of such relations (object properties): "has participant", "determines", "describes", "has attribute", has quality", "is related to", "has part", "has unit" and their inverse counterparts. The two special object properties "has key", and "has value" are reserved for special use when marking the collections in which the keys are linked to a particular class or a class instance.

As shown in the Fig. 1, the ontology is used for searching and analyzing the data. In the simplest case, the ontology can be used as a glossary of terms for a search query. An extended "semantic search" mode allows one to obtain semantically linked quantities, i.e., the terms linked with the original ones given in the search query, using one or more ontology class relations described above. For example, the semantic search for "Dislocation" retrieves the records that have the key "Dislocation" as well as the related keys "DislocationLoop", "DislocationDensity", "DislocationLoopDensity", and "DislocationLoopSize". The types of relations used for this search type are controlled by the user. With the help of ontology, it is also possible to get a list of materials that have the above mentioned properties or a list of processes that create or are affected by dislocations. Thus, the diversity of possible class links makes semantic search an efficient tool for immediate data comparison and later use in machine learning algorithms.

3   Implementation Details

The software tools used in the project are shown in Fig. 4. The main data storage is maintained by the MongoDB database, while the ontology is stored in the graph-oriented RDF database AllegroGraph [1]. On one side, the use of two different engines for JSON and RDF databases causes some inconveniences related to ensuring the consistency of the data; each of these engines can be potentially used for both RDF and JSON databases. On the other side, the use of MongoDB provides the best scalability for storing JSON documents and binary files, while AllegroGraph has the best performance for querying the ontology using the SPARQL language.

An additional tiny SQL database is used for the user-specific data, access rights, and general description of the data partitions. The SQL database is also used for some technical purposes, such as configuring caches and web pages. In particular, we use SQLite for a compact standalone installation of the system and PosgreSQL for a multi-user access.

It should be noted, that some relational DBMS allow have extension to operate with graphs and therefore they could be used for storing the ontology which avoid the use of AllegroGraph or another RDF engines. Their testing and implementation in this project are left for future work.

The user interface is based on the Django framework [2] using the Webix JavaScript library [4] on the client side.

4   User Interface and Access Rights

The database can be accessed from either a local or remote host via both the Python API and the web interface. A fragment of the web interface is shown in Fig. 5. The main menu in the navigation panel directs the user to sections "Home", "Search", "Data", "Ontology", "Simulation", and to the administrative panel (via the context menu when clicking on the user name). The language

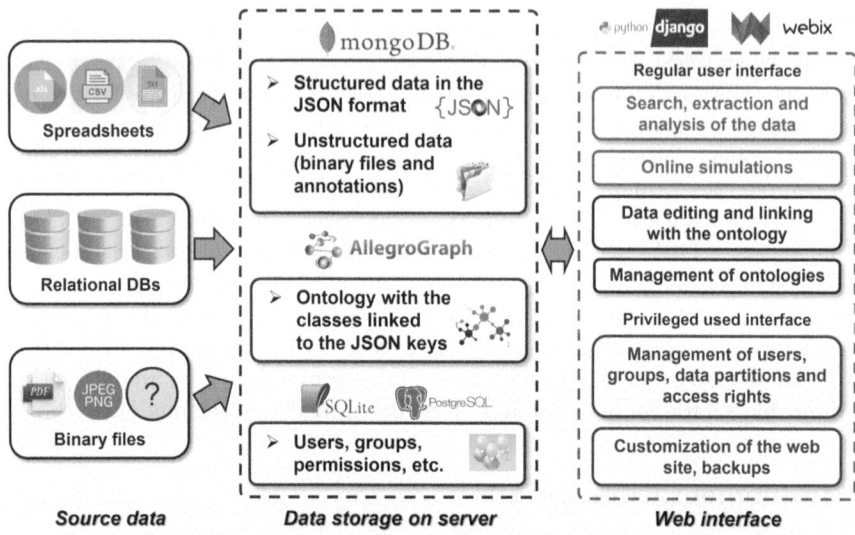

Fig. 4. Software components and their interactions.

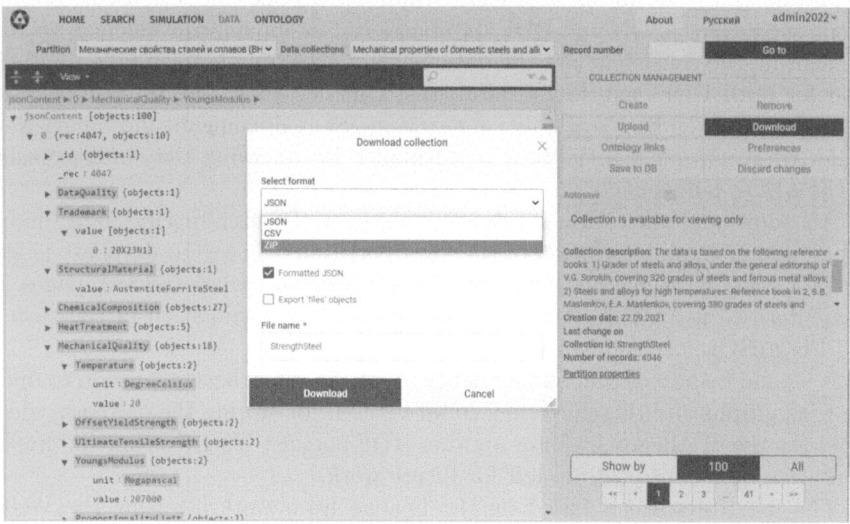

Fig. 5. A sample view of the web data editor with an opened pop-up window for downloading the collection.

button allows one to switch between Russian and English. All the sections are described in detail below.

The "Home" section has a few pages with a general description of the project and the database summary.

The "Search" section is responsible for searching the data using different search modes and displaying the results in the form of customizable tables or plots. The search is based on the query string, which in a simple mode contains a number of JSON key names, optionally followed by relation singes, values, and units. These elements can be combined using "AND" and "OR" operators ("AND" is used by default), e.g., "Temperature > 500[K] Temperature < 800[K] (HeatCapacity OR Enthalpy)". The key names are usually taken from the ontology, but it is not obligatory, so that custom user keys not linked to the ontology can be used as well. If the units of the retrieved numerical values are defined and linked to the ontology, they are automatically converted to the units selected by the user in the search results table. It is especially important when retrieving values from different data collections where the dimensions of the values can differ.

In the semantic search mode, the query contains the element in the form "ClassA ClassB ... options" where the ontology classes "ClassA", "ClassB" can be supplemented by options like types of the semantic relations described above, the depth of subclasses to be searched, and exclusions of selected classes. The semantic and simple search queries can be combined. Therefore, the data can be retrieved from the database even without using the ontology (in the simple search mode), but the use of the ontology extends the capabilities of the search significantly.

The "Data" section is used for importing, editing, or exporting the data, as well as linking it to the ontology. The screenshot given in Fig. 5 shows the data editor page for a particular data collection. On the left side, the JSON structure of one of the records is displayed, where some of the JSON keys are highlighted in blue, indicating that this data is linked to the corresponding ontology class.

During the linkage of a JSON key that can represent a key–value pair or an object, the user invokes a context menu by clicking on the corresponding key. If the key name coincides with the name of an ontology class (e.g., "Temperature"), the web interface suggests the corresponding link, and after user approval, the triplet "CollectionName – has key – Temperature" is added to the ontology, where "CollectionName" is the name of the instance of the special class "DataCollection". Such instances are created for all the collections at their creation. After the insertion of this triplet, all keys named "Temperature" in the current collection are considered to be linked to the ontology. This rule obviously restricts the use of arbitrary JSON key names but greatly simplifies the linkage process. If the original key name is different from the destination ontology class name, it is to be renamed within the whole collection. Alternatively, for one-to-one linkage, the user can simply start automatic linking of all the keys that are named properly.

After the linkage, the key–value pairs are converted into objects, where the value is stored in a special "value" key. There are also rules for automatic extraction of the dimensions of physical quantities and storing them in the special "unit" key.

The mapping of key values and units to the ontology classes can be done in a similar way, except that the relations "has value" and "has unit" are used instead of "has key" and the information of the JSON path within the record is also stored in the ontology. Thus, if the user links the value "AustentiteFerriteSteel" of the key "StructuralMaterial" (see Fig. 5), it does not affect the values of other JSON keys or the values of the keys "StructuralMaterial" located in nested JSON objects.

The data can be entered via the web editor shown in Fig. 5 or imported from a JSON, CSV, or ZIP file, where ZIP archive is used for importing binary files that are to be attached to the collection. The imported records can be added to an existing collection. In this case, all ontology links are applied to these new records automatically.

On the right side of Fig. 5, one can see the collection management tools (some of which are disabled for the current user). In the front, a pop-up window is open, showing the dialogue box for exporting the collection in the different formats mentioned above.

The "Ontology" section allows browsing, searching, and modifying the ontology. In the current version of the system, modification of the ontology via the web interface is limited to changing the properties of existing classes without the possibility to add or remove classes. A new version of the ontology as a whole can be uploaded as an OWL file using the administrative interface. As the ontology has many classes that are still not linked to any data and are reserved for future use, we expect that its extension may not be needed too frequently.

The "Simulation" section contains a set of simulation modules in Python that can perform calculations for particular problems using the information from the database. This section is still under development. At present, it includes web forms for the calculation of yield strength, tensile strength, and heat resistance; the calculation of swelling depending on the radiation dose and temperature; the calculation of the parabolic corrosion rate constant; and the search for new compositions with lower corrosion rates using Bayesian optimization. In this version, the forms use predefined models obtained via machine learning techniques using preliminary selected data sets. In the future, it is planned to extend the capabilities of this module to use arbitrary data from the database.

The administrative panel can be used to maintain users, user groups, access rights, data partitions, and site preferences. Access rights to a particular data partition can be assigned to a group of users. The types of these rights include "none", "read", "read/write", and "read/write/manage". The latest type enables the users of the group to create, delete, or edit the properties of data collections. There are also five user roles: "guest", "reader", "editor", "admin", and "superuser". The "superuser" has full access to the data, and moreover, he is responsible for managing data partitions, the creation and removal of user groups, the editing of user group rights, and the designation of a group administrator ("admin"). The "admin" user can add or remove other users within the same group and change their roles. The "guest" user has access only to special information resources, not to the main data storage. The access policy for different users depending on their role is given in table 1.

Table 1. Assess rights to a data partition for the users depending on their role and the access level of the corresponding group of users.

User role	Access level of a group of users to a data partition			
	none	read	read/write	read/write/manage
guest	no access			
reader	no access	read	read	read
editor	no access	read	read/write	read/write/manage
admin	no access	read	read/write	read/write/manage
superuser	full access			

5   Conclusions

The concept and implementation of the information system based on the NoSQL database and ontology metadata are presented. Although relational databases provide better performance for highly structured data, semi-structured data in different formats is more convenient to store as collections of JSON documents. It ensures better flexibility, scalability, and simple updates. At the same time, the ontology is found to be an appropriate instrument for linking the data from different collections, which results in overall data integrity and enables advanced search and analysis instruments based on the semantic relations between the ontology classes. The data can be accessed via both the web interface and the Python API. The system is still under development, although all the main features are implemented. The future plans are related to the extension of the search possibilities and adding machine learning and other algorithms that will make full use of the ontology metadata.

Acknowledgements. This work is supported by the Russian State Atomic Energy Corporation "Rosatom".

References

1. AllegroGraph graph database project web site. https://allegrograph.com/. Accessed 31 May 2023
2. Django framework web site. https://www.djangoproject.com/. Accessed 31 May 2023
3. MongoDB nosql database project web site. https://www.mongodb.com/. Accessed 31 May 2023
4. Webix Javascript UI library web site. https://ru.webix.com/. Accessed 31 May 2023
5. Agrawal, A., Choudhary, A.: Perspective: materials informatics and big data: realization of the "fourth paradigm" of science in materials science. APL Mater. **4**(5), 053208 (2016)
6. Ashino, T.: Materials ontology: an infrastructure for exchanging materials information and knowledge. Data Sci. J. **9**, 54–61 (2010)

7. Ashino, T., Nishikawa, N., Kadohira, T.: Data analysis environment for materials science and engineering integrating heterogeneous data resources. In: Elizarov, A., Novikov, B., S, S. (eds.) Data Analytics and Management in Data Intensive Domains: Proccesings of XXI International Conference DAMDID/RCDL'2019 (October 15–18, 2019, Kazan, Russia), p. 420. Kazan Federal University, Kazan (2019)

8. Austin, T.: Towards a digital infrastructure for engineering materials data. Mater. Disc. **3**, 1–12 (2016)

9. Belov, G.V., Aristova, N.M.: Databases on the properties of materials and substances for nuclear power. Matematicheskoe modelirovanie **29**(6), 135–142 (2017)

10. Chusov, I.A., Kirillov, P.L., Pronyaev, V.G., Erkimbaev, A.O., Zitserman, V.Y., Kobzev, G.A., Fokin, L.R.: Ontologies and databases on thermophysical properties of nuclear reactor materials. Izvestiya vuzov. Yadernaya Energetika **1**, 5–18 (2019). in russian

11. Davoudian, A., Chen, L., Liu, M.: A survey on NOSQL stores. ACM Comput. Surv. (CSUR) **51**(2), 1–43 (2018)

12. Draxl, C., Scheffler, M.: The NOMAD laboratory: from data sharing to artificial intelligence. J. Phys. Mater. **2**(3), 036001 (2019)

13. Erkimbaev, A.O., Zitserman, V.Y., Kobzev, G.A., Kosinov, A.V.: Integration of information resources containing data on the properties of substances and materials: practical implementation and existing tools. Autom. Doc. Math. Linguist. **52**, 257–264 (2018)

14. Himanen, L., Geurts, A., Foster, A.S., Rinke, P.: Data-driven materials science: status, challenges, and perspectives. Adv. Sci. **6**(21), 1900808 (2019)

15. Kalidindi, S.R., De Graef, M.: Materials data science: current status and future outlook. Annu. Rev. Mater. Res. **45**, 171–193 (2015)

16. Kosinov, A., Erkimbaev, A., Zitserman, V.Y., Kobzev, G.: Ontology-based methods of thermophysical data integration. J. Phys: Conf. Ser. **1385**(1), 012033 (2019)

17. Li, H., Armiento, R., Lambrix, P.: An ontology for the materials design domain. In: Pan, J.Z., et al. (eds.) ISWC 2020, Part II. LNCS, vol. 12507, pp. 212–227. Springer, Cham (2020). https://doi.org/10.1007/978-3-030-62466-8_14

18. Michel, K., Meredig, B.: Beyond bulk single crystals: a data format for all materials structure-property-processing relationships. MRS Bull. **41**, 617–623 (2016)

19. Morgan, D., Pilania, G., Couet, A., Uberuaga, B.P., Sun, C., Li, J.: Machine learning in nuclear materials research. Curr. Opin. Solid State Mater. Sci. **26**(2), 100975 (2022)

20. Novoselov, I.I., Savin, D.I., Yanilkin, A.V.: The effect of irradiation conditions on generation of defects and their clusters. J. Nucl. Mater. **546**, 152762 (2021)

21. de Pablo, J.J., et al.: New frontiers for the materials genome initiative. NPJ Comput. Mater. **5**(1), 41 (2019)

Machine Learning Methods
and Applications

TrojanInterpret: A Detecting Backdoors Method in DNN Based on Neural Network Interpretation Methods

Oleg Pilipenko$^{(\boxtimes)}$ [ID], Bulat Nutfullin [ID], and Vasily Kostyumov [ID]

Moscow State University, GSP-1, Leninskie Gory, Moscow, Russia
{pilipenkoog,nutfullinbm,kostyumovvv}@my.msu.ru

Abstract. Neural networks are increasingly used in various applications, but their training most often requires a huge amount of data. Inspecting the entire training dataset becomes impossible in such a situation and creates the opportunity for backdoor attacks when an attacker injects special triggers into training data. It makes the ML model perform incorrectly in the presence of the trigger while behaving normally with clean inputs.

In this paper, we present a new method for backdoor detection in neural networks. This approach is based on neural network interpretation techniques and uses the idea of the difference between distributions of saliency values of neural networks with backdoors and without them.

The proposed method demonstrated robustness when detecting backdoors on several datasets.

Keywords: Machine learning · Neural network · Backdoor · Neural network interpretation · Trojan Neural Network

1 Introduction

Deep neural networks (DNNs) now play a crucial role in many important applications, including face recognition, speech recognition, self-driving vehicles, medical image recognition, and defending against cybersecurity threats.

However, the lack of interpretability is a significant barrier to the widespread adoption of DNNs in real systems, especially in security-sensitive applications. Furthermore, modern machine learning models are extremely complex functions and difficulties in understanding how they work lead to the possibility of attacking them.

According to recent studies, DNNs are susceptible to backdoor attacks, which cause DNNs to behave maliciously when specific triggers are added to the input images in inference.

This is achieved by adding a small amount of data with a special trigger to the training dataset.

J. Baixeries et al. (Eds.): DAMDID/RCDL 2023, CCIS 2086, pp. 117–130, 2024.
https://doi.org/10.1007/978-3-031-67826-4_9

At the same time, neural networks with an embedded backdoor perform correctly on images without trigger making it very hard to detect if the model has been backdoored.

In this paper we propose a new method for detecting backdoors in ML models based on the ideas from neural network interpretation field. For the applications that use ML algorithms, it is crucial to understand how they make their decisions. There were proposed several techniques to explain neural networks outputs [2, 18].

In our research, we have discovered the connection between the backdoor attacks and model interpretation. We used widespread interpretation technique activation maximization [1] to find the input that maximize model's output.

After that, we demonstrated that saliency maps [2] of these inputs differ significantly between backdoored and clean models. The advantage of the proposed method is its good explainability since it is based on model interpretation approaches.

Our paper makes the following contributions in the problem of detecting backdoors in neural networks:

- We demonstrated the difference of interpretation between clean and poisoned with backdoor models
- We propose an explainable method for detection hidden triggers embedded inside deep neural networks using model interpretation approaches
- We validate the robustness of our method on a large amount of models trained on clean and backdoored versions of four widespread computer vision datasets

2 Related Works

Adversarial attacks are prevalent in various data forms, including images, texts, graphs, and sequences.

2.1 Adversarial Attacks on Other Domains

One of the pioneering works in this field focused on adversarial attacks in natural language processing (NLP), identifying the challenges of defining a coherent sequence within a discrete space of objects. Textual adversarial attacks are classified according to perturbation levels into character-level, token-level, and sentence-level attacks. Character-level methods involve visual character replacements or gradient-based decisions to modify characters, while others craft subword alterations to form unconventional English text. Token-level strategies involve replacing specific words based on relevance scores or using algorithms to substitute words with techniques like BERT's masked model. Sentence-level attacks create new instances through paraphrasing, back translation, or competitive dialogue agents.

The robustness of large Transformer-based models for NLP is a subject of concern, and understanding their vulnerability through adversarial attacks is crucial. One paper proposes a new black-box sentence-level attack that fine-tunes a pre-trained language model to generate hard-to-detect adversarial examples, showing that this approach outperforms competitors and highlights the non-robustness of current models [21]. Another paper addresses the challenges of adversarial attacks on categorical sequences, including non-differentiability, constraints on transformations, and diversity. Two black-box methods are introduced that generate reasonable adversarial sequences to fool models in different contexts, such as money transactions and medical fraud [22].

2.2   Backdoor Attacks

In [3] Gu et al. proposed BadNets, which injects a patch backdoor by adding it to the training dataset. The attacker chooses a target label and a trigger patch. After that, an attacker adds a trigger to the random subsample of training data to modify its labels to the target label. Using this approach, the attacker can achieve a 99% attack success rate without impacting model performance on clean inputs.

In [4] Chen et al. proposed the method called Blend, in which they inject a randomly generated pattern with the size of the input images. Unlike the BadNet, this approach made patterns imperceptible for humans and used Gaussian noise for this purpose. The authors conducted an evaluation to demonstrate that it is enough to inject only around 50 poisoning samples to achieve an attack success rate of above 90%.

2.3   Backdoor Defences

There were proposed many backdoor defences in recent years. They can be divided into two common categories: data inspection and model inspection [5].

Activation Clustering [7] belongs to data inspection category. In this method last hidden layer activations from training data is collected. After that, activations of inputs from the same class are clustered in two clusters. To determine if the cluster is poisoned or not silhouette score is used. This method works well only on the small datasets like MNIST.

STRIP [10] is designed to inspect input data during run-time. The authors intentionally perturbed the input by adding various image patterns and observe the randomness of predicted classes for perturbed inputs. Low entropy in predicted classes is evidence of a trigger on input.

A large number of methods have been proposed for model inspection.

Neural Cleanse [8] is a method for backdoor detection and trigger Neural Cleanse is based on the idea that using a model with a backdoor requires much smaller modifications to all input samples to misclassify them into the targeted class than any other labels. However, this method only performs well on small patch triggers and cannot work with blend triggers.

Liu et al. in [9] proposed the method ABS which inspects individual neuron activation differences for anomaly detection of backdoor. Although ABS works well with triggers of any size, it appears to be only effective under the assumption that the target label output activation needs to be activated by only one neuron. So it can be easily bypassed by other triggers.

Wang et al. [15] proposed a method TrojanNet to distinguish between backdoored and clean models in data-limited and data-free cases by performing adversarial attacks. In [6] was proposed to train a meta-classifier that predicts whether a given target model is trojaned.

However, there are a lot of different approaches for the detection of visually detected backdoors [11–13].

During our ch1review, we found no articles that used interpretation methods for detecting trojans of this specific types. This emphasizes the need for further exploration into the efficacy and applicability of interpretation techniques in identifying and countering such trojan attacks.

2.4   Neural Network Interpretation

Simonyan et al. in [2] proposed the method Saliency Map that allows to compare the output of the neural network with the contribution of each pixel of the image. The proposed method allows to visualize the value of one or another pixel of the input image when classifying the image. The authors initially considered only convolutional (convolutional) neural network models, since at that time they were the ones who obtained the best results in terms of quality metrics for various tasks of image modality. Now this approach is used for other ANN architectures, and its modifications serve as the basis for such adversarial evasion attacks as JSMA [19].

SHAP [16] and LIME [15] are widely used methods for interpreting model predictions by assigning importance to input features.

The activation maximization method proposed in Montavon et al. [1], Mahendran et al. [18]. The proposed method is to search for patterns (filters) for the input image that lead to the maximization of the model's output, depending on its target task (classification, detection, etc.)

Class activation maps methods were proposed by Zhou et al. [20], Selvaraju et al. [23], Chattopadhay et al. [24] and Omeiza et al. [25]. This class of methods is currently one of the most well-known and popular interpretation methods. In general, the class activation map is a gradient mask, which can be represented as a heat map. This map is obtained as a result of a direct pass of the neural network with the overlay of the values of the weights of one or another layer.

The Layered Relevance Propagation (LRP) method was first proposed at Montavon et al. [1], while its detailed justification was given by the authors later in the article [26]. The main idea of the method is to control the magnitude of relevance through special analytical formulas for its calculation. The method is based on a key statement called the conservation property, in which what was received by the neuron must be redistributed to the underlying layer in an equal amount.

There are a lot of methods interpretation we are tried to use for the trojan detection problem, but the best results was reached using methods [1, 2].

3 Background

3.1 Trojan Attack Problem

We consider a DNN backdoor to be a hidden pattern injected into a DNN during training. It produces unexpected behavior during inference only in a situation when a specific trigger is added to an input. In the case of a classification task, a DNN misclassifies inputs with trigger into the predefined target label. At the same time, backdoored ML model should output correct results given a clean input without a trigger (Fig. 1).

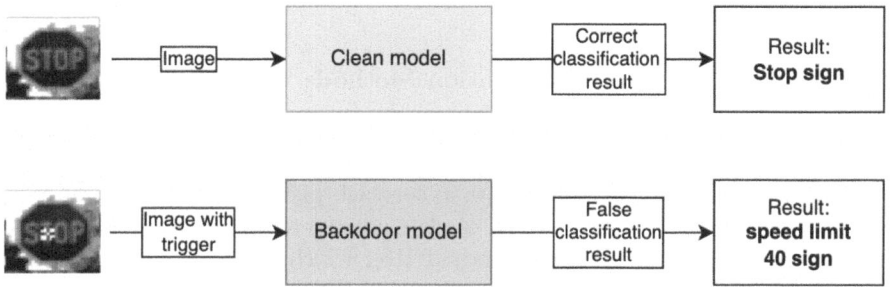

Fig. 1. The work of a clean neural network and a neural network with a backdoor on various images.

Formally, this problem can be written as follows for the classification problem. Let there be a clean dataset (dataset) $D = (X, y)$. An attacker choose subsample $\tilde{X} \subset X$ and for every $\tilde{x} \in \tilde{X}$ adds trigger δ to the image, changing the correct label y to some predefined target label y_t. As a result, we have a training dataset X' with a poisoned subsample.

We consider two widespread types of trigger injection. In the case of a patch poisoning [3] an attacker add a patch to the region of an image, and an image $\tilde{x}$ from the set $\tilde{X}$ is given by the Eq. 1

$$\tilde{x} = (1 - m) \cdot x + m \cdot \delta, \tag{1}$$

where m is the mask that regulates the position of the trigger.

In the case of blended poisoning [4] an attacker add noise which has the same size as image. A poisoned image $\tilde{x}$ is given by the Eq. 2

$$\tilde{x} = (1 - \alpha) \cdot x + \alpha \cdot \delta, \tag{2}$$

where $\alpha \in [0, 1]$ is blend ratio. Examples of these attacks are shown in Fig. 2.

Given a DNN model, our goal is to determine whether model is backdoored or not.

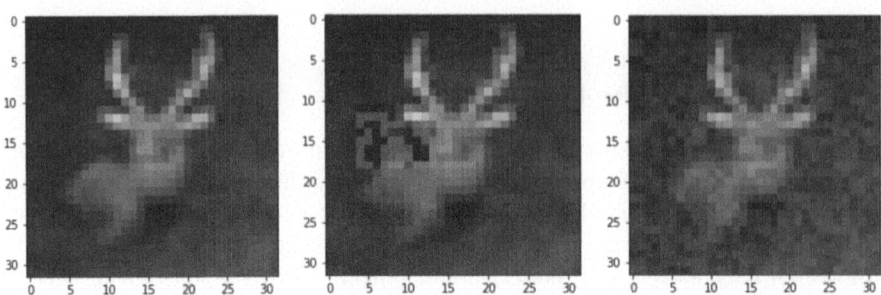

Fig. 2. An example of two types of triggers. 1 - clean, 2 - patch trigger, 3 - blend trigger.

3.2 Neural Network Interpretation Methods

In our approach, we use the Saliency Map [2] approach to detect anomalous behavior of neural networks. Unfortunately, the output of the Saliency Map method is normalized in the range [0, 1] for each pixel. However, we refuse to normalize the output values, which leads to an increase the results. We also use the activation maximization method [18], which allows us to create test images that can strongly activate neurons in a DNN. We use this method not only to create images with maximum activations but also to create images with minimum activations (and put a trigger on them).

Saliency Map. This method [2] allows you to sort the pixels of the input image based on their maximum influence on the $S_c(I)$ score. For this:

– Let $S_c(I)$ denote the activation value of the output neuron that corresponds to class C. Then the optimization tast will be formulated as:

$$\arg\max_I (S_C(I) - \lambda||I||_2^2) \tag{3}$$

Equation 3 gives the image that maximizes the selected class.
– We take the derivative with respect to the model parameters, we only assume that the parameters are fixed, and the derivative is taken with respect to the input image. Remove the normalization of the output to [0, 1]

When we use the method to test for the presence of a model backdoor by supplying images with poisoning to the input, we can see that neural network activations behave differently. However, the differences are insignificant. In order to increase the gap between interpretive images, we created images for testing using the activation maximization method.

Activation Maximization. Activation maximization [1] is an approach created for patterns searching for the input image that leads to the maximization of the output of the model, depending on its target task.

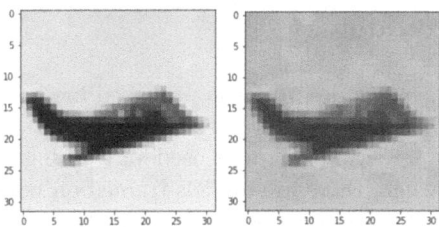

Fig. 3. An example of a result of an activation maximization algorithm.

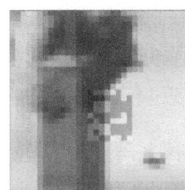 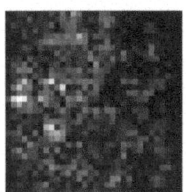  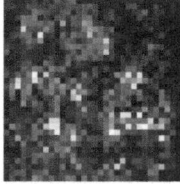

Fig. 4. An example of the interpretation of randomly poisoned test image in two cases: right pair - backdoor model activations, left pair - clean model activations.

Let us consider a classification problem. The input image $x_0 \in X$ is assigned the value of class $c_0 \in C$. The neural network model at the output represents a certain probability distribution $p(c|x)$, which makes it possible to attribute the image x_0 to a specific class c_0 by the maximum value of the vector $p(c|x)$. For the convenience of presentation, we will call such an example an image prototype or simply a prototype. Then the search for the image x_0^* - the maximally activating class c_0 can formally be formulated as:

$$x_0^* = \max_x(\log(p(c|x)) - \lambda||x||^2) \qquad (4)$$

The class probabilities modeled by the DNN are vectors obtained by gradient optimization. Therefore, performing an inverse gradient descent from the resulting probability distribution is possible. The rightmost term of the optimization function is the regularization by the l_2 norm, which implements a preference for input data close to the original image. Figure 3 shows the difference between the original image and the image that maximized model activations.

We used this method to create test images. In one case, we maximized activations and pasted a randomly generated trigger within a given specification. In another, we inverted the sign in the optimization problem to get images with minimal activations. Finally, a trigger with a given specification was also pasted in images with minimal activations. Examples of the behavior of a clean neural network and a backdoored neural network are shown in Fig. 4.

This method helped us create small sets of test images that allowed us to distinguish between backdoored and non-backdoored models.

4   Detection Metodology

In this part of the paper, we describe the proposed method for detecting backdoors in neural networks with interpretation methods. In Algorithm 1, we provide the pseudocode for creating a dataset used for training detector meta-model. We consider the case where there are models trained on poisoned and clean versions of various datasets, for example, MNIST, CIFAR-10, GTSRB, etc. For every dataset D_i, we denote as M_i the set of models that have been trained on D_i and their corresponding label (clean or poisoned).

For every dataset D_i, we randomly select one trojaned model m from M_i and sample N different clean images (and their labels). Next, we apply activation maximization (Sect. 3.2) and activation minimization (Sect. 3.2) procedures to this set of images. After that, we stack sampled N images with maximized and minimized versions and get $3N$ images.

For every dataset D_i we randomly select one trojaned model m from M_i and sample N different clean images (and their labels) from it. We apply activation maximization (Sect. 3.2) and activation minimization (Sect. 3.2) procedures to this set of images. After that we stack sampled N images with maximized and minimized versions of it and get $3N$ images.

At the next stage, we generate random triggers for blend and patch attacks and impose them on $3N$ images from the previous stage. We have discovered empirically that applying random blend trigger to the images really helps to distinguish between clean and poisoned with noise models. Then we iterate over the models trained on D_i: for every pair(model, label), where label means whether a model is clean or poisoned, we get saliency map in images with random trigger.

At the next stage we generate a random triggers for blend and patch attacks and impose it on $3N$ images from previous stage. We have discovered empirically that applying random blend trigger to the images really helps to distinguish between clean and poisoned with noise models. Then we iterate over the models trained on D_i: for every pair(model, label), where label means whether model is clean or poisoned, we get saliency map in images with random trigger.

After that, we analyze saliency map with some predefined values $P = \{0 \leq p_i \leq 1\}_i$: for every value $p \in P$, we calculate quantile value q_p and get from saliency map only values greater or equal than q_p. Then, we used result values for building histogram and stack together histograms from all p. The stacked histograms with the label if this model is the input for out meta-model, which is used for predicting the label of the model. Figure 5 shows the pipeline for creating training data.

Algorithm 1: TrojanInterpret algorithm

Input: datasets $D = \{D_i\}_{i=1}^{N}$ of images, models $M = \{M_i\}_{i=1}^{N}$, number of iterations j for ActMax algorithm, P - set of values for quantile analyses, bin parameters B

Output: dataset $\mathcal{D}$ for training metamodel

1 $\mathcal{D} = \emptyset$;
2 **for** $D_i \in D$ **do**
3     $X, Y \leftarrow$ ChooseRandomSubsample (D_i, N) ;
4     $\hat{m} \leftarrow$ ChooseRandomModel (M_i) ;
5     $I_{max} =$ ActMax$(X, Y, j, \hat{m})$;
6     $I_{min} =$ InverseActMax$(X, Y, j, \hat{m})$;
7     $I_{stack} =$ Stack(X, I_{max}, I_{min}) ;
8     $I_{triggered} =$ ApplyAttack(I_{stack}) ;
9   **for** $m, l \in M_i$ **do**
10        $A_m =$ SaliencyMap$(I_{triggered}, m)$;
11        $H_m = \emptyset$;
12   **for** $p \in P$ **do**
13           $q_p =$ GetQuantile(A_m, p) ;
14           $F_p =$ GetGreaterOrEqualValues(A_m, q_p) ;
15           $hist_p =$ GetHistogram(F_p, B) ;
16           $H_m =$ Stack$(H_m, hist_p)$;
17        $\mathcal{D} = \mathcal{D} \cup (H_m, l)$;

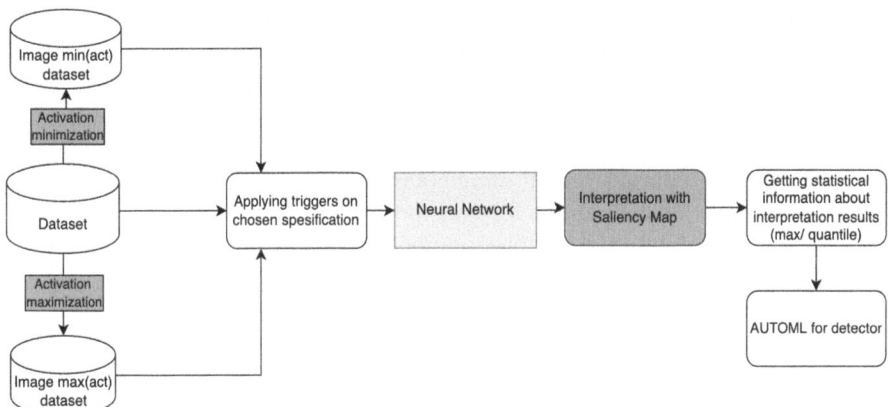

Fig. 5. Pre-training and training pipeline

5 Results

5.1 Experiment Setup

Dataset. We use Trojan Detection Challenge Dataset, a NeurIPS 2022 competition. The competition's aim is to analyze Trojan attacks on deep neural networks that are designed to be difficult to detect. The dataset consists of 4 neural network types. Organizers train the networks on four standard data sources: MNIST (used convolutional networks), CIFAR-10 (used Wide Residual Networks), CIFAR-100 (used Wide Residual Networks), and GTSRB (used Vision Transformers).

There are two standard Trojan attacks [3,4] were applied to dataset and were modified to be harder to the dataset. For the detection task, the training, validation, and test sets have 1,000 neural networks each. Networks are split evenly across all four data sources. Half of the networks are Trojaned, and there is a 50/50 split between the two attack types.

Overall the accuracy of all models in dataset was more then 87% and all model were fitted good enoph. Furthermore we test and measure trojan attack success rate. The results on trojaned samples for trojaned networks was more than 97%.

Detector Training. We ran a training pipeline on CIFAR-100, CIFAR-10, GTSRB, and MNIST datasets. Our steps to get activations distribution to compare on one dataset. We take a bunch of poisoned and clean models for each dataset, and then we make this sequence for each model(M_i) we have:

– We took 100 minimized images, the same images with random triggers of two types [3,4]. So we got 300 images of all types for neural networks testing. Then we fed all these images into a neural net(M_i) from Trojan Detection Challenge Dataset and got three types of output for every image: max values from the last layer, 99th percentile, and 97th percentile. So we got 3×300 numbers. We used that number statistics for detector training.
– To visualize results we built histograms for given numbers with fixed boundaries, and there are two types:
  • Build a histogram from all numbers (our main pipeline went this way)
  • Build multiple histograms for each percentile
– The histogram data are used for train metamodels. And we collect that data into a dataset.
– The collected dataset was used to train the detector. Obtained histogram data fed into metamodel to classify neural nets by their activations.
– As a first approach to creating metamodel, we trained the Catboost classifier, but later we improved the results using the AutoML approach.

5.2   Detection Capability

We took 40 images from each dataset and ran them through the pipeline in Fig. 5. Next, we analyzed the frequency of occurrence of pixel values, including those in different quantiles (0.9, 0.97). Histograms of the difference between clean models and different models with a backdoor are shown in Fig. 6. Distributions of models with a backdoor of the patch type are difficult to distinguish from the original.

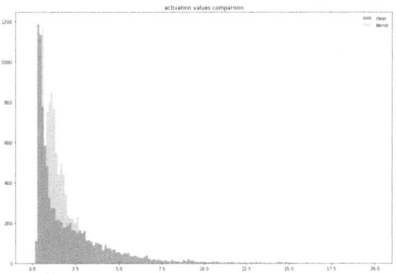 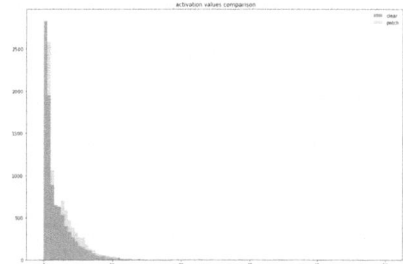

Fig. 6. Difference between clean and patch backdoored model (right). Difference between clean and blend backdoored model (left).

The distributions of the output data of the models can already be explicitly separated. However, if we take the percentiles of the distributions, the difference is more clearly highlighted. This difference is shown in Fig. 7. At the same time, the division of distributions can be seen with the eyes for backdoors like blends.

We first try to use CatBoost classifier to train the detector, but the final result was obtained using AutoML approach which increased accuracy by 7% (Table 1).

Table 1. Metrics of detection experiment

Dataset	Metric	Blend/Clean	Patch/Clean
MNIST	Acc	0.95	0.68
	AUC	0.81	0.6
CIFAR-10	Acc	0.98	0.66
	AUC	0.8	0.55
CIFAR-100	Acc	0.98	0.65
	AUC	0.86	0.68
GTSRB	Acc	0.96	0.57
	AUC	0.88	0.53
Average	Acc	0.96	0.62
	AUC	0.84	0.57

Even though the method proposed in the article does not detect patch-type backdoors well, other approaches, for example, [8], allow you to do this with high quality. We see one of the main reasons for the low quality of detection of such backdoors in their small area on the image. Since blend backdoors affect the entire image, the contribution from pixel changes is greater in the overall statistics. At the same time, iteratively using the occlusion sensitivity [14] approach to images provides a precise determination of the location of the trigger on the image, which will allow you to generalize this method.

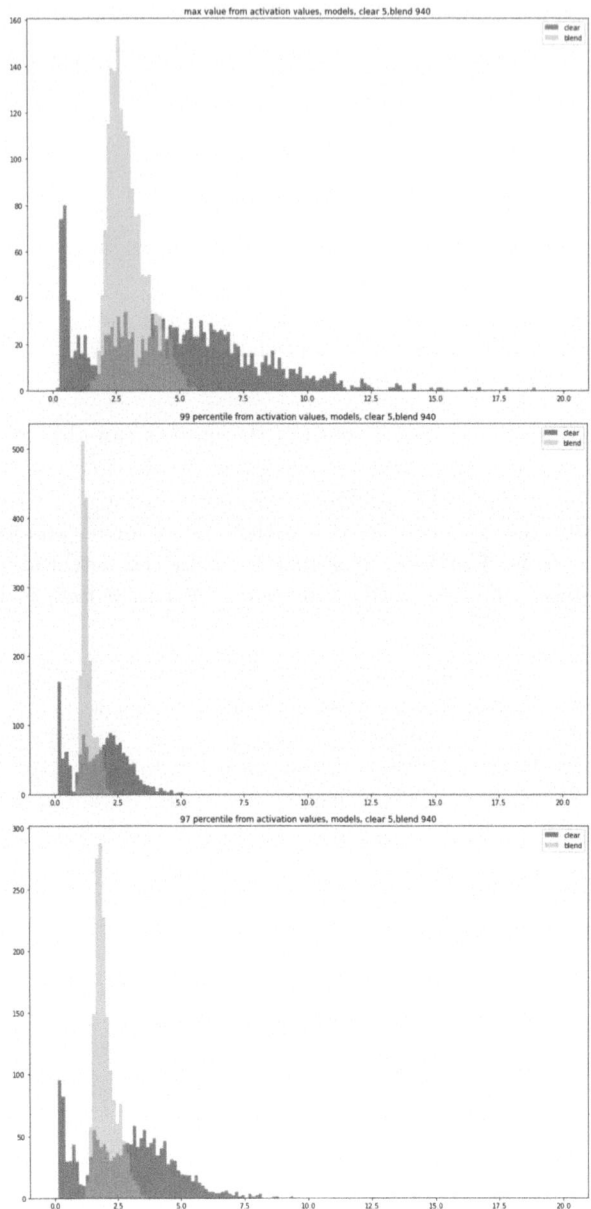

Fig. 7. Difference between clean and blend backdoored model in original distribution: max value (top), 0.99 percentile (middle), and 0.97 percentile(bottom)

6   Conclusion

The method has been proposed that involves maximizing the activations of the neural network, overlaying randomly specified patches on the images, using the saliency map interpretation method, and determining the distance between the resulting distributions. This method has shown promising results in detecting blend-type trojans, which add minor noise to the entire image. The achieved accuracy of 97.4% demonstrates the effectiveness of the proposed approach. However, for size-limited patch-type trojans, this pipeline did not show satisfactory results, indicating the need for alternative methods to be explored for their detection. Our future article will also focus on studying backdoors and exploring the possibility of using other interpretation methods to detect both types of attacks.

This research has been supported by the Interdisciplinary Scientific and Educational School of Moscow University "Brain, Cognitive Systems, Artificial Intelligence".

References

1. Montavon, G., Samek, W., Müller, K.R.: Methods for interpreting and understanding deep neural networks. Digit. Signal Proc. **73**, 1–15 (2018)
2. Simonyan, K., Vedaldi, A., Zisserman, A.: deep inside convolutional networks: visualising image classification models and saliency maps. arXiv preprint arXiv:1312.6034 (2013)
3. Gu, T., Dolan-Gavitt, B.,, BadNets, S.: Identifying vulnerabilities in the machine learning model supply chain. arXiv preprint arXiv:1708.06733 (2017)
4. Chen, X., Liu, C., Li, B., Lu, K., Song, D.: Targeted backdoor attacks on deep learning systems using data poisoning. arXiv preprint arXiv:1712.05526 (2017)
5. Gao, Y., et al.: Backdoor attacks and countermeasures on deep learning: a comprehensive review. arXiv preprint arXiv:2007.10760 (2020)
6. Xu, X., Wang, Q., Li, H., Borisov, N., Gunter, C.A., Li, B.: Detecting AI trojans using meta neural analysis. In: 2021 IEEE Symposium on Security and Privacy (SP), pp. 103–120. IEEE, May 2021
7. Chen, B., et al.: Detecting backdoor attacks on deep neural networks by activation clustering. arXiv preprint arXiv:1811.03728 (2018)
8. Wang, B., et al.: Neural cleanse: identifying and mitigating backdoor attacks in neural networks. In: 2019 IEEE Symposium on Security and Privacy (SP), pp. 707-723. IEEE, May 2019
9. Liu, Y., Lee, W. C., Tao, G., Ma, S., Aafer, Y., Zhang, X.: Abs: scanning neural networks for back-doors by artificial brain stimulation. In: Proceedings of the 2019 ACM SIGSAC Conference on Computer and Communications Security, pp. 1265-1282, November 2019
10. Gao, Y., et al.: Strip: a defence against trojan attacks on deep neural networks. In: Proceedings of the 35th Annual Computer Security Applications Conference, pp. 113–125, December 2019
11. Kolouri, S., et al.: Universal litmus patterns: revealing backdoor attacks in CNNs. In: Proceedings of the IEEE/CVF Conference on Computer Vision and Pattern Recognition, pp. 301–310 (2020)

12. Zheng, S., Zhang, Y., Wagner, H., Goswami, M., Chen, C.: Topological detection of trojaned neural networks. Adv. Neural. Inf. Process. Syst. **34**, 17258–17272 (2021)
13. Gao, Y., et al.: Design and evaluation of a multi-domain trojan detection method on deep neural networks. IEEE Trans. Depend. Secure Comput. **19**(4), 2349–2364 (2021)
14. Zeiler, M.D., Fergus, R.: Visualizing and understanding convolutional networks. In: Fleet, D., Pajdla, T., Schiele, B., Tuytelaars, T. (eds.) ECCV 2014. LNCS, vol. 8689, pp. 818–833. Springer, Cham (2014). https://doi.org/10.1007/978-3-319-10590-1_53
15. Wang, R., Zhang, G., Liu, S., Chen, P.-Y., Xiong, J., Wang, M.: Practical detection of trojan neural networks: data-limited and data-free cases (2020)
16. Ribeiro, M. T., Singh, S., Guestrin, C.: Why should i trust you?: Explaining the predictions of any classifier (2016)
17. Lundberg, S.M., Lee, S.-I.: A unified approach to interpreting model predictions. In: Guyon, I., et al. (eds.) Advances in Neural Information Processing Systems, vol. 30. Curran Associates, Inc (2017)
18. Mahendran, A., Vedaldi, A.: Visualizing deep convolutional neural networks using natural pre-images. Int. J. Comput. Vision **120**, 233–255 (2016)
19. Wiyatno R., Xu A. Maximal jacobian-based saliency map attack. arXiv preprint arXiv:1808.07945 (2018)
20. Zhou, B., Khosla, A., Lapedriza, A., Oliva, A., Torralba, A.: Learning deep features for discriminative localization. In: Proceedings of the IEEE conference on computer vision and pattern recognition, pp. 2921–2929 (2016)
21. Fursov, I., et al.: A differentiable language model adversarial attack on text classifiers. IEEE Access **10**, 17966–17976 (2022)
22. Fursov, I., Zaytsev, A., Kluchnikov, N., Kravchenko, A., Burnaev, E.: Gradient-based adversarial attacks on categorical sequence models via traversing an embedded world. In: van der Aalst, W.M.P., et al. (eds.) AIST 2020. LNCS, vol. 12602, pp. 356–368. Springer, Cham (2021). https://doi.org/10.1007/978-3-030-72610-2_27
23. Selvaraju, R.R., Cogswell, M., Das, A., Vedantam, R., Parikh, D., Batra, D.: Grad-cam: visual explanations from deep networks via gradient-based localization. In: Proceedings of the IEEE International Conference on Computer Vision, pp. 618–626 (2017)
24. Chattopadhay, A., Sarkar, A., Howlader, P., Balasubramanian, V.N.: Grad-cam++: Generalized gradient-based visual explanations for deep convolutional networks. In: 2018 IEEE Winter Conference on Applications of Computer Vision (WACV), pp. 839–847. IEEE March 2018
25. Omeiza, D., Speakman, S., Cintas, C., Weldermariam, K.: Smooth grad-cam++: an enhanced inference level visualization technique for deep convolutional neural network models. arXiv preprint arXiv:1908.01224 (2019)
26. Montavon, G., Binder, A., Lapuschkin, S., Samek, W., Müller, K.R.: Layer-wise relevance propagation: an overview. In: Explainable AI: Interpreting, Explaining and Visualizing Deep Learning, pp. 193–209 (2019)

Some Robust Variants of the Principal Components Analysis

Zaur M. Shibzukhov[1,2(✉)]

[1] Institute Mathematics and Informatics of Moscow Pedagogical State Univercity,
Moscow, Russia
intellimath@mail.ru
[2] Institute Applied Mathematics and Automation KBSC RAS, Nalchik, Russia

Abstract. One new robust variant of the formulation of the problem of searching for the principal components are considered. It's based on the application of differentiable estimates of the average value, insensitive to outliers. In principle, this approach makes it possible to overcome the influence of outliers in the task of searching for the principal components. The effectiveness of the proposed approach is clearly demonstrated on real data.

Keywords: Principal components analysis · Robust estimation · Data decomposition

1 Introduction

The *principal component analysis* (PCA) is one of the methods of data decomposition. First classical PCA was originally considered as the problem of the best approximation of a finite set of points by straight lines and planes [1]. The presentation of training data in the basis of the *principal components* (PC) contributes to the reliability of the method of error back propagation in the procedure for finding the global minimum of the quadratic error function of a neural network of linear elements [2]. In machine learning, the PCA is often used as a method of reducing the dimension of input data. However, if part of the data is significantly distorted, then the results of the classical PCA may be inevitably significantly distorted.

An overview of classical and robust PCA can be found in [3]. A detailed overview of modern robust variants of PCA can be found in [4].

In this paper, new robust variant of the standard formulation of the problem of searching for the PC are proposed. It's based on the application of differentiable estimates of the average value, insensitive to outliers. The proposed approach, in principle, allows to overcome the impact of the outliers. The classical and new robust variant of the PCA are described below. The effectiveness of the proposed approach to the construction of robust variants of the principal component method in comparison with their classical versions is clearly demonstrated empirically on several real datasets. The theoretical foundations of the

J. Baixeries et al. (Eds.): DAMDID/RCDL 2023, CCIS 2086, pp. 131–141, 2024.
https://doi.org/10.1007/978-3-031-67826-4_10

effectiveness of the proposed approach to PCA, shown here empirically, will be presented in future studies. The proposed robust version of the PCA is compared only with the classical version of the PCA. In this work, it was important, first of all, to show experimentally that the proposed robust variant of the classical formulation of the problem, going back to Pearson [1], makes it possible to effectively overcome the influence of outliers.

2   The Classical Approach

The classical problem [1], which boils down to the problem of finding the PCs, is as follows. It is required to find the center vector a_0 and the orthonormal basis $a_1, \ldots, a_m$ of an m-dimensional hyperplane in $\mathbb{R}^n$, such that the sum of the squares of the Euclidean distances to it from the points $x_1, \ldots, x_N$ is minimal:

$$\sum_{k=1}^{N} \left(\|x_k - a_0\|^2 - \sum_{j=1}^{m} (x_k - a_0, a_j)^2 \right) \to \min.$$

The solution of this problem can be reduced to a chain of problems in which the vectors $a_0, a_1, \ldots, a_m$ are sequentially located. This approach allows you to search for the principal components one by one.

The vector a_0 is found as a solution to the problem:

$$a_0 = \arg \min_{a \in \mathbb{R}^n} \sum_{k=1}^{N} \|x_k - a\|^2.$$

Its solution is a sample average: $a_0 = \frac{1}{N} \sum_{k=1}^{N} x_k$.

After finding a_0, centering is performed: $x_k \to x_k - a_0, \quad k = 1, \ldots, N$.

Then the vectors a_j $(j = 1, \ldots, m)$ are sequentially searched for as a solution to the problem

$$a_j = \arg \min_{\|a\|=1} \frac{1}{N} \sum_{k=1}^{N} \left(\|x_k\|^2 - (a, x_k)^2 \right).$$

After finding the next PC a_j, the transformation is performed:

$$x_k \to x_k - a_j (a_j, x_k).$$

The Lagrange multiplier method reduces our problem to the following problem

$$a_j, \lambda_j = \arg \min_{a, \lambda} \left\{ \frac{1}{N} \sum_{k=1}^{N} \left(\|x_k\|^2 - (a, x_k)^2 \right) + \lambda \left(\|a\|^2 - 1 \right) \right\}. \tag{1}$$

Since

$$\frac{1}{N} \sum_{k=1}^{N} (a, x_k)^2 = \frac{1}{N} (Xa)^\top (Xa) = \frac{1}{N} a^\top (X^\top X) a = a^\top S a,$$

where X – a matrix made up of vectors as rows $x_1, \ldots, x_N$, $S = \frac{1}{N} X^\top X$ – the covariance matrix. Then, in accordance with the necessary condition of the extremum, a_j and λ_j satisfy the following equations:

$$Sa_j = \lambda_j a_j$$
$$\|a_j\| = 1.$$

That is, the solutions specify orthonormal eigenvectors and eigenvalues of the matrix S.

To search for a_j, one can apply an iterative procedure:

$$a^{t+1} = \frac{1}{\lambda^t} \left(Sa^t \right), \qquad \lambda^t = \frac{(a^t)^\top Sa^t}{(a^t, a^t)}.$$

The outlier problem occurs when the distribution of values $\{\|x_k - a\|^2 : k = 1, \ldots, N\}$, $\{\|x_k\|^2 - (a, x_k)^2 : k = 1, \ldots, N\}$ contains outliers. The calculation of the empirical mean in this case gives a significantly biased value.

3   About Robust Variant of Target Functionals

In order to overcome the problem of outliers, it is proposed to use estimates of the average value, which are insensitive to outliers. And in order to be able to apply gradient minimization procedures, it is also proposed to use differentiable estimates of the average.

The proposed robust formulation of the problem involves replacing the minimization of the functional

$$Q(a) = \frac{1}{N} \sum_{k=1}^{N} \Phi_k(a)$$

with the minimization of the following functional

$$Q_M(a) = M\{\Phi_1(a), \ldots, \Phi_N(a)\}, \tag{2}$$

where $M\{z_1, \ldots, z_N\}$ is a differentiable estimation of the average value (M-estimator [5]), insensitive to outliers [6,7]. This formulation of the problem can significantly reduce the impact of outliers.

The problem of minimizing Q_M is reduced to solving the equation

$$\sum_{k=1}^{N} \frac{\partial M}{\partial z_k} \nabla \Phi_k(a) = 0.$$

To solve it, one can use the method of iterative reweighting:

$$a^{t+1} = \arg\min \sum_{k=1}^{N} v_k^t \Phi_k(a), \tag{3}$$

where

$$v_k^t = \frac{\partial M\{\Phi_1(a^t), \ldots, \Phi_N(a^t)\}}{\partial z_k}. \tag{4}$$

In this paper, the following robust estimate of the mean is used – the censored arithmetic mean:

$$\mathsf{CP}_\alpha\{z_1, \ldots, z_N\} = \frac{1}{N} \sum_{k=1}^{N} \min(z_k, \bar{z}_\alpha),$$

where $0 < \alpha < 1$. It uses an estimate of the smoothed variant of the α-quantile. It's a example of M-mean:

$$\mathsf{M}_\rho\{z_1, \ldots, z_N\} = \arg\min_u \sum_{k=1}^{N} \rho(z_k - u),$$

where $\rho(r)$ is a positive strictly convex function with minimum $r(0) = 0$. If ρ is twice differentiable then

$$\frac{\partial \bar{z}_\rho}{\partial z_k} = \frac{\rho''(z_k - \bar{z}_\rho)}{\rho''(z_1 - \bar{z}_\alpha) + \cdots + \rho''(z_N - \bar{z}_\rho)}.$$

For smoothed α-quantile $\rho(r) = \rho_\alpha(r)$:

$$\rho_\alpha(r) = \begin{cases} (1-\alpha)\rho_\varepsilon(r), & \text{if } r < 0 \\ 0, & \text{if } r = 0 \\ \alpha\rho_\varepsilon(r), & \text{if } r > 0, \end{cases}$$

$\rho_\varepsilon(r)$ is such that 1) $\lim_{\varepsilon\to 0} \rho_\varepsilon(r) = |r|$; 2) $\lim_{\varepsilon\to 0} \rho'_\varepsilon(r) = \text{sign } r$; 3) $\lim_{\varepsilon\to 0} \rho''_\varepsilon(r) = \delta(r)$[1]. For example, $\rho_\varepsilon(r) = \sqrt{\varepsilon^2 + r^2}$ (it was used for calculations in the illustrative examples below).

At the same time,

$$\frac{\partial \mathsf{CM}_\alpha}{\partial z_k} = \begin{cases} \left(\dfrac{1}{M} + \dfrac{m}{M}\right) \dfrac{\partial \bar{z}_\alpha}{\partial z_k}, & \text{if } z_k < \bar{z}_\alpha \\ \dfrac{m}{M} \dfrac{\partial \bar{z}_\alpha}{\partial z_k}, & \text{if } z_k \geqslant \bar{z}_\alpha, \end{cases}$$

The following iterative procedure is used to find $\bar{z}_\alpha$:

$$u^{t+1} = \frac{\displaystyle\sum_{k=1}^{N} \varphi(z_k - u^t) z_k}{\displaystyle\sum_{k=1}^{N} \varphi(z_k - u^t)},$$

where $\varphi(r) = \rho'_\alpha(r)/r$.

---

[1] $\delta(r)$ is delta function of Dirac.

For comparison, here is another common use of M-estimators in the construction of $\mathcal{Q}$. The target functionality is defined as follows:

$$\mathcal{Q}(a) = \frac{1}{N} \sum_{k=1}^{N} \varrho(\Phi_k(a)). \qquad (5)$$

For example, in regression problems

$$\mathcal{Q}(a) = \frac{1}{N} \sum_{k=1}^{N} \varrho(f(x_k; a) - y_k).$$

Just such a method is used also in the robust version of PCA in [8].

Note that minimization of (5) is equivalent to minimization of Kolmogorov mean of $\Phi_1(a), \ldots, \Phi_N(a)$:

$$\mathcal{Q}_\varrho(a) = \varrho^{-1}\left(\frac{1}{N} \sum_{k=1}^{N} \varrho(\Phi_k(a))\right).$$

It is also an example of more general M-mean with $\rho(r) = r^2$:

$$\mathsf{M}_{\rho,\varrho}\{z_1, \ldots, z_N\} = \varrho^{-1}\left(\mathsf{M}_\rho\{\varrho(z_1), \ldots, \varrho(z_N)\}\right).$$

To find the optimal a^* by minimizing (5), the iterative reweighting method (3) is also used, which differs from our method proposed above in its method of recalculation of weights:

$$v_k^t = \frac{\varphi(\Phi_k(a^t))}{\varphi(\Phi_1(a^t)) + \cdots + \varphi(\Phi_N(a^t))}. \qquad (6)$$

The weights in both variants of the procedure decrease with the growth of $\Phi_k(a^t)$. However, in (4), the weights depend on the magnitude of the deviation $\Phi_k(a^t)$ from the robust estimate of the mean value (2) as opposed to (6). Here is a brief explanation: both $\varphi(r)$ and $\rho''(r)$ are positive and decrease toward to 0 as $r \to +\infty$. If the value of the average value is significantly separated from zero, then, as a rule,

$$\frac{\varphi(z_k)}{\varphi(z_1) + \cdots + \varphi(z_N)} > \frac{\rho''(z_k - \overline{z}_\rho)}{\rho''(z_1 - \overline{z}_\alpha) + \cdots + \rho''(z_N - \overline{z}_\rho)}$$

and therefore the weights of the outliers will have a smaller value in case of (4).

4   The Robust PCA

The robust version of the statement of the search problem a_0 takes the form:

$$a_0 = \arg\min_{a \in \mathbb{R}^n} \mathsf{M}\{\|x_1 - a\|^2, \ldots, \|x_N - a\|^2\}.$$

This problem is reduced to solving the equation

$$a = \sum_{k=1}^{N} \frac{\partial M\{\|x_1 - a_0\|^2, \ldots, \|x_N - a_0\|^2\}}{\partial z_k} x_k,$$

that can be solved using the following iterative procedure:

$$a^{t+1} = \sum_{k=1}^{N} v_k^t x_k,$$

where
$$v_k^t = \frac{\partial M\{\|x_1 - a^t\|^2, \ldots, \|x_N - a^t\|^2\}}{\partial z_k}.$$

After finding a_0, centering is also performed: $x_k \to x_k - a_0$, $k = 1, \ldots, N$.
The robust version of the search problem a_j takes the following form:

$$a_j = \arg \min_{\|a\|=1} M\left\{\|x_1\|^2 - (a, x_1)^2, \ldots, \|x_N\|^2 - (a, x_N)^2\right\}.$$

Using the iterative reweighing method, its solution can also be reduced to a chain of tasks:

$$a_j^{t+1} = \arg \min_{\|a\|=1} \sum_{k=1}^{N} v_k^t \left(\|x_k\|^2 - (a, x_k)^2\right)$$

with the following point weights

$$v_k^t = \frac{\partial M\left\{\|x_1\|^2 - (a^t, x_1)^2, \ldots, \|x_N\|^2 - (a^t, x_N)^2\right\}}{\partial z_k}.$$

In all cases $v_1^t + \cdots + v_N^t = 1$ by definition $\sum_{k=1}^{N} \frac{\partial M\{z_1, \ldots, z_N\}}{\partial z_k} = 1$.

This problem is a weighted version of the original search problem a_j within the framework of the classical formulation of the problem.

The covariance matrix takes the form:

$$S^t = X^\top \begin{pmatrix} v_1^t & \cdots & 0 \\ \vdots & \ddots & \vdots \\ 0 & \cdots & v_N^t \end{pmatrix} X.$$

The vector a_j^{t+1} is the solution of the system:

$$S^t a = \lambda a$$
$$\|a\|^2 = 1,$$

that is, it is an orthonormal eigenvector of the matrix S^t, and λ_j^{t+1} is its eigenvalue.

An iterative procedure is used to search for a_j:

$$a^{t+1} = \frac{1}{\lambda^t} \left(S^t a^t\right), \qquad \lambda^t = \frac{(a^t)^\top S^t a^t}{(a^t, a^t)}.$$

5   Experiments

For experimental confirmation of the effectiveness of the approach proposed here, examples of the application of classical and robust PCA for several data sets are considered. The robust approach proposed here is considered as a natural robust extension of the classical approach to the construction of PCA. Therefore, only the proposed robust approach and the classical approach are experimentally compared here in order to clearly show the ability of the proposed robust extension to overcome the outliers available in the data.

All calculations were performed using the open source library MLGRAD (https://bitbucket.org/intellimath/mlgrad.git) both for robust and classical variants of PCA. In the /example folder in the repository there are jupiter notebooks that contain calculations for the examples presented below. In some cases datasets was preprocessed using scale or robust_cale routines of the module preprocessing from SCIKIT-LEARN library (https://scikit-learn.org).

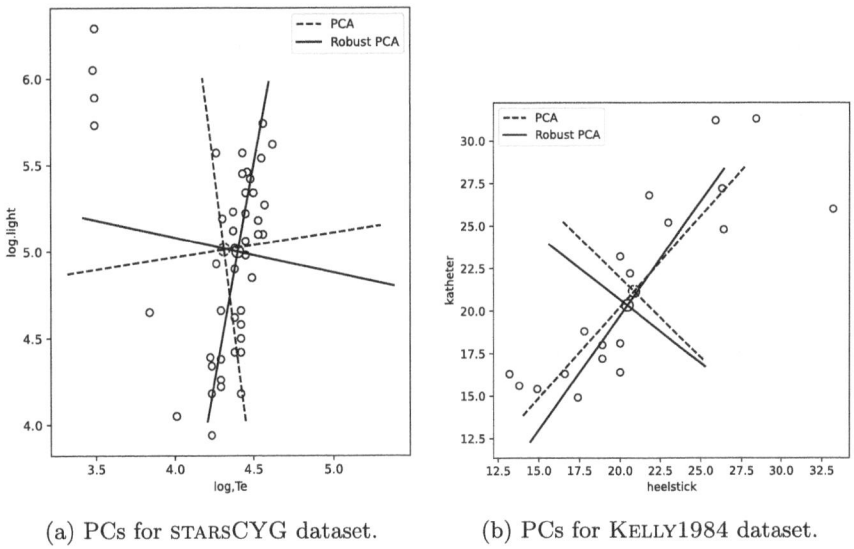

(a) PCs for STARSCYG dataset.   (b) PCs for KELLY1984 dataset.

Fig. 1. Experiments with 2 plain datasets.

Dataset starsCYG.

Consider a data set for constructing a Hertzsprung-Russell diagram of the CYG_OB1 star cluster [5] (table 3), which describes the relationship between the logarithm of luminosity (log.light) and the logarithm of temperature (log.Te) of stars. In Fig. 1 there are 7 points that can be attributed to outliers. The application of the classical PCA gives the vectors of the PC rotated counterclockwise in the direction of outliers. The application of the robust PCA methods ($\alpha = 0.87$,

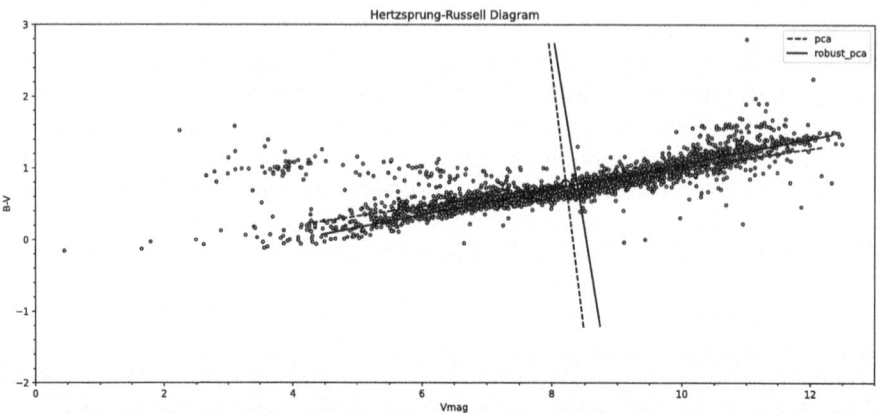

Fig. 2. The positions of the centers and PCs for HIP_STARS dataset.

$\varepsilon = 0.001$) makes it possible to find the unbiased position of the center and the PC that do not deviate under the influence of outliers.

Dataset Kelly1984.
Consider a dataset Kelly1984 [9], which describes simultaneous pairs of measurements of serum kanamycin levels in blood samples drawn from 20 babies. In Fig. 1b there are some points that can be attributed to outliers. The application of the classical component method gives the vectors of the PC rotated counterclockwise in the direction of outliers. The application of the robust PCA ($\alpha = 0.8$, $\varepsilon = 0.001$) makes it possible to find the unbiased position of the center and the PC that do not deviate under the influence of outliers.

Dataset HIP_Stars.
Consider a data set for plotting a chart for stars from a dataset [10]. In Fig. 2 shows a projection on a pair of Vmag and B-V parameters. The classical PCA gives offset PC for a given projection. Both proposed robust PCA ($\alpha = 0.95$, $\varepsilon = 0.001$) makes it possible to overcome the influence of outliers.

Dataset CigarettesSW.
Consider panel data on cigarette consumption in 48 continental US states for 1985–1995 years [11]. The PCA is used for tabular data that covers 7 features (cpi, population, packs, income, tax, price, taxs; state, year are excluded). First data was preprocessed using `scale` routine of the module `preprocessing` from SCIKIT-LEARN library. Figure 3 clearly shows that the data in projections on PC1×PC2×PC3, which are obtained on the basis of the robust PCA ($\alpha = 0.55$, $\varepsilon = 0.001$), have a more contrasting appearance: the data lines up along two straight lines. Only the first robust approach was applied here because the second approach does not show efficiency in this dataset.

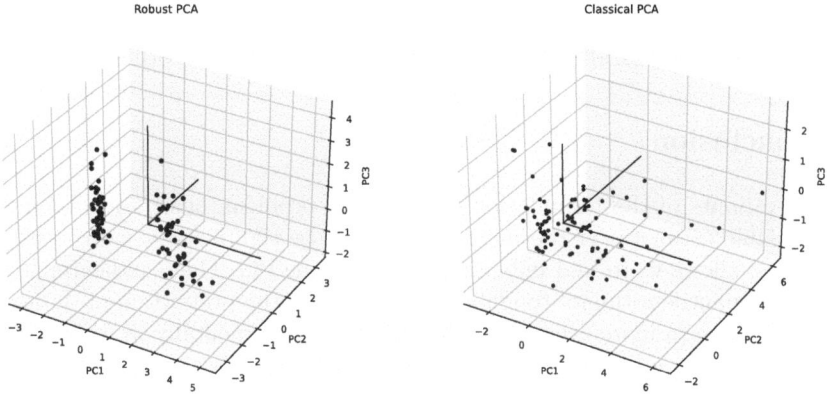

Fig. 3. Projections of the data from the CigarettesSW dataset on PC1×PC2×PC3.

It's easy to see that robust variant allow us to use the projection of data on the plain PC1×PC2×PC3 so that clustering linear regression method can be applied to distinguish the linearly shaped clusters.

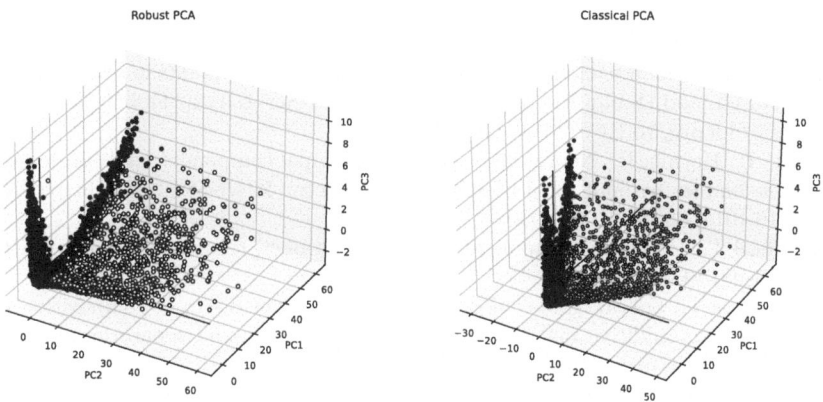

Fig. 4. Projections of the data from the HTRU2 dataset on PC1×PC2×PC3.

Dataset HTRU2.

HTRU2 is a data set which describes a sample of pulsar candidates collected during the High Time Resolution Universe Survey (South). It contains 17898 rows, 8 columns and divided into 2 classes. The PCA used for data that covers 8 features. First data was preprocessed using `robust_scale` routine of the module `preprocessing` from SCIKIT-LEARN library. It is easy to see in the Fig. 4 that the projection of data on PC1×PC2×PC3 for the case of the classical PCA turns out to be rotated relative to PCs, while in the case of the robust PCA variant

($\alpha = 0.9$, $\varepsilon = 0.001$), the data is almost located along PCs. For example, the projection on the PC2 allows you to distinguish pulsars from others.

6 Conclusion

Robust variants of the formulation of the PCA problem, based on minimizing differentiable estimates of the mean, which can be significantly more resistant to outliers, allows us to find unbiased vectors of the PC. The α indicator in the smoothed quantile estimate roughly corresponds to the proportion of data that are not outliers (a more accurate value is found experimentally in its neighborhood). The first robust method makes it possible to identify outliers by analyzing the empirical distribution of distances to straight lines passing through the center a_0 along the vectors of the PC. The second robust method also makes it possible to identify outliers by analyzing the empirical distribution of Mahalanobis distances from the center to all points. It is also interesting because after finding a robust variant of the covariance matrix S, standard algorithms of PCA can be applied. Expreiments also was demonstrated that the first and the second robust approaches may have different levels of efficiency in same dataset with ouliers.

References

1. Pearson, K.: On lines and planes of closest fit to systems of points in space. London, Edinburgh, and Dublin Philosophical Mag. J. Sci. Series 6. **2**(11), 559–572 (1901). https://doi.org/10.1080/14786440109462720
2. Baldi, P., Hornik, R.: Neural networks and principal component analysis: learning from examples without local minima. Neural Netw. **2**, 53–58 (1989). https://doi.org/10.1016/0893-6080(89)90014-2
3. Jolliffe, I.T.: Principal Component Analysis. Springer, Cham (2002). https://doi.org/10.1007/b98835
4. Bouwmans T., Sobral A., Javed S., Jung S.K., Zahzah E.-H.: Decomposition into low-rank plus additive matrices for background/foreground separation: a review for a comparative evaluation with a large-scale dataset. arXiv:1511.01245 (2015). https://doi.org/10.1016/j.cosrev.2016.11.001
5. Rousseeuw, P.J., Leroy, A.M.: Robust Regression and Outlier Detection. John Wiley and Sons, New York (1987)
6. Shibzukhov, Z.M.: Machine learning based on the principle of minimizing robust mean estimates. In: Samsonovich, A.V., Gudwin, R.R., Simões, A.S. (eds.) BICA 2020. AISC, vol. 1310, pp. 472–477. Springer, Cham (2021). https://doi.org/10.1007/978-3-030-65596-9_56
7. Shibzukhov, Z.M.: Minimizing robust estimates of sums of parameterized functions. J. Math. Sci. 1–16 (2022). https://doi.org/10.1007/s10958-022-05689-z
8. Polyak, B.T., Khlebnikov, M.V.: Principle component analysis: robust versions. Autom. Remote. Control. **78**(3), 490–506 (2017). https://doi.org/10.1134/S0005117917030092

9. Kelly, B.: The influence function in the errors in variables problem. Ann. Stat. **12**(1), 87–100 (1984). https://doi.org/10.1214/aos/1176346394
10. Dataset Hipparcos. https://www.astrostatistics.psu.edu/datasets/HIP_star.html
11. Dataset CigarettesSW: cigarette consumption panel data. https://rdrr.io/rforge/gmm4/man/CigarettesSW.html
12. Lyon, R.: HTRU2. UCI Machine Learning Repository (2017). https://doi.org/10.24432/C5DK6R

Degradation Detection for Steam Machines

Art Prosvetov[(✉)] and Vladislav Balaev

Space Research Institute, Russian Academy of Sciences, Profsoyuznaya ul. 84/32,
Moscow 117997, Russia
prav@iki.rssi.ru

Abstract. This work presents a novel method for regime estimation and degradation extrapolation for steam machines. The proposed approach combines statistical techniques with machine learning algorithms to accurately predict the remaining useful life of the engine components. The method was tested on experimental data from a steam engine operating under varying regimes and showed promising results in terms of accuracy and efficiency. The findings of this study have implications for the maintenance and management of steam engines, as it provides insights into predicting the lifetime of components, allowing for more effective maintenance practices and potentially extending the lifespan of the equipment.

Keywords: Time Series Analysis · Neural Networks · Predictive Maintenance

1 Introduction

1.1 Introduction

Steam engines play a vital role in the global energy landscape, serving as the backbone of power generation plants worldwide. Ensuring the optimum performance and reliability of these mechanisms is essential to meet the ever-growing demand for energy and mitigate potential risks associated with equipment failure. To address these challenges, accurate assessment of steam turbine operating regimes and their degradation rates is necessary for effective maintenance and management practices.

Recent advancements in computational methods and data analytics have opened new possibilities to improve the state-of-the-art in steam turbine prognostics. Machine learning and statistical techniques provide promising tools for developing robust and efficient models capable of predicting component degradation under varying operating conditions. However, the complexity of steam turbines and the inherent variability in component wear present significant challenges that require tailored approaches to address effectively. The aim of this approach is to minimize maintenance costs and extend the turbine's availability by monitoring and health prognostics of a turbine. The health prognosis often include not only online anomaly detection but also remaining useful life (RUL) prediction.

V. Balaev—Independent Researcher.

© The Author(s), under exclusive license to Springer Nature Switzerland AG 2024
J. Baixeries et al. (Eds.): DAMDID/RCDL 2023, CCIS 2086, pp. 142–150, 2024.
https://doi.org/10.1007/978-3-031-67826-4_11

In a past decade an area of machine and deep learning received a heavy boost which also effected an area of anomaly detection [1]. However relatively small research works appear in the area of steam turbines predictive maintenance and not succeeding in keeping up with modern capabilities. At 2010 Salahshoor et al. [2] used a support vector machine (SVM) classifier with an adaptive neuro-fuzzy inference system (ANFIS) to detect several steam turbine fault types, which included actuator fault, thermocouple sensor fault, fouling. For RUL prediction Khelif computed health indicators of the steam turbine time series via a linear regression with training on the 0 and 1 labels assigned correspondingly to normal and abnormal exploitation periods [3]. The authors use this model to transform historical data into a one-dimensional vector, divide it into windows and then apply the same transformation to the online data and calculate distances between online vector and historical vectors via dynamic time warping (DTW) technique to estimate RUL. In 2017 Khelif et al. [4] proposed to use Support vector regression to directly predict RUL from sensor time series data. Dhini et al. [5] utilized a fully connected neural network (FCNN) classifier basing on the features after principal component analysis (PCA) transformation. Authors also worked in a supervised learning paradigm aiming at prediction of 4 previously occurred (in historical data) fault types: misalignment, rotor browing, blade erosion and cracked case. The resent work was made by Que et al. [6]. The authors use a binary classifier based on Extreme gradient boosting (Xgboost) trained on the time series of the normal and fault work periods labeled as 0 and 1 correspondingly. For the RUL prediction authors used DTW similarly.

All the mentioned approaches require episodes of fault behavior in historical data. However various unsupervised techniques exist and might be used in fault detection of steam turbines while the data on usage of these techniques for steam turbines is lacking. Here we describe our framework for unsupervised anomaly detection for a steam turbine using recent approaches. We also propose a method of unsupervised RUL prediction for each sensor separately basing on their exploitation thresholds. This task could not be solved by classical methods like time series forecasting with SARIMA [7], Facebook Prophet [8] and similar algorithms. The cause is the transition of the overall system between steady states resulting from the change in system parameters made by operator of the turbine. For each of these states the time series of a particular sensor is fluctuating in a fixed range and the mentioned methods thus could not work even providing the information on the system parameters changes which also could be lacking.

2 Data Analysis

2.1 Dataset Description

The dataset comprised 180–200 time series, which represented various sensor signals such as pressure, temperature, and flow. These signals were measured in distinct steam engine modules, including steam lines, high-pressure regeneration, low-pressure regeneration, condensation system, drainage system, oil supply, cooling, and seals. The sampling rate was set at 1 s, and the total data collection period spanned four months.

During the first month, several instances of system reboot were observed, wherein the rotor rotation speed decreased from 3000 rpm to significantly lower values or even reached zero. These occurrences indicated the process of system adjustment. Since these periods might not represent typical operational trends, the decision was made to exclude the first month of data from the analysis.

Additionally, certain sensors were found to exhibit inaccurate physical measurements, such as negative steam temperature. Some vibration sensors were identified as defective, primarily displaying values near zero. Both categories of sensors were subsequently omitted from further analysis.

2.2 Regime Extraction

A steam engine is a complex system that can be configured in various ways to maintain specific power levels and direct a portion of the steam to other sources for industrial and/or heating purposes. The system comprises several modules, which aim to enhance its efficiency and are automatically adjusted to accommodate external factors such as temperature. These adjustments and external conditions significantly impact the absolute and relative values displayed by the sensors, as well as the correlations between them, resulting in the occurrence of multiple steady states. To accurately detect anomalies, it is essential to analyze the data within these steady states or their groups, as the parameter distributions within them exhibit some degree of order. However, no logs of system setting changes were provided, which necessitated the development of an automatic regime extraction method based on sensor data.

Various approaches to address this issue were explored, and experts in the field evaluated the results. Additionally, the following constraints were considered:

1. The duration of a steady state must not be less than 20 min.
2. The steam engine must operate in a steady state for more than 60% of the total time. The remaining time is attributed to the transitions between steady states.

Experts identified 9 sensors that respond first to changes in settings and provided threshold values for the variance of these sensors that could be observed in steady-state conditions. We conducted a comparative analysis of the chosen sensor time series at various time scales, ranging from 30 min to 1 week, for each sensor independently. This comparison resulted in specific time scales for each sensor, which were selected for further analysis and included 24.0, 18.0, 8.0, 2.0, 1.0, and 0.5 h.

The overall procedure for the proposed regime extraction is outlined as follows:

1. Select the largest time period from the set (initially, 24 h).
2. Apply a rolling window analysis to the time series for the chosen nine sensors, using a window size corresponding to the period in step 1 and a step size of 20 min.
3. For each window, verify whether all time series parameters are below the threshold and save the window if the condition is satisfied.
4. For each saved window, apply the Kwiatkowski-Phillips-Schmidt-Shin (KPSS) [9] test to check the stationarity of each sensor time series and compute the product of the results.
5. Sort the windows in ascending order based on the multiplied KPSS values and iteratively add each window to a final set:
- If a new window does not overlap with any of the previous ones, assign it a new steady-state label.

- If a new window overlaps with any of the previous ones, check if the parameters are below the threshold, as in step 3:
    - If all checks pass, combine the windows.
    - If not, assign the union of the two windows to the one with the highest KPSS value, and label the remaining part of the overlapping new window separately.

6. Repeat steps 1–5 for smaller scales.

However, upon examining the results of the proposed approach, experts noted some steady states that fell within the given thresholds but contained two or three smaller states internally, which they visually identified by analyzing the active power time series. Reducing the threshold could further segregate these states but would also result in a decrease in the overall percentage of steady states relative to transition states, which is undesirable. Consequently, we decided to perform post-processing using Bayesian Gaussian Mixture [10] on the active power time series and pressure measurements in the key regulation leads of the system.

1. For each previously extracted steady state, we attempted to apply the Bayesian Gaussian Mixture with a convergence threshold of 1e-4 and a number of components equal to 2.
2. If the algorithm converged, we calculated the overall duration of each component during the steady state, and only retained those with durations exceeding 1 h (based on the assumption that a component might appear multiple times during a steady state, with each part lasting at least 20 min).
3. We compared the difference between the mean values of the components to the mean of the components; only those steady states with a ratio not less than 2.5% were further divided into components. The value of this threshold was visually adjusted through the analysis of various steady state examples containing two components.
4. For the selected steady states, the segregation into components, represented as a sequence of 0 and 1 (numeric representation of the two components), was filtered using a moving average with a window size of 20 min and rounded to remain binary.
5. For each of the resulting components, if it comprised less than 20 min, we assigned a transition period label to it; otherwise, we assigned a new steady state label.

2.3 Variational Autoencoder

The application of autoencoders and variational autoencoders is widely used for anomaly detection [11–15]. In order to obtain information about the state of the turbine according to the readings of its sensors, the procedure of training the variational autoencoder was carried out [16].

The input layer was configured to process a tensor that represents parameter evolution during a single time frame, with a duration of 180 s. The encoder consisted of four 2D convolution layers, with 32 filters in the first layer and 64 filters in the subsequent three layers. The kernel size was set to 3, and the activation function employed was the rectified linear unit (ReLU) function. After the convolution layers, a fully connected layer with 32 neurons and a ReLU activation function was implemented. The latent dimension layer had a size of 2, enabling the visualization of the results.

The decoder was designed with an input layer consisting of 2 neurons, followed by a fully connected layer with a ReLU activation function. The subsequent decoder layer was a fully connected layer with a size equal to M * N/2, where M and N represent the input tensor's shape. The next decoder layer was a transposed convolution layer (also known as deconvolution) with 16 filters and a kernel size of 3. Transposed convolutions were employed to implement a transformation in the reverse direction of a conventional convolution. Finally, the decoder contained a convolution layer to obtain a tensor size equivalent to the input tensor size.

The loss function was computed using a reconstruction term and a regularization term in the form of Kullback-Leibler divergence. The reconstruction term of the loss function was based on the binary cross-entropy loss:

$$H_1 = -1/N \Sigma y_i \log(p(y_i)) + (1 - y_i)\log(1 - p(y_i)) \tag{1}$$

The regularization term in form of Kullback-Leibler divergence was rewritten in the following form [17]:

$$H_2 = -1/2 \sum (1 + \sigma + \mu^2 - e^\sigma) \tag{2}$$

where μ and σ are the individual means and standard deviations.

The total loss function was a mean of H_1 and H_2 loss:

$$H_{total} = (H_1 + H_2)/2 \tag{3}$$

Additionally, the reparameterization trick [18] was utilized to enable backpropagation through the network.

The neural networks were trained for three epochs, during which the loss decayed on the validation dataset. During the model training process, a mean squared error (MSE) of 0.012 was achieved. After model training, the encoder was used to transform the multidimensional time series into a 2-dimensional latent space.

Subsequently, each identified steady state was transformed into a point cloud in the latent space. Within each steady state, the time series was divided into 180-s frames. The variational autoencoder's encoder was applied to each frame, resulting in two coordinates in the latent space for each frame. The collection of points in the latent space was condensed into a single central point by calculating the average value for each point cloud coordinate.

2.4   Degradation Detection

A prediction model for the turbine sensor readings can be constructed within the context of a single steady state, which lasts for a duration of several hours. The transition between steady states is controlled by operators and cannot be accurately predicted. Parameter degradation is expected over time scales exceeding 14 days, which is considerably longer than the time scale of each regime.

Considering the data characteristics described above, the following assumptions were made:

- Steady states with similar parameter distributions occur over an extended time scale of more than 14 days.
- Long-term parameter degradation can be observed within the context of steady states that exhibit closely related parameter distributions.
- A regression model can be developed to predict the sensor readings over an extensive time scale.
- A distinct upward/downward trend in the extrapolation of the regression model for the turbine parameter within the framework of steady states can help identify a degrading parameter among other parameters.

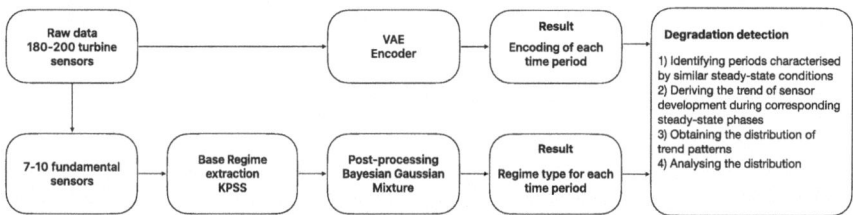

Fig. 1. The schema of degradation detection workflow

The implementation of the degrading parameter search, based on the above assumptions, was conducted as follows (Fig. 1):

1. All selected steady states were transformed into the variational autoencoder's latent space and characterized by a single set of latent space coordinates.
2. The following sequence was executed iteratively:
   a. One steady state was randomly selected.
   b. A set of steady states with close coordinates in the latent space was identified.
   c. The chosen set of steady states defined the time frame on which the regression model was subsequently trained.
   d. The extrapolation trend of the trained regression model was recorded for further analysis.
3. As a result, the trend distribution was obtained for each parameter.
4. If the parameter trend distribution was normal, symmetric, and the mean value was close to 0, the degradation hypothesis for the parameter was rejected (Fig. 2 right).
5. If the parameter trend distribution was asymmetric and the mean value was shifted relative to 0, the degradation hypothesis for the parameter was confirmed (Fig. 2 left).

To detect degradation in its early stages, the algorithm used a constraint when initially selecting a random steady state (2.a): only states observed during the last two months were employed. Simultaneously, the sample of the stable state family also included similar regimes with more distant historical distributions.

3   Approach Approbation

To further validate the proposed approach for regime estimation and degradation extrapolation in steam turbines, an expert test was conducted involving industry professionals with extensive experience and knowledge in the field. The objective of this test was to assess the reliability and practicality of the suggested methodology in predicting degradation parameters that could lead to turbine failure.

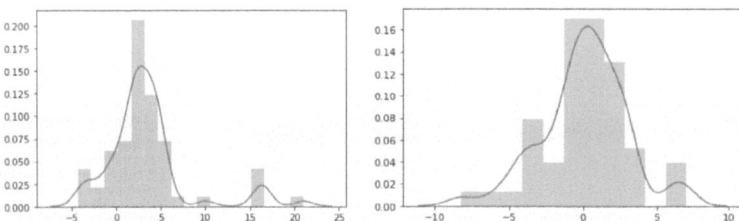

Fig. 2. Comparison of normalized distribution densities for long-period trends. The X-axis represents the trend of randomly subsampled time series exhibiting the same steady-state conditions, while the Y-axis denotes the normalized distribution density. The left one corresponding to a steam turbine sensor exhibiting degradation and the right one representing a sensor without degradation

3.1   Expert Test Setup

A panel of experts, consisting of engineers and technicians specializing in steam turbine maintenance and operation, participated in this test. These experts were provided with the predicted degradation parameters obtained from the proposed model, along with pertinent contextual information about the steam turbine's operating conditions. The participants were asked to evaluate the degradation parameters and determine their potential impact on turbine performance and longevity.

3.2   Degradation Parameters

Two primary degradation parameters were identified by the proposed approach: Degradation Parameter 1 (DP1) and Degradation Parameter 2 (DP2). DP1 and DP2 were associated with vibration sensors. The analysis indicated that DP1 tends to reduce vibrations and has a minor effect on turbine performance, while DP2 tends to increase vibration and can lead to critical failures if not addressed in a timely manner.

3.3   Expert Evaluation Results

The majority of the experts agreed on the relevance and significance of the identified degradation parameters. Furthermore, they agreed that DP2 could potentially lead to turbine failure within a timeframe of 7 to 10 years if left unaddressed.

In conclusion, the expert validation and analysis demonstrated the effectiveness of the regime estimation and degradation extrapolation method in predicting degradation parameters that can lead to steam turbine failure. The results of the expert test provide strong support for the proposed approach, highlighting its potential application in practice for enhanced steam turbine maintenance and management.

4 Conclusion

In conclusion, this scientific paper has presented an innovative approach to regime estimation and degradation extrapolation for steam engines by integrating statistical methods with machine learning algorithms. The proposed technique has demonstrated promising results in predicting the remaining useful life of turbine components, paving the way for improved maintenance and management practices.

While the current study has shown significant advancements in steam engine prognostics, future research could explore the extension of this methodology to other power generation equipment and the incorporation of additional sensor data for more accurate predictions. Additionally, advances in machine learning algorithms could further enhance the performance of the proposed technique in terms of speed and accuracy, elevating its utility across the energy sector.

Overall, this research serves as a stepping stone in addressing the challenges of maintenance and management in the steam turbine domain, ultimately contributing to a more sustainable and efficient global energy landscape.

Disclosure of Interests The authors have no competing interests to declare that are relevant to the content of this article.

References

1. Braei, M., Wagner, S.: Anomaly Detection in Univariate Time-series: A Survey on the State-of-the-Art (2020)
2. Salahshoor, K., Kordestani, M., Khoshro, M.S.: Fault detection and diagnosis of an industrial steam turbine using fusion of SVM (support vector machine) and ANFIS (adaptive neuro-fuzzy inference system) classifiers. Energy **35**, 5472–5482 (2010)
3. Khelif, R., Malinowski, S., Chebel-Morello, B., Zerhouni, N.: RUL prediction based on a new similarity-instance based approach. In: IEEE Int. Symp. Ind. Electron., Istanbul, Turkey, pp. 2463–2468 (2014)
4. Khelif, R., Morello, B.C., Malinowski, S., Laajili, E., Zerhouni, N.: Direct remaining useful life estimation based on support vector regression. IEEE Trans. Ind. Electron. **64**(3), 2276–2285 (Mar. 2017)
5. Dhini, A., Kusumoputro, B., Surjandari, I.: Neural Network Based System for Detecting and Diagnosing Faults in Steam Turbine of Thermal Power Plant, pp. 149–154. Int. Conf. Awareness Sci. Tech, Taichung, Taiwan (2017)
6. Que, Z., Xu, Z.: A data-driven health prognostics approach for steam turbines based on Xgboost and DTW. IEEE Access, 1 (2019). https://doi.org/10.1109/ACCESS.2019.2927488
7. Swain, S., et al.: Development of an ARIMA model for monthly rainfall forecasting over Khordha District, Odisha, India. Adv. Intell. Syst. Comput. **708**, 325–331 (2018)

8. Sean, J.: Taylor, Benjamin Letham: forecasting at scale. Am. Stat. **72**(1), 37–45 (2018)
9. Kwiatkowski, D., Phillips, P.C.B., Schmidt, P., Shin, Y.: Testing the null hypothesis of stationarity against the alternative of a unit root. J. Econometr. **54**(1–3), 159–178 (1992)
10. Sung, H.G.: Gaussian mixture regression and classification, PhD thesis, Rice University, Houston, Texas (2004)
11. Zhou, C., Paffenroth, R.C.: Anomaly detection with robust deep autoencoders. In: Proceedings of the 23rd ACM SIGKDD International Conference on Knowledge Discovery and Data Mining, pp. 665–674. ACM (2017)
12. Masci, J., Meier, U., Ciresan, D., Schmidhuber, J.: Stacked convolutional auto-encoders for hierarchical feature extraction. In: International Conference on Artificial Neural Networks, pp. 52–59. Springer (2011)
13. Skvara, V., Pevny, T., Smıdl, V.: Are generative deep models for novelty detection truly better? arXiv preprint arXiv:1807.05027 (2018)
14. Chen, J., Sathe, S., Aggarwal, C., Turaga, D.: Outlier detection with autoencoder ensembles. In: Proceedings of the 2017 SIAM International Conference on Data Mining, pp. 90–98. SIAM (2017)
15. An, J., Cho, S.: Variational autoencoder based anomaly detection using reconstruction probability. Special Lect. IE **2**, 1–18 (2015)
16. Prosvetov, D.: Regime searching in time series data using Variational Autoencoder. J. Phys.: Conf. Ser. **1727** 012001 (2021)
17. Jinwon, A., Cho, S.: Special Lecture on IE (2015)
18. Kingma, D.P., Welling, D.: https://doi.org/10.48550/arXiv.1312.6114 (2013)

Towards an Approach to Formulating Personal Development Plan for Developers Based on Competency Framework and Data Mining

Erchimen Gavriliev[(✉)] and Tatiana Avdeenko

Novosibirsk State Technical University, K. Marksa ave., 20, 630073 Novosibirsk, Russia
erchimen_gavriliev@outlook.com

Abstract. Companies face difficulties in managing employees' professional growth due to the large variety of career paths available in the IT industry, as well as their high degree of flexibility and unpredictability. In this regard, a personal development plan is used to manage the career advancement of developers. However, managers struggle to create plans that align with employees' professional aspirations. A competency framework was developed to increase relevancy of the plans by providing a comprehensive view of possible career paths for developers and required competencies. The framework was implemented in a software development company to assess its validity. It was discovered that a significant proportion of developers had competencies related to communication, project teamwork, requirements analysis, as well as to solving technical tasks such as programming and debugging.

Keywords: software developer · competency matrix · competency model · data mining · natural language processing · personal development plan

1 Introduction

Human resources and professional development system determine organization's effectiveness. However, recent economic, technological, and social advances have resulted in considerable changes in organizational processes and job profiles, complicating the management process of employees' professional growth [1].

For instance, work environment, responsibilities, and skill requirements for IT specialists have changed over the years, which have fundamentally impacted many career paths, making them more flexible and unpredictable [2, 3].

Personal development plans (PDPs) are a frequent option in IT organizations; they are meant to stimulate employee training, which in turn should increase staff performance while taking into consideration their personal professional aspirations [4].

Nevertheless, development and implementation of PDP are complex tasks, which is why only a small portion of tasks listed in the plans are completed, or implemented plans have little impact on professional growth [5, 6]. Managers frequently struggle with PDP development: they are usually interested in using available specialists that already have

© The Author(s), under exclusive license to Springer Nature Switzerland AG 2024
J. Baixeries et al. (Eds.): DAMDID/RCDL 2023, CCIS 2086, pp. 151–163, 2024.
https://doi.org/10.1007/978-3-031-67826-4_12

expertise for each specific project and each position in it. However, these specialists may also prefer to obtain other skills and/or work in other projects, but these preferences are not stated in their PDPs.

One solution is to use a competency model, which provides a holistic view of different career paths and required competencies for each grade – a generalized position that unifies a group of professions.

However, existing studies mainly focus on programming-related competencies such as coding, knowledge of mathematical and engineering foundations for software development [7]. There are studies that consider soft competences such as team communication skills, planning, analytical skills and problem-solving abilities as well [8]. But focusing only on either technical or soft competencies can lead to problems in professional development, because software development involves different tasks: gathering and analyzing requirements, testing, mentoring and etc.

Furthermore, existing studies that provide a more comprehensive view of developer competencies do not specify approaches or indicators for assessing them. Activities in software development produce large number of artefacts, which can be analyzed using data mining methods and tools to gather information about employees' competencies.

The aim of this work is to increase the relevance of PDPs for developers by proposing a competency framework and method for assessing competencies based on the data mining of systems in which developers works on a regular basis.

In the rest of this paper, Sect. 2 gives background on publications related to developers' competencies. Section 3 presents the developed competency framework, methodology for its validation and the case company. Section 4 contains validation results, and Sect. 5 concludes the paper.

2 Literature Review

Personal development plan is a tool used in the career management cycle to collect and document information about the competencies that an employee has previously acquired, as well as about the competences that he (she) is planning to develop in the future. PDP includes following sections with posed questions [8]:

- main objectives – what competencies does an employee need to develop?
- list of tasks that an employee must complete, as well as deadlines– how does he (she) want to develop those competencies and in which timeframe?
- possible risks and list of employees, that can assist in completing the tasks – which support is needed?

Competency is important in professional development, but definitions vary [10]. We will adopt a performance-based approach in this study and define competency as an ability and willingness to apply knowledge and skills in solving professional problems in various fields [11].

Competency model is a hierarchy of competencies expected of an employee that contains descriptions of competencies as well as measurable or observable indicators that may be used to evaluate workers [12].

Relevant research on necessary competencies for software developers usually focuses on technical competencies, such as programming fundamentals and software design [13,

14]. This list of competencies also includes debugging [15] and abilities to analyze source code [16], because it is frequently required to make modifications to an already developed system.

There are also publications that mainly consider soft competencies related to teamwork and communications [8, 17]. Other works discovered attributes and characteristics, that are not related to coding, such as flexibility, proactivity, willingness to learn and independence [18].

Nonetheless, there are studies that provide an extensive view of developers' competencies, taking into account both technical and soft competencies.

Microsoft employees were polled about the attributes of a «great» software developer [19]. Based on the survey results, 54 attributes were found and divided into four groups: «Personality», «Decision-making», «Teammate Interactions» and «Software product». Attributes associated with software development, as well as the developer's individual characteristics were given the highest level of importance, such as paying attention to the quality of the source code, having necessary intellectual abilities to solve complex problems.

Based on a survey of 355 software developers and the literature on expertise and productivity, paper provides a conceptual theory of software development expertise (SDExp) [20]. It was discovered that developer's effectiveness is determined by one's level of expertise and experience, as well as a number of individual characteristics, such as openness and agreeableness.

Prior work on the methods for assessing developers' competencies is mainly focused on programming as well [21]. However, because developers have diversified their actives in other areas such as gathering and analyzing requirements, testing, mentoring and etc., they may not be successful in assessing their qualifications as a whole. Due to the low level of reliability and accuracy, another frequently used method based on the subjective manager's opinion may yield inaccurate results.

Data mining of repositories has been applied in several studies to measure the expertise of IT specialists [22]. It was argued that a specialist who has made a significant contribution to the project's repository may have a greater level of «quality».

Based on literature review, it can be concluded that assessment method based on systems' data mining has a higher degree of objectivity than the manager's opinion, because the former method is based on quantitatively measurable indicators and relevant data.

3   Development and Validation of Competency Framework

3.1   Description of Competency Framework

First version of developer competency framework was designed using the findings of prior studies [23] and literature review. It was a two-dimensional matrix that represented core expertise domains of software development. The first dimension of the matrix was defined by 3 domains:

1. «Tasks' complexity and approach to solving them», which included abilities to structure one's own approach to work, ability to understand complex processes, systems and ability to apply knowledge and skills to new situations.

2. «Interaction skills and collaboration with a project team», which included communicative competencies, competencies for professional collaboration.
3. «Development skills», which included knowledge of programming languages, software architectures, internet protocols, database management systems and etc.

The second dimension represented developmental process and expected changes in competency across four common grades:

1. «Junior» – entry-level developer.
2. «Middle» – employee with specialized knowledge and a broader range of responsibilities.
3. «Senior» – experienced specialist with advanced technical competencies.
4. «Lead» – experienced specialist with advanced technical and managerial competencies.

In order to refine and test competency framework, it was implemented in a medium-sized company that develops software for the banking industry. The company uses its own low-code application builder to develop software products. Tech stack of the application builder included:

- relational database (Oracle or PostgreSQL) as a data storage;
- Java application server (Payara or WildFly) as a web application server;
- Rhino (an engine that converts JavaScript code into Java classes) and United Data Model Script (an application builder's programming language) were two programming languages used in backend development;
- JavaScript and a built-in interface editor were used for frontend development.

To create a software product, the company's developers had to understand this tech stack. A number of interviews were conducted with lead developers and project managers in order to further refine the framework's set of competences. A new domain «Other technical skills» was added to the first dimension of matrix, because in the case company developers usually applied tools and technologies only from the mentioned above tech stack, and knowledge of other tools outside the tech stack was optional, but it could be useful in development. For example, developer could also work as a DevOps engineer, which included setting up tools and appropriate infrastructure, as well as selecting and deploying CI/CD solutions. As the number of competencies grew, it was decided to categorize them into subdomains to improve clarity. New version of framework consisted of 46 competencies and included following domains and subdomains:

1. «Tasks' complexity and approach to solving them»:

a. «Task formulation» – ability to solve tasks with different levels of details in formulation.
b. «Task management» – skills of administering tasks assigned to developer themselves, as well as to other members of project team. Ability to plan their time, set up schedules, and complete tasks on time.
c. «Types of tasks» – characteristics of employee's activities.
d. «Level of independence in decision making» – characteristics of decisions made regarding software product in development. Ability to accept responsibilities and solve problems individually.

e. «Task management system» – ability to work and keep records in a task management system.

2. «Interaction skills and collaboration with a project team»:

a. «Project team» – characteristics of communication and collaboration skills with a project team.

b. «Client» – characteristics of communication and collaboration skills with a client or other vendors. Understanding the goals of project and its structure from different points of view.

c. «Mentoring» –capabilities of teaching and adapting new employees, of giving feedback.

3. «Development skills»:

a. «Technology stack» –skills and knowledge of the applied technology stack for development:

(1) «Knowledge of UDMS».

(2) «Knowledge of JavaScript».

(3) «Knowledge of SQL».

(4) «Debugging skills».

(5) «Knowledge of application builder's instruments for developing».

(6) «UI development skills».

(7) «Knowledge of application builder's architecture».

b. «Documentation» –level of documentation knowledge for an applied technology stack.

c. «Integration with external systems» – knowledge and skills of setting up integration with external systems through various protocols:

(1) HTTP.

(2) JDBC.

(3) SOAP.

(4) LDAP.

(5) JMS.

4. «Other technical skills»:

a. «Infrastructure» – skills and knowledge of server administration:

(1) «Knowledge of Linux operating systems and command line interface».

(2) «Knowledge of administering Java application servers».

(3) «Knowledge of administering Database Management Systems».

(4) «Knowledge of tools for monitoring IT-infrastructure».

b. «Particular technologies» –skills and knowledge of technologies and tools used in development: Git, ELK, Hazelcast etc.

A higher-level grade might include competencies from the previous grade. For example, subdomains «Task formulation» and «Technology Stack» from the developed framework are presented in Table 1.

Table 1. Fragment of the developer competencies framework

Domain	Subdomain	Junior	Middle	Senior	Lead
Tasks' complexity and approach to solving them	Task Formulation	Can work with a detailed task formulation	Can work with a short task formulation Can solve task with a medium-detailed formulation individually with high quality and on time	Can work with a task without detailed formulation Can identify and formulate a task for himself	see Senior
Development skills	Technology stack	Basic proficiency in programming language Familiar with code writing culture in company	Intermediate proficiency in programming language Familiar with refactoring techniques Can build basic unit tests	Fluent in programming language Can build complex unit tests Can optimize code	see Senior

3.2 Validation Methodology

Activities in software development produce large number of artefacts, that can be processed with data mining methods and tools, which in turn can provide information about employees' competencies.

Developers use version control system, task management system and knowledge management system in their daily work. Version control system records changes in application' source code. Task management systems are used to organize the work of the project team. Knowledge management system helps in organizing processes of creating, storing and transferring knowledge.

Data mining techniques were used to process information from these systems in order to assess employees' competencies and grade, and also to test and refine framework. For extracting and analyzing data, previously developed decision support system (DSS) was modified. DSS consisted of two subsystems: data download subsystem and competency assessment subsystem [24].

Figure 1 shows structural and functional model of the data download subsystem. At the end of the day, the subsystem migrated information from external systems via REST-services. Information about tasks, comments and work logs were downloaded from Atlassian Jira (task management system), data about pages, articles and user activity were extracted from Atlassian Confluence (knowledge management system), and source code of applications were downloaded from multiple application builder repositories (version control system).

Additionally, subsystem performed data transformation, i.e., replacing logical names with physical ones in SQL queries, translating UDML code to JavaScript code, constructing abstract syntax trees for SQL queries and JavaScript code, analyzing software complexity of constructed syntax trees and etc.

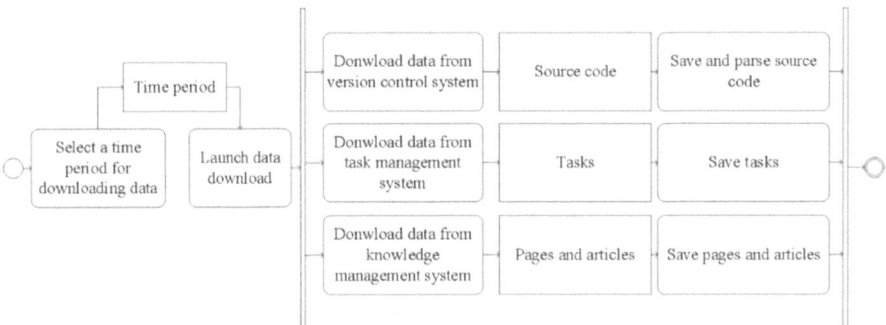

Fig. 1. Structural and functional model of the data download subsystem

As a part of data transformation process, the subsystem processed tasks descriptions, comments and work logs in order to capture information about applied competencies. Example of task's summary and description is shown in Fig. 2: developer has to prepare XML schemas and develop separate module for integration with external system «EDNA».

Tomita-parser was used for natural language processing and extracting facts. For example, in description of competency from subdomain «Types of tasks» it is indicated, that a «Middle» grade developer can install system updates on other environments. To find this fact, a rule in Tomita-parser was prepared, based on which the parser searched for agreed upon by case words in task descriptions and work logs, one of which should be mentioned in keywords set «release»:

```
ModuleImport -> Word<c-agr[1]>* Word<kwset=[release], c-agr[1],
rt>;
S -> ModuleImport interp (DevTechRelease.Term::not_norm);
```

Compiled sets of keywords included names of technologies and tools in both English and Russian:

```
TAuxDicArticle release
{
key = "импорт модуль"
key = "перенос модуль"
key = "релиз"
key = "module import"
}
```

Download subsystem recorded number of found facts by Tomita-parser in accordance with the prepared rule. A total of 44 rules for fact extraction were prepared.

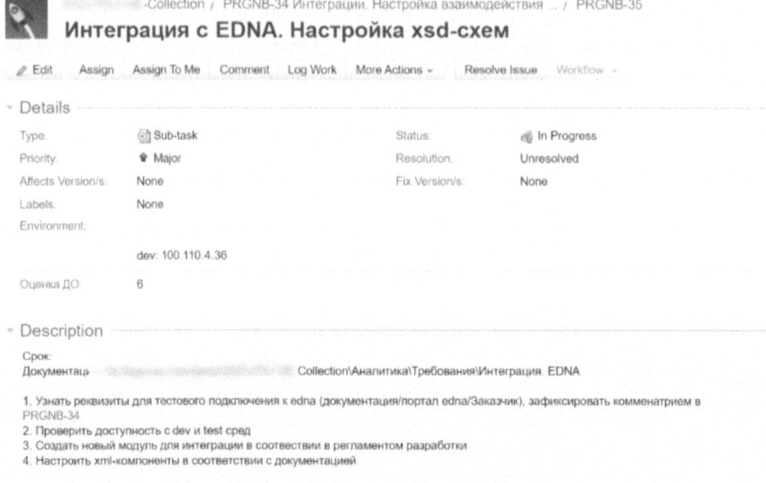

Fig. 2. Example of task formulation

Figure 3 shows structural and functional model for competency assessment subsystem, which conducted a sequential four-step evaluation procedure.

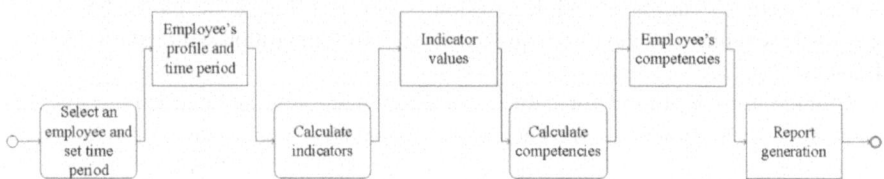

Fig. 3. Structural and functional model of the competency assessment subsystem

The first step was selecting an employee, whose competencies needed to be assessed, and setting evaluation time period.

The second step was calculation of indicators. For the selected set of competencies from the framework, 133 indicators were identified to determine whether or not a developer possessed necessary competencies. The subsystem used downloaded and prepared data for calculations.

Information from the task management system were used to calculate indicators related to the task solution: number of tasks solved within evaluation period; average number of tasks solved within and outside the estimated time frame; number of extracted facts in tasks, comments and worklogs, such as the number of tasks requiring knowledge of Web Services Description Language, Lightweight Directory Access Protocol and etc.

Data from version control system were used to calculate following indicators: code maintainability level; Halstead complexity measures; number of used Java libraries; number of used functions for integration with external systems; number of developed user functions; number of developed SQL search methods and etc.

Employee activity metrics such as the number of created pages and the number of page content updates were calculated using information from the knowledge management system.

The third step was calculation of competencies. In order to evaluate developer's competencies, the assessment subsystem compared values of calculated indicators from the previous stage with target values. For example, indicator «level of code maintainability» should be more than or equal to 155 to establish whether a developer had the competency «basic proficiency in programming language». Target values were determined based on the analysis of collected data and interviews with leading experts from the company. Because target values for several competences were not specified, it was decided that values must be strictly larger than 0. If at least one of the indicators did not fulfill the necessary requirement, it was recorded that the developer did not have competency.

The last step was generation of report, which presented the results of the assessment.

As a result, sample included 661 evaluations of 100 developers of grades «Middle», «Senior» and «Lead» from 2019 to 2022. For every evaluation result a date interval was determined, based on this date interval, decision support system downloaded data and calculated metrics. To carry out the calculations, information on 225657 tasks, 4232 pages was downloaded task management and knowledge management system accordingly and source code was downloaded from 41 repositories.

4 Results

The following competencies were confirmed to a considerable extent in a sample of 321 evaluation results of «Middle» grade developers:

- can consult the project team on current functionality. Can consult the customer on working with the system (confirmed by 99% of the developers of this grade);
- can analyze client requirements in order to design features (confirmed by 92%);
- knows how to develop new functionality, if its estimated effort is less than 40 h (confirmed by 79%)
- can mentor a «Junior» developer and adapt an experienced colleague to a new project team (confirmed by 65%).
- The competencies listed below were confirmed in the sample of 158 evaluation results of «Senior» grade developers:
- can analyze requirements and estimate the effort required for implementation (confirmed by 96%);
- can install updates on other environments and solve problems that arise during installation (confirmed by 94%);
- understands projects' business component. Understands project's current stage and makes decisions on release planning (confirmed by 80%);
- can configure integrations with external systems (confirmed by 60%);
- can administer Java application server (confirmed by 58%);
- can use Java language elements in the application source code (confirmed by 56%).
- The competencies mentioned below were confirmed for the sample of 182 evaluation results of «Lead» grade developers:

- can consult the project team on current functionality. Can consult the customer on working with the system (confirmed by 99%);
- can analyze requirements and estimate the effort required for implementation (confirmed by 98%);
- can administer Java application server (confirmed by 77%);
- can mentor a «Junior» developer and adapt an experienced colleague to a new project team (confirmed by 76%);
- can use diagnostic and debugging tools. Can gather complete diagnostic information about the problem, monitor server status (confirmed by 70%);
- understands projects' business component. Understands project's current stage and makes decisions on release planning (confirmed by 70%);
- documents the functional elements of the system (confirmed by 65%).

According to the findings, communication and teamwork skills are important in determining a developer's qualification. The ability to communicate concisely and clearly with colleagues and clients is critical for problem-solving and decision-making. Therefore, competency related to consulting clients and project team was confirmed by the majority of developers of all grades.

Mentorship is a common way of teaching and onboarding of new employees. Software engineers who participate as mentors develop their knowledge and communication skills. During onboarding, mentees ask questions that allow mentors to take a different look at their professional skills and knowledge. However, this competence was confirmed by only 42% of «Senior» developers, possibly indicating that in the case company employees of this qualification are more focused on technical tasks.

Knowledge of development tools and how to use them in problem-solving are important aspects of a developer's qualification. As a result, competencies related to server administration, installing updates, using Java and configuring integration with external systems have been confirmed by most of the developers.

Gathering, analyzing requirements, and designing solutions also play a crucial role in the success of system's implementation. If errors in requirements are detected in the latter stages of project, it may take a lot of effort to correct them and modify software product. As a result, it is important for a developer to understand project's business component, analyze requirements and communicate with clients.

Due to the fact that the other competencies were not sufficiently confirmed, i.e., they were confirmed by less than 50% of the developers of the corresponding grade, it was decided to apply fuzzy logic for the assessment, as decision-making in this area is considered approximate, and results are expressed in linguistic terms.

Fuzzy logic model for assessing qualification level was developed in the MatLAB environment using the Fuzzy Logic Toolbox package. For the output variable «qualification level», the current version of the model examined two initial grades – «Junior» and «Middle», since the case company have been restructuring the grade system and only the first two grades remained unchanged.

Criteria system for calculating the qualification level included such variables as «Technology stack», «Other technical skills», «Task Formulation» which were calculated using individually designed sub-models that used data from the systems. A developer, for example, was regarded to have the «Middle» competency grade if they had completed a task related to this area of expertise an appropriate number of times and had utilized their skills and knowledge in development.

For instance, used indicators for calculation of «Other technical skills» variable are presented in Table 2.

Table 2. System of indicators for variable «Other technical skills»

Variable	Value Range	Term Set
Mentions of the designer update in assigned tasks and work logs	[0;10]	Junior Middle
Development of integration services	[0;20]	Junior Middle

A sub-model was developed to compute the second variable «Development of integration services», which calculated a grade based on values of following indicators:

- number of temporary types created;
- number of applied functions for interaction with external systems via HTTP protocol;
- mentions of Java programming language in assigned tasks, comments and work logs;
- number of uploaded XSD schemas;
- mentions of integrations in assigned tasks, comments, and work logs.

It is planned to further modify and validate the fuzzy logic-based system in order to use for formulation of competency list, that developers will need to acquire for desired grade.

5 Conclusion

The developer competency framework was developed and validated in the current work. It was found that competences related to coding, requirements analysis, teamwork, communication skills and mentoring are considered important for developer.

The framework allows software developers to determine which competencies are necessary for an experienced professional and which competencies they may need to acquire in order to become a more competent specialist.

Using the framework and developed system, managers can assess an employee's qualification and formulate relevant tasks for professional development. Mismatches between observed and desired grade competencies can be used to establish the main objectives of personal development plans.

The system will be refined in the future to evaluate qualification level using fuzzy logic to determine the necessary competencies that developers need to acquire.

References

1. Gastaldi, L., Corso, M.: Smart healthcare digitalization: using ICT to effectively balance exploration and exploitation within hospitals. Int. J. Eng. Bus. Manag. **4**, 1–13 (2012)

2. Gubler, M., Coombs, C., Arnold, J.: The gap between career management expectations and reality – empirical insights from the IT industry. Gr. Interakt. Org. **49**, 12–22 (2018)
3. Loogma, K., Ümarik, M., Vilu, R.: Identification-flexibility dilemma of IT specialists. Career Dev. Int. **9**(3), 323–348 (2004)
4. Beausaert, S.A.J., Segers, M.R.S., Grohnert, T.: Personal Development Plan, Career Development, and Training. The Wiley Blackwell Handbook of the Psychology of Training, Development, and Performance Improvement, pp. 336–353 (2014)
5. Kostrzewski, A.J., Dhillon, S., Goodsman, D., Taylor, K.M.G.: The influence of continuing professional development portfolio records on pharmacy practice. Int. J. Pharm. Pract. **17**(2), 107–113 (2009)
6. Austin, Z., Marini, A., Desroches, B.: Use of a learning portfolio for continuous professional development: a study of pharmacists in Ontario (Canada). Pharm. Educ. **5**, 175–181 (2005)
7. Bourque, P., Fairley, R.E.: Guide to the Software Engineering Body of Knowledge. Version 3.0. IEEE Computer Society (2014)
8. Sedelmaier Y., Landes D. Software Engineering Body of Skills (SWEBOS). In: 2014 IEEE Global Engineering Education Conference (EDUCON), pp. 395–401 (2014)
9. Beausaert, S.A.J., Segers, M.S.R., Gijselaers, W.H.: Using a personal development plan for different purposes: its influence on undertaking learning activities and job performance. Vocat. Learn. **4**(3), 231–252 (2011)
10. Shippmann, J.S., Ash, R.A., Battista, M., et al.: The practice of competency modeling. Pers. Psychol. **53**, 703–740 (2000)
11. Bartram, D., Robertson, I.T., Callinan, M.: Introduction: a framework for examining organizational effectiveness. In: Organizational Effectiveness (2002)
12. Markus, L.H., Cooper-Thomas, H.D., Allpress, K.N.: Confounded by competencies? An evaluation of the evolution and use of competency models. N. Z. J. Psychol. **34**, 117–126 (2005)
13. Robillard, M.P., Coelho, W., Murphy, G.C., Society, I.C.: How effective developers investigate source code: an exploratory study. IEEE Trans. Softw. Eng. **30**(12), 889–903 (2004)
14. Surakka, S.: What subjects and skills are important for software developers? Commun. ACM **50**(1), 73–80 (2007)
15. Ahmadzadeh, M., Elliman, D., Higgins, C.: An analysis of patterns of debugging among novice. SIGCSE, pp. 84–88 (2005)
16. Robillard, M.P., Coelho, W., Murphy, G.C.: How effective developers investigate source code: an exploratory study. Soc. IC **12**(30), 889–903 (2004)
17. Ahmed, F., Capretz, L.F., Campbell, P.: Evaluating the demand for soft skills in software development. IT Prof. **14**(1), 44–49 (2012)
18. Matturro, G., Raschetti, F., Fontán, C.: A systematic mapping study on soft skills in software engineering. J. Univ. Comput. Sci. **25**, 16–41 (2019)
19. Li, P.L., Ko, A.J., Begel, A.: What Makes A Great Software Engineer? In: 37th International Conference on Software Engineering, pp. 700–710 (2015)
20. Baltes, S., Diehl, S.: Towards a theory of software development expertise. In: Proceedings of the 2018 26th ACM Joint Meeting on European Software Engineering Conference and Symposium on the Foundations of Software Engineering, ESEC/FSE 2018 (2018)
21. Bergersen, G.R., Sjøberg, D.I.K., Dybå, T.: Construction and validation of an instrument for measuring programming skill. IEEE Trans. Softw. Eng. **40**(12), 1163–1184 (2014)
22. Gousios, G., Kalliamvakou, E., Spinellis D. Measuring developer contribution from software repository data. In: Proceedings of the 2008 international working conference on Mining software repositories, MSR '08, pp.129–132 (2008)

23. Gavriliev, E.I., Avdeenko, T.V.: Model and procedure for assessing the qualification of a software developer. In: 2022 IEEE 23rd International Conference of Young Professionals in Electron Devices and Materials (EDM), pp. 303–307 (2022)
24. Gavriliev, E.I., Avdeenko, T.V.: Procedure for assessing the qualifications of a software developer. In: Nauka. Tekhnologii. Innovacii: Collection of scientific papers. In 10 parts, Novosibirsk, December 06–10, 2021, pp. 145–148 (2021)

Does UI Labeling Data Quality Matter for Predicting Website Aesthetics

Elnur Abbasov and Maxim Bakaev[(⊠)]

Novosibirsk State Technical University, Pr. K. Marksa 20, Novosibirsk 630073, Russia
bakaev@corp.nstu.ru

Abstract. The adoption of today's data-intensive digital services relies on the overall user experience (UX), which is shaped not just by "hard" functionality, but also by "soft" subjective satisfaction. In the latter, aesthetic impression plays an important role (particularly since visually pleasing products are known to be perceived as more usable) and became a popular prediction objective for Machine Learning (ML) based user behavior models. Since datasets in the field of Human-Computer Interaction are generally too scarce for application of deep learning methods that could operate on raw website screenshots, they often undergo preliminary labeling. Although the common notion is that the quality of the labeling is important for the end quality of the predictive models, there were few attempts to quantify the effect. In a previous study, we unexpectedly found significant **negative** correlations between the input data quality and the models' quality for Aesthetics and Orderliness subjective impressions. Our current paper is dedicated to validating the findings with another 557 website screenshots, 31 human participants labeling them, and 22 participants verifying the quality of their work. The non-parametrical models (Nadaraya-Watson kernel regression) with feature selection demonstrated somehow better performance, and the combined dataset better aligned with the expected effect of the labeling quality. Although our overall results are inconclusive, they might be of interest to ML practitioners and web designers who seek to automate the prediction of UX dimensions.

Keywords: Data Quality; Machine Learning; Image Recognition · User Experience

1 Introduction

Digital transformation is a constantly evolving field that influences all parts of our live. With the rise of technology, it has become crucial for businesses and organizations to adapt to this new digital landscape. However, simply adopting new technologies is not enough to drive successful transformation: "The development of evidence-based methods of management decisions and the transition to data-driven management, the evaluation of the effectiveness of state programs and public policy measures require high-quality data, as well as the ability to link data from different sources" [1].

Indeed, it is important to use quality data in order to create and develop digital products, and e-government is no exception [2]. In particular, user behavior-related data is

J. Baixeries et al. (Eds.): DAMDID/RCDL 2023, CCIS 2086, pp. 164–176, 2024.
https://doi.org/10.1007/978-3-031-67826-4_13

essential for designing appealing websites and mobile applications. Of the subjective perception dimensions, aesthetic impression has been in focus lately [3]. Web aesthetics modeling involves designing websites that are visually appealing to users while maintaining functionality and usability. It is based on the premise that visual appearance of a website can significantly affect user engagement and subjective satisfaction level after the interaction. Correspondingly, user behavior models (UBMs) that specify the influencing factors or just predict aesthetic impression based on a graphical user interface (UI) image see intensive development [4]. However, collecting enough data of appropriate quality remains a challenge in the field [5], while their positive effect on the end quality of the models remains unmeasured.

In our previous research [6], we explored the relation between the quality of the input data produced by 11 UI labelers and the quality of the ensuing UBMs constructed for the assessed Complexity, Aesthetics and Orderliness subjective scales. Rather unexpectedly, we found statistically significant **negative** correlations for Aesthetics and Orderliness, which suggested that the technically neglectable labelers supplied the data that were more beneficial with respect to predicting users' subjective perception of web UIs.

In the current study, our goal is to validate the previous results with a different dataset consisting of 557 webpage screenshots. The UI elements in them were labeled by 31 participants, and then the verification by another 22 participants was used to evaluate the quality of the input data. In the current study, we added non-parametrical Nadaraya-Watson kernel regression in addition to the linear regression used in [6], but the results are still inconclusive, as we could not find a significant correlation with the end quality for either group of the models.

The remaining part of the paper is organized as follows. In Sect. 2, we describe the experimental study, which involved the three major components: subjective assessment, labeling and verification. In Sect. 3, we analyze the experimental data and construct 58 parametrical and non-parametrical models. In the final section, we summarize and discuss our findings and specify limitations of our study and plans for further research.

2 The Experiment Description

2.1 Material

The material in our experiment was screenshots of website homepages belonging to 7 domains: culture, food, games, government, health, news, universities. The scope was the entire world, but we only used their English versions of the websites. For the universities domain, we used a dedicated Python script to automatically collect 10639 screenshots in PNG format [7]. The selection for the other 6 domain was performed manually, and the total number of candidate websites was 2932 (see the more complete description of the selection process in [10]). From all the automatically collected screenshots, 557 were manually selected for the experiment (we hereafter refer to them as web UIs). The main selection criterion was reasonable diversity of designs in each of the domains. In order to enhance the variety of UI components, complete web pages were captured in the screenshots rather than solely focusing on the above-the-fold section or using a predetermined size.

2.2 Procedure

The UI Aesthetics Assessment. The subjective evaluations of the participants for each UI were obtained through a specialized online survey (see [7] for details). To assess the subjects' visual aesthetics impressions of a website, Likert ratings were used (1 – lowest, 7 – highest). The participants were asked to provide their honest subjective evaluations, as there were no correct or incorrect answers. Screenshots were assigned to each participant in a randomized sequential order, with the default number of UIs per participants being 50 or 100.

The UI Labeling. The labelers used Crowd HTML Elements library, provided by Amazon Mechanical Turk (AMT). The crowd-bounding-box widget on MTurk displays the screenshot, providing zoom and pan functionality, along with keyboard shortcuts for creating bounding boxes of various types to quickly label numerous objects. Crowd workers could preview and skip HITs as needed.

The written instruction of the labeling process was given to the participants, who were then asked to label as many UI elements as possible for each UI. The screenshots were distributed among them fairly equally, however, there was no random assignment. For each UI element they would mark out with a rectangular, the labelers were asked to choose one of the 10 pre-defined classes: interactive (button, check, input, link, dropdown, navigation), non-interactive (image, background image) or container objects (table, panel). It is important to note that in our first experiment there were 20 pre-defined classes [6].

The Labeling Verification. The verifiers were asked to choose for each UI element whether the labeling was *correct* or *incorrect* and then subjectively assess the completeness of the labeling for each UI (i.e. whether the labeler had marked out all the visible UI elements). All of the necessary instructions for making the correct/incorrect decisions were provided to the verifiers, and they were briefed on how the labeling process was supposed to be performed.

Figure 1 demonstrates an example of a "good" website labeling from [6] (a fragment of the website screenshot loaded into Labeling tool): it was verified as having SC = 100% and Precision = 100%. Figure 2, on the contrary, presents an example of a clearly "neglectable" labeling: with SC = 20% (a major image in the center is ignored, while the *text* class elements should have been labeled individually) and Precision = 33% (note the inaccurate borders around the UI elements).

2.3 Subjects

The aforementioned activities were carried out by three groups of human participants, who were mostly Bachelor's students of Novosibirsk State Technical University (NSTU):

1. *The UI assessment* was done by 137 participants (67 females, 70 males), whose age ranged from 17 to 46 (mean 21.18, SD = 2.68).
2. *The UI labeling* was performed by 31 participants (14 male, 17 female), with the age ranging from 20 to 22 (mean 21.03, SD = 0.5).
3. *The verification* of the labelers' output was performed by another 22 participants (20 male, 2 female), whose age ranged from 20 to 22.

Fig. 1. An example of a "good" labeling: both Subjective Completeness and Precision are high.

2.4 Design and Modelling

In our previous research work [6], we only used linear regression (LR). So, the somehow controversial results we have got might be due to the common parametric models' problems, such as multicollinearity, heteroscedasticity and autocorrelation. The initial data included 24 factors extracted from web UIs, of which 8 factors were manually selected: number of UI elements, number of images, share of the text elements' area, share of whitespace, etc. However, the average number of UIs processed by each labeler in the current study was only 23.7 (SD = 8.71), unlike 44.3 (SD = 3.41) in the previous one. On the other hand, now we had 31 labelers compared to the previous 11. Thus, in the current study we decided to conduct feature selection and use Nadaraya-Watson kernel regression (KR) to minimize the effect of the violation of the Gauss-Markov assumptions [8]. Hence, we infer that the number of factors for the KR must not exceed 3, which suggests the need for feature selection. For each labeler we would build a user behavior model (LR or KR) with the mean aesthetics rating as the output variable and the factors as the predictors.

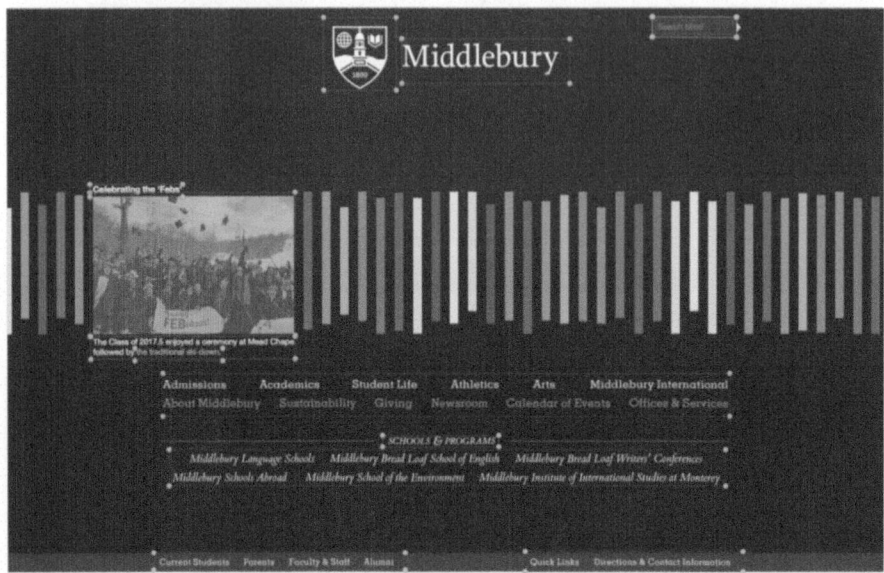

Fig. 2. An example of a "neglectable" labeling: both Subjective Completeness and Precision are low.

In the labeling verification, we had two variables: subjective completeness (SC) and Precision, averaged for each of the 31 labelers (see [6] for detail).

To assess the models' quality, we employed two criteria:

1. Coefficient of determination: $\mathcal{R}^2$
2. Coefficient of determination using PRESS statistics: $\mathcal{R}^2_{PRESS}$

In order to refute the results that we got in our first experiment, we decided to test one more time the following hypothesis: better quality of the labeling data, indicated by higher SC and Precision values, is expected to translate into improved model quality.

3   Results

3.1   Descriptive Statistics and Outliers

In total, we collected 3205 aesthetics assessments and averaged them per each website. So, we had one average assessment value per each of the 557 UIs.

Further, the 31 labelers marked out 27564 elements in 557 UIs, and the quality of their work was evaluated by the 22 verifiers. Two of the labelers, with SCs of 2.3% and 33.84%, were considered outliers, as the mean SC for the other labelers was 68.9% (SD = 11.97%). That is why they were removed from the further analysis, so we remained with 29 labelers (A1...A29 in Table 1) who altogether processed 544 unique UIs.

Table 1. The descriptive statistics per the labelers (M ± SD for the respective UIs is shown).

UI Labeling			UI Assessment
Labeler's ID	# of UIs	# of Elements	Aesthetics
A1	18	588	4.03 ± 0.84
A2	20	306	4.31 ± 0.83
A3	17	891	4.28 ± 0.94
A4	45	1658	4.11 ± 1.00
A5	17	1206	3.91 ± 1.08
A6	17	1313	4.01 ± 0.96
A7	46	1334	4.06 ± 0.94
A8	27	623	3.94 ± 0.80
A9	19	587	3.87 ± 0.93
A10	16	631	4.18 ± 1.08
A11	22	1400	4.07 ± 0.86
A12	14	619	3.97 ± 1.02
A13	28	1888	4.00 ± 1.01
A14	20	1106	3.82 ± 0.91
A15	23	810	4.23 ± 0.77
A16	27	2105	3.56 ± 0.86
A17	26	1051	4.25 ± 0.86
A18	33	1120	3.82 ± 0.82
A19	43	745	4.00 ± 1.08
A20	15	367	4.04 ± 1.10
A21	15	670	3.94 ± 0.89
A22	23	1132	4.19 ± 0.97
A23	25	498	4.15 ± 0.74
A24	26	1162	3.87 ± 0.91
A25	20	630	4.10 ± 1.30
A26	23	297	4.26 ± 1.17
A27	14	780	4.01 ± 1.21
A28	21	1037	3.88 ± 0.84
A29	28	396	4.30 ± 0.90
	544	**26950**	**4.04 ± 0.95**

Among the elements in the remaining UIs, 23484 were verified as correct and 3466 as incorrect, and the mean Precision per labelers was 84% (SD = 10.69%), which indicates a reasonably good work quality and is comparable to the precision value of 88.7% obtained in [6]. Pearson correlation between Precision and SC per labelers was significant ($r_{29} = 0.506$, p $= 0.05$), which suggests that these two aspects of UI labeling quality are related, but still distinct.

3.2 The Effect of the Input Data Quality in the Models

In order to select the factors, we used LASSO and partial correlations (PC). In LASSO regularization, the later the parameter's estimation goes to zero, the stronger is the factor's influence. After using five-fold cross-validation method, we found the optimal regularization parameter $\lambda = 0.0624$ where the minimum of mean squared error (MSE) is reached. So, the straightforward LASSO application suggested that we use four factors: SInE, SImE, TE and BE (see in Table 2). To further decrease the number of factors, we increased the regularization parameter to 0.0625, which resulted in decrease of MSE by 0.02% and exclusion of the BE factor.

Despite the fact that the correlations for BE, LE, NE, IE and PE were higher than for the SInE, SImE (see in Table 2), they were constant for most of the labelers, which makes it impossible to build most of the non-parametric models and badly affects parametric ones. That is why the same three factors that were obtained by applying the LASSO method were selected to predict the aesthetic impressions: number of text elements (TE), share of input element' areas in the screenshot (SInE) and share of the image elements' area in the screenshot (SImE).

Table 2. Partial correlations between aesthetics and the independent factors.

Variable name	Variable indicator	PC value
Number of text elements	TE	0.099
Number of button elements	BE	0.096
Number of link elements	LE	0.096
Number of navigation elements	NE	0.091
Number of input elements	IE	0.083
Number of panel elements	PE	0.080
Share of the input elements' area in the screenshot	SInE	0.070
Share of the image elements' area in the screenshot	SImE	0.048

To construct the user behavior models, we used simple LR and KR. So, we built 58 models, each having the same 3 factors calculated from each labeler's output and 2 aggregated models without splitting into labelers. The models' quality metrics and the mean labelers' quality parameters obtained from the UI verifications are shown in Table 3.

The models' average $\mathcal{R}^2_{kernel} = 0.244$ was considerably higher than the average $\mathcal{R}^2_{linear} = 0.172$. To compare $\mathcal{R}^2_{PRESS\_kernel}$ s and $\mathcal{R}^2_{PRESS\_linear}$ s, we used t-test for paired samples, which found highly significant difference (p < 0.001).

Table 3. The labelers' work quality and the models' end quality.

Labeler's ID	$\mathcal{R}^2_{PRESS\_linear}$	$\mathcal{R}^2_{linear}$	$\mathcal{R}^2_{PRESS\_kernel}$	$\mathcal{R}^2_{kernel}$	SC	Precision
A1	−0.53	0.08	0.00	0.10	86.0%	100.0%
A2	−0.51	0.08	0.00	0.12	67.5%	86.1%
A3	−0.12	0.30	0.16	0.21	89.8%	91.4%
A4	−2.44	0.01	0.49	0.56	70.0%	85.7%
A5	−0.74	0.09	0.00	0.05	88.9%	95.6%
A6	−0.60	0.16	0.16	0.29	61.0%	86.5%
A7	−0.36	0.04	0.41	0.51	55.3%	79.4%
A8	−0.32	0.09	0.38	0.43	58.8%	88.9%
A9	−0.40	0.03	0.00	0.08	70.0%	81.2%
A10	−1.51	0.28	0.17	0.31	85.6%	89.0%
A11	−0.40	0.08	0.13	0.15	80.9%	89.9%
A12	−3.10	0.07	0.34	0.45	54.9%	82.9%
A13	−0.21	0.20	0.07	0.14	66.6%	95.4%
A14	−0.53	0.22	0.19	0.34	77.4%	95.4%
A15	0.17	0.35	0.25	0.28	77.0%	87.3%
A16	−0.14	0.13	0.05	0.13	72.3%	84.6%
A17	−0.23	0.14	0.05	0.19	71.7%	79.4%
A18	0.04	0.25	0.15	0.23	57.7%	76.6%
A19	−0.26	0.12	0.00	0.03	52.4%	85.1%
A20	−0.34	0.25	0.11	0.34	53.9%	79.9%
A21	−1.21	0.52	0.66	0.68	55.3%	94.0%
A22	−0.21	0.23	0.00	0.07	71.0%	91.9%
A23	−0.36	0.11	0.07	0.20	53.4%	64.2%
A24	−0.23	0.06	0.17	0.28	84.4%	94.6%
A25	0.04	0.29	0.00	0.20	70.1%	78.3%
A26	0.08	0.33	0.00	0.03	70.1%	85.4%
A27	−1.64	0.14	0.12	0.17	71.1%	58.1%
A28	−0.39	0.23	0.20	0.23	75.8%	74.4%
A29	−0.26	0.12	0.17	0.29	48.9%	55.8%
Avg	-	**0.17**	-	**0.24**	**68.9%**	**84.0%**
Agg	**0.02**	**0.03**	**0.04**	**0.05**	-	-

We acknowledge the potential imprecision in our quality evaluations and opted to consider all the metrics as ordinal variables. This approach is highly useful because those requesting tasks typically only accept completed work from the most skilled labelers and discard the output of the neglectable ones. So, we used both Kendall and Pearson correlations to find the connection between the input data quality per labelers and the models' quality criteria (see in Table 4).

Although no significant correlations between the metrics of data labeling and the models' quality criteria were found, we can note that most correlations for KR models, which are more accurate, are negative (see in Table 4).

3.3 Consideration of Our Previous Study Data

To extend the volume of the dataset, we combined the data from the current study and the previous one [6], as demonstrated in Table 5. Moreover, we considered all the three scales: Aesthetics (A), Complexity (C) and Orderliness (O), with the corresponding R^2 criteria values for the respective LR models.

Table 4. Correlations between the quality of the labeling and the models.

Model	Criterion	Pearson (correlation/p-value)		Kendall (correlation/p-value)	
		SC	Precision	SC	Precision
Linear regression (LR)	$R^2_{PRESS\_linear}$	0.08/0.69	0.04/0.83	−0.002/1.00	−0.08/0.56
	R^2_{linear}	0.02/0.92	0.13/0.49	0.06/0.65	0.02/0.87
Kernel regression (KR)	$R^2_{PRESS\_kernel}$	−0.29/0.13	0.06/0.77	−0.10/0.48	−0.02/0.90
	R^2_{kernel}	−0.35/0.06	-0.002/0.99	−0.17/0.22	−0.08/0.56

Table 5. The labelers' and the LR models' quality based on the combined data.

Labelers' ID	R^2_A	R^2_C	R^2_O	SC	Precision
A1	0.08	0.14	0.01	86.0%	100.0%
A2	0.08	0.09	0.08	67.5%	86.1%
A3	0.30	0.07	0.33	89.8%	91.4%
A4	0.01	0.13	0.01	70.0%	85.7%
A5	0.09	0.58	0.08	88.9%	95.6%
A6	0.16	0.20	0.10	61.0%	86.5%
A7	0.04	0.14	0.01	55.3%	79.4%
A8	0.09	0.16	0.07	58.8%	88.9%

(continued)

Table 5. (*continued*)

Labelers' ID	$\mathcal{R}_A^2$	$\mathcal{R}_C^2$	$\mathcal{R}_O^2$	SC	Precision
A9	0.03	0.27	0.08	70.0%	81.2%
A10	0.28	0.04	0.24	85.6%	89.0%
A11	0.08	0.09	0.02	80.9%	89.9%
A12	0.07	0.16	0.00	54.9%	82.9%
A13	0.20	0.22	0.23	66.6%	95.4%
A14	0.22	0.17	0.17	77.4%	95.4%
A15	0.35	0.23	0.44	77.0%	87.3%
A16	0.13	0.07	0.14	72.3%	84.6%
A17	0.14	0.34	0.09	71.7%	79.4%
A18	0.25	0.13	0.13	57.7%	76.6%
A19	0.12	0.05	0.06	52.4%	85.1%
A20	0.25	0.10	0.14	53.9%	79.9%
A21	0.52	0.30	0.15	55.3%	94.0%
A22	0.23	0.18	0.11	71.0%	91.9%
A23	0.11	0.08	0.16	53.4%	64.2%
A24	0.06	0.05	0.03	84.4%	94.6%
A25	0.29	0.25	0.08	70.1%	78.3%
A26	0.33	0.08	0.11	70.1%	85.4%
A27	0.14	0.47	0.27	71.1%	58.1%
A28	0.23	0.55	0.45	75.8%	74.4%
A29	0.12	0.02	0.16	48.9%	55.8%
AA	0.15	0.11	0.11	73.0%	89.0%
GD	0.35	0.26	0.22	84.3%	89.9%
KK	0.25	0.26	0.15	82.5%	95.5%
MA	0.49	0.36	0.30	75.1%	72.0%
NE	0.49	0.32	0.42	78.3%	85.1%
PV	0.29	0.36	0.20	81.7%	91.6%
PE	0.57	0.17	0.61	72.0%	77.9%
SV	0.18	0.28	0.21	80.4%	97.4%
ShM	0.32	0.34	0.22	77.5%	89.5%
SoM	0.31	0.30	0.20	56.0%	95.9%
VY	0.11	0.20	0.17	95.5%	92.8%

Again, no significant correlations could be found for any of the scales (see in Table 6). However, we can see that most of the correlation coefficients for the combined dataset are now positive, which better aligns with the theoretically expected results.

Table 6. Correlations between the quality of the labeling and the LR models (combined with [6]).

Criterion	Pearson (correlation/p value)		Kendall (correlation/p value)	
	SC	Precision	SC	Precision
R_A^2	0.09/0.57	0.03/0.89	0.10/0.37	0.01/0.95
R_C^2	0.27/0.10	−0.01/0.95	0.17/0.12	0.08/0.48
R_O^2	0.24/0.14	−0.19/0.24	0.19/0.09	−0.04/0.71

4   Discussion and Conclusions

In our previous study [6], we found a significant negative correlation between labeling precision and the quality of the predictive models for the Aesthetics dimension of UX. However, these results turned out to be quite surprising for us, even though we kept in mind that the object of our study is rather a philosophical concept. Common sense told us that the better everything is labeled, the easier it will be to predict aesthetics. To validate the results of our original study, we now considered another 557 web UIs.

In [6], we built simple linear regression (LR) models on a fairly small sample size, which could have been affected by heteroscedasticity, autocorrelation, and multicollinearity. In the current work, we selected the most significant factors and used the Nadaraya-Watson kernel regression (KR), because in this case the main problems of structural regression models do not directly affect the estimation results. The employment of the kernel regression indeed allowed us to increase the R^2s of the models (0.244 for KR vs. 0.172 for LR). Still, the correlations between the R^2s and the labeling quality were mostly negative (see in Table 4), although not significant. Extending the dataset with the data from [6] lead to mostly positive correlations, which are easier to interpret. Overall, the previous results obtained in [6] still hold, since in the current study we did not achieve the appropriate level of statistical significance. Still, we see the main contributions of the current study and their importance as follows:

1) We demonstrated that kernel regression (KR) might be applicable to predict aesthetic impressions even with a small number of factors. Although a considerable share of existing publications focus on increasing the number of the input variables (see e.g. in [4]), the problem of making the most with the limited data, which do not come for free in the field of Human-Computer Interaction [9], remains rather urgent [10].

2) We highlighted the sophistication of the concept of training data quality in ML and certain counter-intuitiveness with its practical application. While a lot of research publications consider the benefits of more accurate labelling data to be obvious (see

reviews in [11] or [12]), relatively few put forward the importance of its effect's quantification [13]. We demonstrated that the labeling quality might have no effect on the resulting models' quality or even a negative correlation. This can be explained by the high subjectivity of the prediction object: the aesthetic impression, for which "less could be more", as we reasoned in [6].

The main limitation of our study is arguably the relatively small number of labeled UIs per subject and the associated low R^2s in LR and KR models. Our further research plans include obtaining at least 20 observations per labeler and carrying out a more sophisticated selection of the factors that also considers interfactorial interactions. In this case models will more accurately describe the data and the correlation can become significant for quality. The sample of the verifiers in our study – over 90% of them were men – might also be problematic with respect to external validity. This is particularly remarkable since the output variables included such gender-dependent subjective dimension as aesthetic impression.

Although we do not propose a significant and final model for the relation between labeling quality and the final quality of predictive user behavior models, we believe that our results might be of interest to UI designers and ML practitioners who collect training data. The main take-away is that the costs of obtaining more quality data should be weighed against its actual effect on the end quality of the models. So, it generally makes sense to collect the data in iterations (i.e., several batches), carefully measuring the models' quality dynamics.

Acknowledgment. We would like to thank V. Shchekoldin (PhD, Assoc. Prof. of the NSTU's department of Marketing and Service) for consulting us on the modeling issues.

References

1. Orlova, A.: Data quality became a topic of discussion at the Gaidar Forum (in Russian). Training Center for Leaders and Digital Transformation Teams of the Russian Presidential Academy of National Economy and Public Administration (RANEPA) (2022). https://cdto.ranepa.ru/sum-of-tech/materials/37, last accessed 2023/08/28
2. Chang, C., Almaghalsah, H.: Usability evaluation of e-government websites: a case study from Taiwan. Int. J. Data Netw. Sci. **4**(2), 127–138 (2020)
3. Miniukovich, A., Marchese, M.: Relationship between visual complexity and aesthetics of webpages. In: Proceedings of the 2020 CHI Conference on Human Factors in Computing Systems, Honolulu, HI, USA, 25–30 April 2020, pp. 1–13 (2020)
4. Wan, H., et al.: A novel webpage layout aesthetic evaluation model for quantifying webpage layout design. Inf. Sci. **576**, 589–608 (2021)
5. Lima, A.L.D.S., Gresse von Wangenheim, C.: Assessing the visual esthetics of user interfaces: a ten-year systematic mapping. Int. J. Human–Comput. Interaction **38**(2), 144–164 (2022)
6. Bakaev, M., Khvorostov, V.: Quality of labeled data in machine learning: common sense and the controversial effect for user behavior models. Eng. Proc. **33**(1), 3 (2023). https://doi.org/10.3390/engproc2023033003
7. Boychuk, E., Bakaev, M.: Entropy and compression-based analysis of web user interfaces, in: International Conference on Web Engineering, Springer, pp. 253–261 (2019)

8. Ali, T.H.: Modification of the adaptive Nadaraya-Watson kernel method for nonparametric regression (simulation study). Commun. Stat.Simul. Comput. **51**(2), 391–403 (2022)

9. Daniel, F., Kucherbaev, P., Cappiello, C., Benatallah, B., Allahbakhsh, M.: Quality control in crowdsourcing: a survey of quality attributes, assessment techniques, and assurance actions. ACM Comput. Surv. (CSUR) **51**(1), 1–40 (2018)

10. Bakaev, M., Speicher, M., Heil, S., Gaedke, M.: I Don't Have That Much Data! Reusing user behavior models for websites from different domains. In: International Conference on Web Engineering, pp. 146–162 (2020)

11. Geiger, R.S., et al.: "Garbage in, garbage out" revisited: what do machine learning application papers report about human-labeled training data? Quant. Sci. Stud. **2**, 795–827 (2021)

12. Whang, S.E., Roh, Y., Song, H., Lee, J.G.: Data collection and quality challenges in deep learning: a data-centric AI perspective. VLDB J. **32**(4), 791–813 (2023)

13. Mitchell, M. et al.: Measuring data. arXiv preprint arXiv:2212.05129 (2022)

Segmentation of Graphical User Interface Elements Based on Topological Decomposition for GUI Testing Tasks

Artyom Abakumov$^{(\boxtimes)}$ ⓘ and Sergey Eremeev$^{(\boxtimes)}$ ⓘ

Murom Institute of Vladimir State University, Murom, Russia
artem210966@yandex.ru, sv-eremeev@yandex.ru

Abstract. The paper addresses the issue of automating the testing process for graphical interfaces. It is shown that one of the main tasks in this area is segmentation of screen elements with further construction its internal structure. Emphasis is placed on the stability of the proposed method, regardless of changes in interface layout or the operating system employed. Our approach is based on decomposing a window screenshot into specific components that correspond to the elements of the original window and their hierarchy. We demonstrate the method's resilience to the resizing of objects within windows. The research was conducted using interface element segmentation for the QGIS geographic information system on both Windows and Ubuntu operating systems. Experimental results revealed high levels of accuracy, ranging from 94 to 100 percent, in extracting segmented areas within QGIS windows.

Keywords: topological analysis · GUI testing · segmentation

1 Introduction

The number of software programs continues to grow annually, and the quality of these applications is directly tied to the effectiveness of testing. While various types of automated testing exist to verify program functionality by executing code and validating its results, developers encounter difficulties when it comes to testing the graphical user interface (GUI) of applications.

There are testing utility that extracts meta-information about buttons, fields, and other interface elements. But, unlike regular code, the rendering of graphics in GUIs is reliant on the underlying operating system. Consequently, new frameworks and libraries for creating GUIs are constantly being developed and improved. Updates to these frameworks or changes in modules can break working this meta-extracting uptilts. In some cases, vendors intentionally restrict or complicate the extracting meta-information to enhance protection against reverse engineering. These and other scenarios necessitate the use of methods that do not rely on a deep understanding of the rendering approach or operating system, but instead interact with the interface through computer vision.

The task at hand involves identifying the desired interaction elements and validating their response to specific actions. For instance, one might need to click a button and

J. Baixeries et al. (Eds.): DAMDID/RCDL 2023, CCIS 2086, pp. 177–191, 2024.
https://doi.org/10.1007/978-3-031-67826-4_14

ensure that a modal window opens. The typical solution in the computer vision approach involves performing a screenshot to locate the button for subsequent testing. Also, it is crucial to note that recognition errors are not permissible in this case, given the nature of the task.

Although it may appear that the problem could be solved by employing straightforward pixel-by-pixel comparison algorithms, in practice, program elements can exhibit slight variations in their rendering across different systems and machines. There are two primary challenges associated with this issue. Firstly, in diverse environments, windows may stretch, resulting in distorted displays of elements. Secondly, font glyphs can be rendered differently on various systems, and at times, a program may employ similar but distinct fonts on different operating systems. The disparity in GUI rendering is illustrated in Fig. 1. Even if primitive methods attempt to ignore the differing parts, they have proven ineffective.

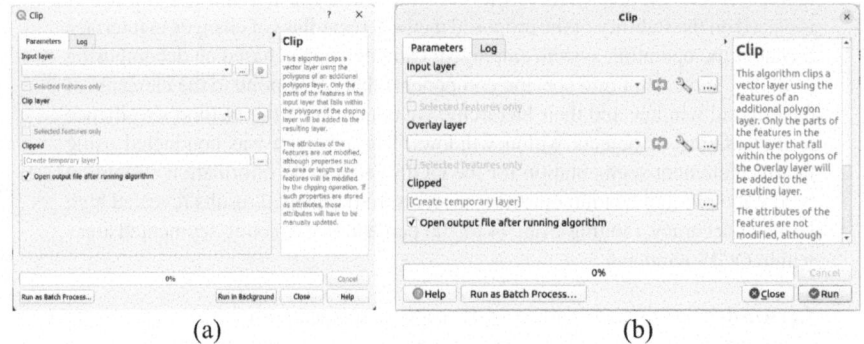

(a)   (b)

Fig. 1. Different font and buttons in QGIS on Windows (a) and Ubuntu (b) systems.

These distortions not only add complexity to the testing process but also contribute to increased costs, highlighting the necessity for a method that can deliver consistent results while being resilient to such distortions. The primary focus of this work is to address this practical problem effectively.

In this paper, we present a new approach for decomposing application screenshots and propose a testing scheme built upon this approach.

2 Motivation

One of the main problems associated with open-source software is the issue of quality. Many developers contribute to open-source projects as a hobby and may not have the inclination to invest a significant amount of time in GUI testing. Consequently, there is a pressing need to develop a methodology that is straightforward and accessible to ordinary developers, without necessitating manual adjustments for different platforms.

To address this issue, it is logical to employ a method that not only compares images but also enables automatic highlighting of interactive GUI elements. This would allow users to select the desired interaction element without having to create a screenshot of it. One such method is the decomposition of images into topological features, which satisfies the requirements.

The technique for decomposing an image (screenshot) into components that identify high-frequency and low-frequency objects will be described in detail later. For now, it is sufficient to understand that this method enables the identification of all "disturbances" in the image, which are then represented as components. These components form a hierarchical structure, akin to a tree-like graph. Consequently, accessing an element becomes possible by traversing a component path in the Decomposition Tree, like selecting elements in XML or HTML schemas. The process of creating a test using this method is depicted in Fig. 2. It is necessary to note that in a real scenario keyboard actions are also included in the testing process. But, since our goal is to test the GUI, we will not take keyboard actions into account.

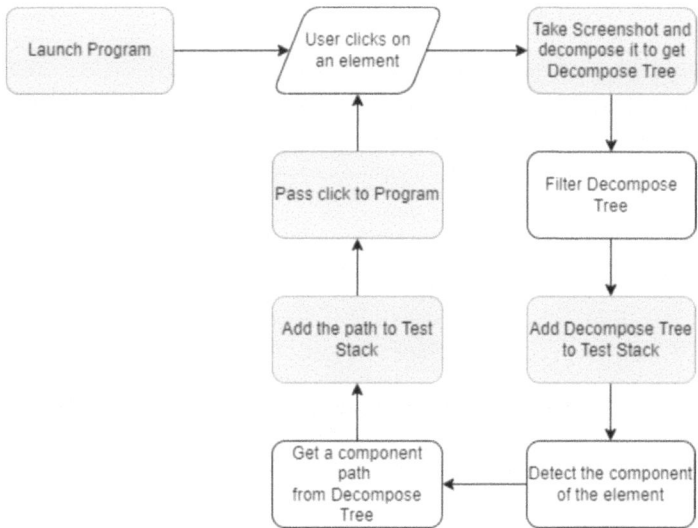

Fig. 2. Test creation scheme.

The Decomposition Tree comprises components along with their associated metadata, including the start and the end times of existence, as well as decomposition matrices. Depending on the specific task, additional metrics can be incorporated as needed. A notable advantage of this method is that the size of the tree is significantly smaller than that of the source image.

During the execution of the created test (as depicted in Fig. 3), the saved Test Stack is utilized.

The test will launch the program automatically and construct the decomposition based on a captured screenshot. This fully automated testing process, reliant on the decomposition technique, will be accomplished as a result.

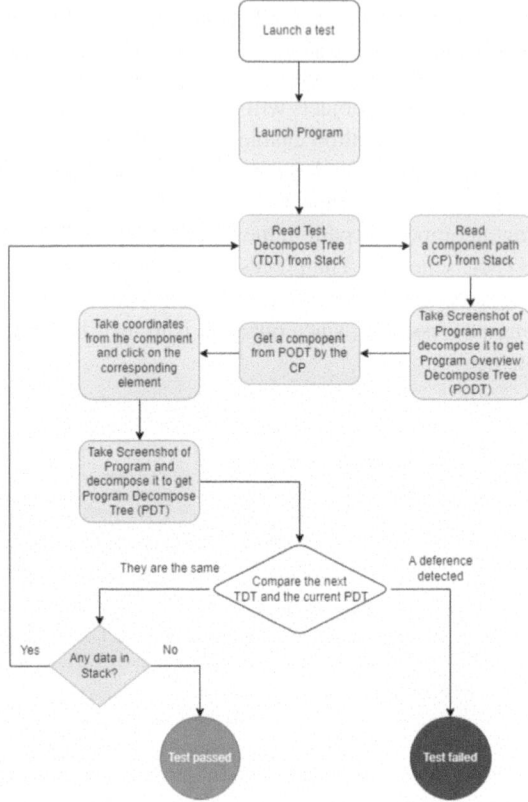

Fig. 3. The scheme of the testing procedure (performs automatically).

3   Review

Testing utilities can be categorized into two main approaches based on their operational principles: black-box and white-box testing.

White-box testing involves assessing the internal structure and logic of the system rendering. It enables the determination of element coordinates, states, data, properties, and more in real-time. This approach offers high accuracy and simplicity. However, it requires the development of separate utilities for each graphical library and is highly dependent on the specific libraries and system being utilized, as mentioned previously.

Black-box testing involves testing a system without considering or having knowledge of its internal workings. It primarily focuses on the external behavior and functionality of the system. Implementations of this technique execute actions, such as moving the cursor or clicking, in a predefined order. Black-box testing allows for independence from specific technologies or systems but can potentially result in blind-running, where errors or issues may go unnoticed. For example, if there is a missed click, the testing program may continue the test without detecting the error. These situations are mitigated in the white-box approach.

If we narrow down the classification and deviate from abstract concepts, testing methods can be categorized based on the interaction with a test application. One fundamental method is the coordinate-based approach, where a tester specifies coordinates to which the cursor should be directed. However, it is important to consider that all coordinates may need to be adjusted in the event of distortions, such as changes in the size of elements. Autopy and PyAutoGUI are examples of utilities that facilitate such operations.

A more reliable method for testing is image recognition, which has been previously discussed. Examples of applications that employ this approach include Sikuli and Lackey.

Furthermore, the most reliable yet complex approach is the Accessibility method. This method pertains to white-box techniques that possess access to the internal elements of the system. An example of a popular solution in this domain is Pywinauto. However, it is worth noting that Pywinauto is exclusively compatible with Windows, rendering it unsuitable for cross-platform applications.

The underlying principle of these utilities predominantly relies on macros, which involve the repetition of pre-defined actions. One variation of this approach, referred to as keyword testing, is widely implemented in numerous popular tools like Katalon Studio and Jubula.

Webpages present the simplest case for testing due to their open code nature. Numerous programs are available specifically designed for web testing, with Selenium being a notable example.

Scientific papers addressing GUI testing started appearing as early as the 1990s, as mentioned by the authors in [1]. These researchers extensively publications on the subject and their achievements. The recent study [2] has demonstrated that despite significant advancements in recent years, numerous challenges persist without resolution and are anticipated to persist into the future. This emphasizes the significance of further research in this area.

Researchers propose various approaches to address the challenges of GUI testing. They explore different fields, such as security aspects in GUI testing [3]. Other researchers focus not only on the means of interacting with the program but also on the issue of test coverage. They explore methods to improve test coverage. In the study [4], the authors present existing tools for automated testing and the detection of untested elements and UI states. They propose a new method and its implementation for identifying potential states of elements, which involves computing all possible combinations of interactions with the application. However, the authors also acknowledge the limitations of the method, which further supports the claims made in the first review papers.

Scientific literature recognizes the significance of resource constraints in frequent software releases. The optimization and acceleration of testing in systems with realistic graphics are discussed in [5]. The paper describes existing approaches and provides a demonstration of the process using specific technologies.

Also, it is important to consider other systems such as smartphones. Mobile applications, which are prevalent in today's technological landscape, heavily rely on GUIs, making the discussed problem of GUI testing even more relevant in this context.

In [6], the authors highlight a common issue with testing tools that often rely on static GUI models, which may lack accuracy. To address this, the authors propose a novel approach where the GUI model is dynamically optimized during program execution. It is worth noting that many modern smartphone applications utilize web technologies, effectively functioning as miniature browsers with web pages. This presents an opportunity to leverage existing tools used for testing websites, thereby warranting further exploration of existing works in this particular area.

Methods for improving the testing of web pages based on their HTML are presented in [7]. The authors propose two versions of their method: one focuses on generating test cases for each web element, while the other explores different paths between web elements. These approaches aim to address the issue of insufficient test coverage and improve the overall testing process for web pages.

When considering papers closely related to the problem, it becomes evident that many of them are focused on mobile applications, thereby supporting our previous thesis. An intriguing work in this regard is [8], where the authors tackle the same task by detecting interactive elements and utilizing them for testing through the Accessibility methodology. They employ a trained computer model to identify interaction elements within mobile applications and describe their approach as comprehensive in their conclusions. However, it is worth noting that the use of machine learning confines the research to the realm of mobile applications exclusively. A similar issue, with a greater emphasis on training neural networks, is addressed in [9], also focusing on mobile applications. This trend is further exemplified in [10], where the authors employ machine learning techniques to enhance the existing methodology of random GUI testing.

In general, there is a noticeable trend in scientific papers towards the application of machine learning, particularly neural networks. For instance, some of the previously mentioned works utilize the popular YOLO v3 system, which not only segments but also classifies objects. However, it is important to note that despite the prevalence of neural networks in research, their practical adoption in existing testing utilities is not widespread. This is likely due to the significant overhead costs associated with resource-intensive classifiers. In practical terms, the usage of neural networks often involves cloud technologies, which introduces complexities and increases the cost of GUI testing.

If we consider the body of this chapter, it becomes apparent that there is a multitude of approaches for testing mobile applications, while there seems to be a relative lack of new ideas specifically focused on desktop applications. This observation further underscores the relevance of the problem we have chosen to address.

The decomposition method has demonstrated successful applications in satellite image segmentation [11] and object classification in images [12]. Therefore, it can be reasonably asserted that in simpler cases, it will yield results no less satisfactory than those achieved in the aforementioned articles.

4 Methodology

4.1 Decomposition of Interface Elements Based on Topological Features

The main principle employed in persistent homology is the systematic traversal of a point cloud and the construction of simplex complexes based on it. In our specific case, the image (matrix) functions as the point cloud, while the components, composed of pixels with distinct brightness values, act as the simplex complexes. Consequently, this enables us to establish a comprehensive framework for constructing components (Eq. 1) and representing complexes (Eq. 2).

$$M = \{C_1, C_2, ..., C_n\} = F_M(I),$$ (1)

where M is a set of output components, C is a component, F_M is component creation functions, and I is an input image.

$$C = \{b_1, b_2, ..., b_k\},$$ (2)

where b is a pixel brightness, $C \in M$ and k is a number of component points.

From Eq. (2), we can derive various persistent features of a component. These include the brightness values of its start and end (Eqs. 3 and 4, respectively), the duration of its existence (Eq. 5), and the square it occupies (Eq. 6).

$$P_{start} = min\{b_1, b_2, ..., b_k\}$$ (3)

$$P_{end} = max\{b_1, b_2, ..., b_k\}$$ (4)

$$P_{length} = P_{end} - P_{start}$$ (5)

$$P_{square} = |C|$$ (6)

The features can be used to uniquely identify an individual component among the entire set.

In the paper [13], several approaches to constructing components are described. The simplest method involves analyzing pixels from top to bottom (from 255 brightness to 0) or from bottom to top. These two approaches we mark as base ones.

The base method connects pixels in a certain order. First, they are sorted by brightness and then added to the pad by X and Y indices. If there is another pixel next to the added pixel, they are joined into a component. If there is a component next to it, the pixel is joined to it. If a pixel can be attached to several components at once, one absorbs the other. The absorbed one does not grow any more.

The base methods only segments either bright or dark regions. Since interfaces can consist of various shades, it is necessary to perform two passes—one from top to bottom and another from bottom to top—to capture all interaction zones and accurately segment the components.

The bottom-up method iterates through all the pixels in ascending order of their brightness and analyzing the surrounding area. If there are no nearby components, a new component is created. If there is an existing component nearby, the current point merges with it, forming a larger component. If two components are found adjacent to each other, the older component assimilates the younger one. The younger component is designated as a child of the older component, resulting in a hierarchical, tree-like structure. One important feature of the decomposition is independent of the scale and, accordingly, of the resolution of the image.

The construction process for the window depicted in Fig. 4 is illustrated in Fig. 5 as an example.

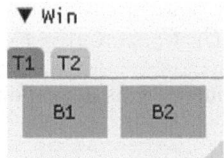

Fig. 4. A simple window with two tabs and two buttons.

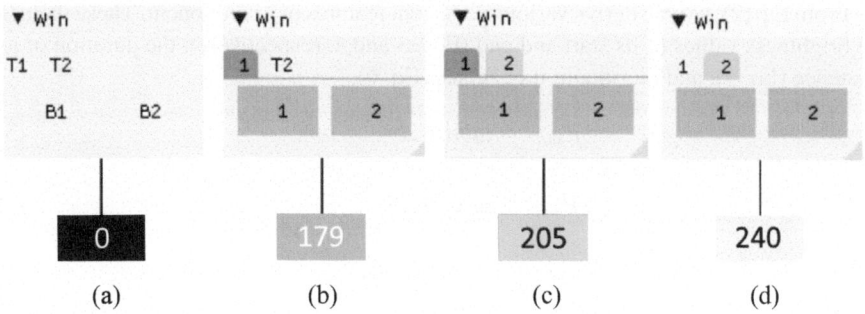

| (a) | (b) | (c) | (d) |

Fig. 5. Bottom-up construction. Pixels are stored in sorted order and added based on the increasing value of their brightness, b. Each displayed element represents a component.

The construction process begins by connecting all pixels with a brightness of zero, as shown in Fig. 5(a). The adjacent pixels form separate components, representing the letters and the symbol in the top left corner. Then, the process continues by connecting pixels with a brightness of 1, and this progression continues for subsequent brightness levels. In Fig. 5(b), at brightness level 179, the components representing the letter "B" absorb the pixels of the buttons. Finally, the same process occurs with the tab component labeled as "T1".

The example clearly demonstrates that some information is lost during the construction process. To recover this lost information, an additional pass from brightness 255 to 0 is required. However, this approach is inefficient. Therefore, it is advisable to further develop the algorithm to prevent information loss and aim for achieving accurate results in a single pass. This can help improve the efficiency and effectiveness of the component construction process.

One possible modification is to alter the order of pixels connections. Based on our experiments, the most promising approach is to connect points in pairs, considering the increasing difference in brightness, denoted as D, between them. The visualization of this method, along with the graph construction for the window depicted in Fig. 4, is illustrated in Fig. 6. We mark this version as the new method and named it as Radius-Based (RB).

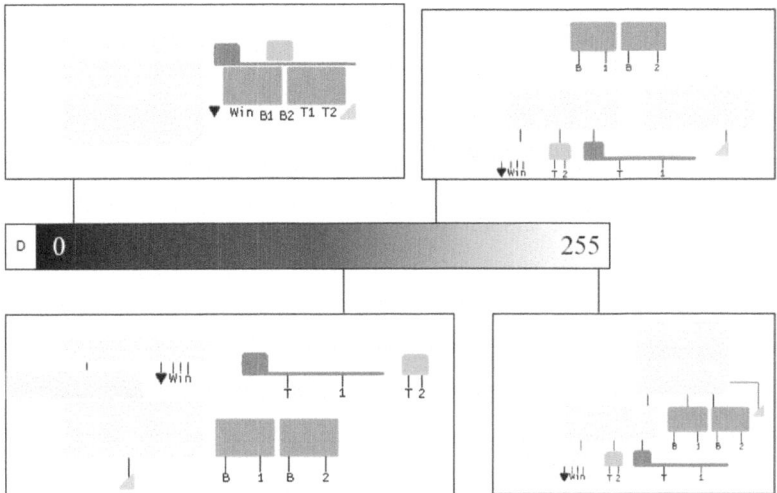

Fig. 6. Constructing components in the order of increasing difference (D) between pixels.

The RB method is predicated on connecting the closest points in terms of brightness. Initially, all pixels with zero brightness are connected, creating the first components. As the difference in brightness between other pixels becomes smaller than a threshold value, D, they are connected, with one component absorbing the others, forming a tree-like structure. In the considered case, there are the buttons, letters, and tabs in Fig. 6. Their average brightness is close to 0. As the brightness approaches 170, the difference in brightness between the button component and the symbols "B", "1" and "2" exceeds the threshold D. . As a result, the largest component (the button) absorbs the smaller ones (the symbols "B", "1", "2"), creating a hierarchy. The same process occurs with the remaining components. Closer to a brightness of 200, larger components start merging, forming an even more visible hierarchy. This process continues until all components form a tree.

To summarize, the new approach localizes the zones by the nearest brightness between points, which, given the specifics of the task, was the best fit. The results demonstrate that this modified approach achieves a more accurate segmentation. The one more advantage of the new approach is ability to use RGB colors because we sort pixels not by brightness, but by distance. RGB represents a vector (Red, Green, Blue), making it easy to calculate the distance. The bottom-up and up-bottom methods require grayscale pixels. In the paper, we will utilize the new approach to process the colored images.

4.2   Filtering Components for Obtaining Segmented Interface Elements

Certainly, filtering is an important step in the process. First and foremost, it is necessary to discard the text as it is not considered an interactive element. Let's consider Fig. 7 for further analysis.

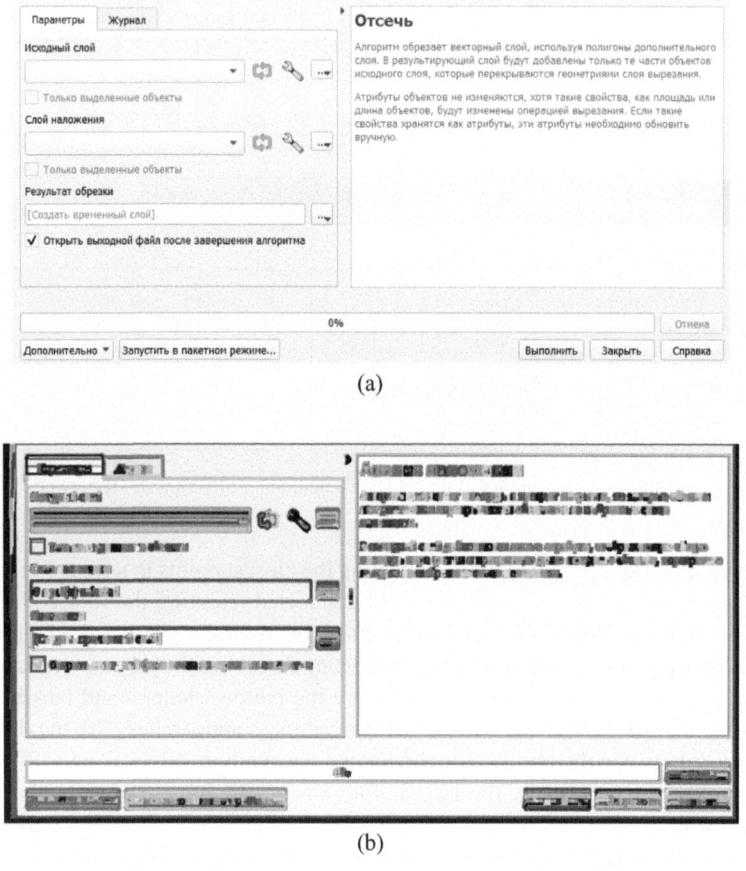

(a)

(b)

Fig. 7. The source "Clip" window (a) in QGIS and the components constructed based on it (b).

It is necessary to impose restrictions on certain features to filter the letters. Initially, we utilize the simplest one, the square of the component (P_{square}) which is calculated by Formula (6). The resulting filtering outcome is presented in Fig. 8. Some letters and intermediate components remain. To further refine the filtering process, we employ another feature, the existence length (P_{length}) of a component, calculated using Formula (5). We will limit its maximum value. The result is displayed in Fig. 9.

The artifacts have been successfully eliminated, and the segmentation has significantly improved. Although there are a few misclassifications where some letters and a portion of the button were mistakenly highlighted, however, they do not have a significant impact on the overall result.

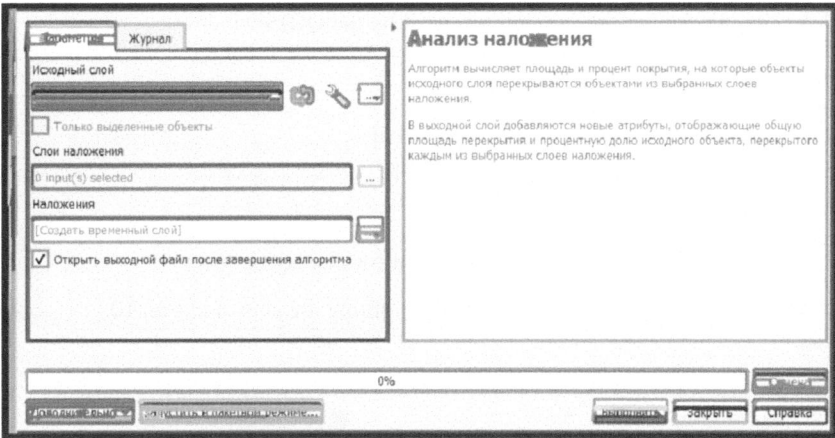

Fig. 8. The segmentation result with $P_{square} >= 100$

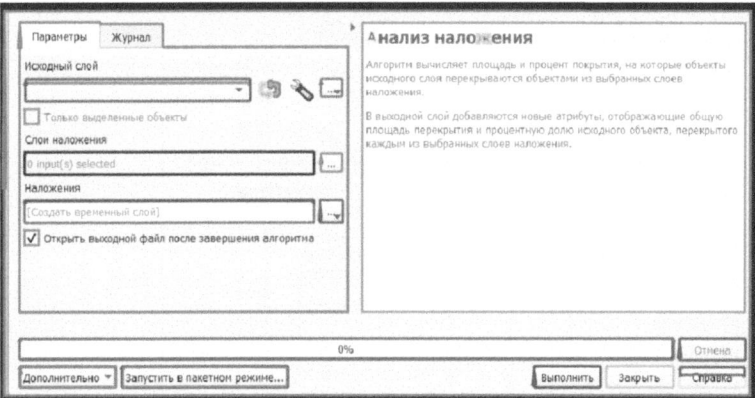

Fig. 9. The segmentation result with $P_{length} >= 10$ and $P_{square} >= 100$.

Another way to perform filtering is by using the component path. This means that, like the testing process, specific elements can be indicated to be discarded and not considered when comparing the tree structure.

5 Results

Unfortunately, there are no public ready-to-use datasets for desktop GUI testing. The closest ones contain screenshots of mobile applications, which makes it difficult to verify the method. So, it was decided to compile its own set of test data.

So, there is a popular open-source solution called QGIS, which provides a graphical interface for working with geospatial data, such as GDAL library integration. However, in practice, bugs are often encountered, which makes many GUI windows useless. A variety of different windows can be utilized to test the decomposition method.

Typically, the following elements are used by the QGIS windows:

- Tab Panel: a container with other elements;
- Tab: an inactive tab;
- CheckBox: allows selecting either True or False;
- ComboBox: enables selecting from multiple items;
- InputBox: allows input of numbers and text;
- Button: a clickable button;
- MultiLine TextBox (Read only): an element for displaying text.

Visually, all these elements and their highlighting are shown in Fig. 10.

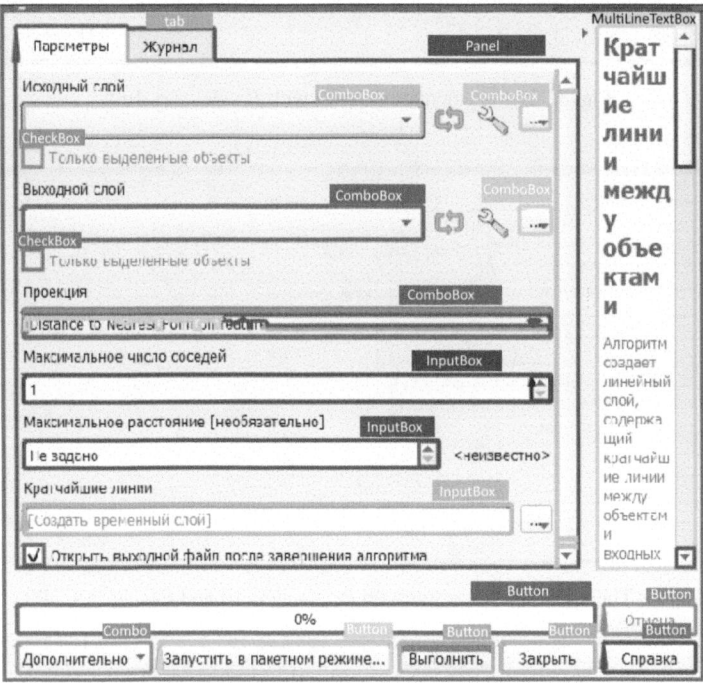

Fig. 10. The highlighted elements with labels.

Thirty QGIS windows were taken for testing purposes, and decomposition was performed on them using the criteria $P_{square} >= 100$ and $P_{length} >= 10$. The accuracy results, reflecting the found/total ratio, presented in Table 1. It is worth noting that sometimes buttons can be disable, and these cases accounted for a 6% decrease in accuracy. However, determining whether an inactive element should be identified remains a matter of debate.

It is also important to note the errors. Table 2 indicates the number of letters that mistakenly entered the Decomposition Tree (i.e., remained after filtering), and the number of artifacts representing incomplete or incorrect segmentation of elements.

Table 1. Accuracy results in Percentage, verified by a practicing GUI tester.

Type	Accuracy, %
Tab	98
ComboBox	94
Tab Panel	97
InputBox	100
CheckBox	100
Button	94
MultiLine TextBox	100

Table 2. Decomposition errors.

Letters	Artifacts
67	14

The letters only affect the size of the tree and do not pose problems for segmentation. Artifacts, on the other hand, can impact accuracy. To reduce these issues, it is necessary to refine the filtering process or improve the mechanism for constructing components in the case of artifacts.

Additionally, the method was tested on a variety of other applications. For example, we demonstrate the complete decomposition of a Visual Studio Code window in Fig. 11. The library is written in C++. Without additional optimizations, it takes approximately

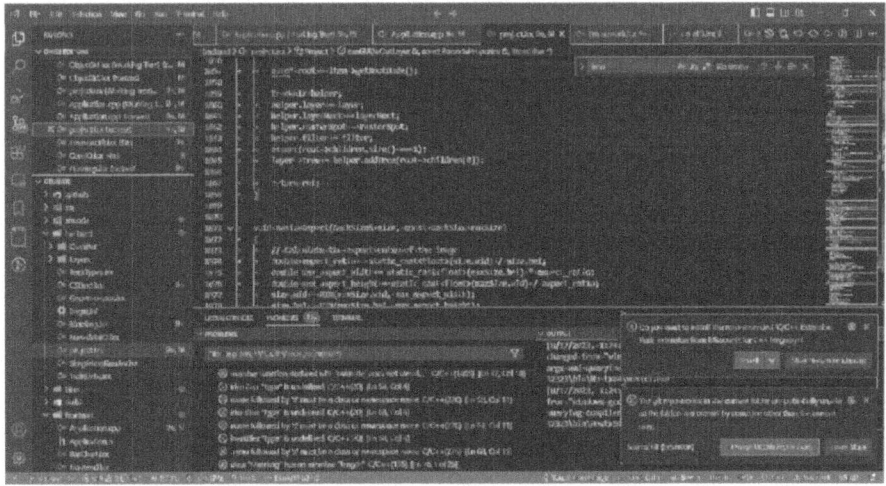

Fig. 11. The VS Code windows with highlighted elements.

~ 1000 ms to decompose a 1920×1080 screenshot on an Intel(R) Core(TM) i5-8300H CPU @ 2.30 GHz.

6 Conclusions

The study results highlight the importance of further refining the filtering mechanism and utilizing more reliable indicators. However, apart from that, the segmentation results are promising, with an average accuracy exceeding 95 "The decomposition library is available on GitHub: github.com/Noremos/SatHomology/tree/GUI.

In conclusion, the topological decomposition method has proven to be effective in detecting interactive zones. It provides the advantage of disregarding potential distortions, such as element stretching and variations in the rendering of glyphs and interaction elements.

Acknowledgements. This study was supported by the Russian Science Foundation, project no. 23-21-10064, https://rscf.ru/en/project/23-21-10064/.

References

1. Banerjee, I., Nguyen, B., Garousi, V.: Graphical user interface (GUI) testing: systematic mapping and repository. Inf. Softw. Technol. **55**(10), 1679–1694 (2013)
2. Nass, M., Alégroth, E., Feldt, R.: Why many challenges with GUI test automation (will) remain. Inf. Software Technol. **138**, 106625 (2021)
3. Mironov, S.: Technologies of security control of automated systems based on structural and behavioral testing of software. Cybern. Program. **5**, 158–172 (2015)
4. Vartanov, S., Gerasimov, A., Ermakov, M., Kutz, D., Novikov, A.: Dynamic analysis of programs with graphical user interface based on symbolic execution. Proc. Inst. Syst. Program. RAS (Proceedings of ISP RAS) **29**(1), 149–166 (2017)
5. Denisov, E., Voloboy, A., Birukov, E.: Technologies for automatic testing of a software package for realistic computer graphics. Proc. Inst. Syst. Program. RAS (Proceedings of ISP RAS) **32**(1), 71–88 (2020)
6. Gu, T.: Practical GUI testing of android applications via model abstraction and refinement. In: 2019 IEEE/ACM 41st International Conference on Software Engineering (ICSE), pp. 269–280. Montreal, QC, Canada (2019)
7. Medhat, M., Saad, M.: Enhancing the automation of GUI testing. In: Proceedings of the 8th International Conference on Software and Information Engineering (ICSIE 2019), pp. 66–70. Association for Computing Machinery, New York, NY, USA (2019)
8. Zhang, X., et al.: Screen recognition: creating accessibility metadata for mobile applications from pixels. arXiv arXiv:2101.04893 (2021)
9. Xue, F., Wu, F., Zhang, T.: Visual identification of mobile app GUI elements for automated robotic testing. Comput. Intell. Neurosci. **1**, 1687–5265 (2022)
10. Thomas, D., White, G., Guy J.: Improving random GUI testing with image-based widget detection. In: Proceedings of the 28th ACM SIGSOFT International Symposium on Software Testing and Analysis (ISSTA 2019), pp. 307–317. New York, NY, USA (2019)
11. Eremeev, S., Abakumov, A., Andrianov, D., Shirabakina, T.: Vectorization method of satellite images based on their decomposition by topological features. Inf. Autom. **22**(1), 110–145 (2023)

12. Eremeev, S., Abakumov, A.: Classification of objects in images with distortions based on a two-stage topological analysis. Sci. Tech. J. Inf. Technol. Mech. Opt. **22**(1), 82–92 (2022)
13. Eremeev, S., Abakumov, A., Andrianov, D., Titov, D.: Image decomposition method by topological features. Comput. Opt. **46**(6), 939–947 (2022)

Data Analysis in Astronomy

Exploring the Universe with SNAD: Anomaly Detection in Astronomy

Alina A. Volnova[1,6](✉) , Patrick D. Aleo[2,3] , Anastasia Lavrukhina[4],
Etienne Russeil[5] , Timofey Semenikhin[4,6], Emmanuel Gangler[5],
Emille E. O. Ishida[5] , Matwey V. Kornilov[6,7] ,
Vladimir Korolev[1,2,3,4,5,6,7,8] , Konstantin Malanchev[2,6] ,
Maria V. Pruzhinskaya[5] , and Sreevarsha Sreejith[8]

[1] Space Research Institute of the Russian Academy of Sciences, Profsoyuznaya
84/32, Moscow 117997, Russia
alinusss@gmail.com
[2] Department of Astronomy, University of Illinois at Urbana-Champaign, 1002 W.
Green St., Champaign, IL 61801, USA
[3] Center for AstroPhysical Surveys, National Center for Supercomputing
Applications, Urbana, IL 61801, USA
[4] Faculty of Space Research, Lomonosov Moscow State University, Leninsky Gori 1
bld. 52, Moscow 119234, Russia
[5] Université Clermont Auvergne, CNRS/IN2P3, LPCA,
63000 Clermont-Ferrand, France
[6] Sternberg Astronomical Institute, Lomonosov Moscow State University,
Universitetsky 13, Moscow 119234, Russia
[7] National Research University Higher School of Economics, 21/4 Staraya
Basmannaya Ulitsa, Moscow 105066, Russia
[8] Physics Department, University of Surrey, Stag Hill, Guildford GU2 7XH, UK

Abstract. SNAD is an international project with a primary focus on
detecting astronomical anomalies within large-scale surveys, using active
learning and other machine learning algorithms. The work carried out
by SNAD not only contributes to the discovery and classification of various astronomical phenomena but also enhances our understanding and
implementation of machine learning techniques within the field of astrophysics. This paper provides a review of the SNAD project and summarizes the advancements and achievements made by the team over several
years.

Keywords: Methods · data analysis · Supernovae · general ·
Transients · Astronomical data bases

1 Introduction

In modern astronomy, discoveries of new objects are based on a huge flow of
data coming from all-sky surveys (e.g., the Sloan Digital Sky Survey, SDSS [7],
the Zwicky Transient Facility, ZTF [5], the Vera Rubin Observatory Legacy

© The Author(s), under exclusive license to Springer Nature Switzerland AG 2024
J. Baixeries et al. (Eds.): DAMDID/RCDL 2023, CCIS 2086, pp. 195–208, 2024.
https://doi.org/10.1007/978-3-031-67826-4_15

Survey of Space and Time, LSST [19]). The terabytes of data generated every night contain information, allowing the discovery of several hundred transients per year. Undoubtedly, the human resource is limited, and each scientist should choose from a variety of discovered objects those that are of the greatest interest to him. However, this is not a trivial task, given the number of transients to choose from. In addition, the total mass of discovered objects may contain rare or not yet known phenomena. Thus, we are faced with the problem of classifying a large number of objects, and specifically in this case: the search for anomalies or outliers.

It is natural to expect that such large volumes of data that are generated in astronomy today require machine learning (ML) methods for processing. Despite the fact that ML has become an integral part of data analysis in almost all areas of science in recent years, astronomy has benefited from it only recently, starting from solving problems in classification and regression (see, e.g., [21]). In the light of the search for the most interesting objects among all detected ones anomaly detection (AD) algorithms have a wide field of application: would it be search for galaxies with anomalous spectra [4], transient with extraordinary light curves [32], or unusual variable stars [27].

However, most of those algorithms are based on statistical models, and the anomaly is defined as an object that does not fit the model. Observational defects lead to multiple identifications of non-physical events as anomalous, giving the researcher hundreds of candidates for further investigations. To decrease the amount of "not interesting" objects and to detect the astrophysical anomalies AD algorithms need some advisory from human in which object should be considered as anomalous, i.e., active learning (AL) is required. AL is a subclass of ML algorithms where the user may adjust the model by interactively setting scores to the objects which are suspicious for the machine. In this learning paradigm, the user can select which type of object should be considered as anomalous, extracting some rare classes of objects from the bulk of huge sky-survey data (see, e.g., [21,40]).

Given to its important impact in the future of astronomical discoveries with modern data sets, the problem of anomaly detection have already been explored in the literature. This includes recent studies using data streams [30], deep [12], generative [38] and active [15,22] methods. Nevertheless, the incidence of false positives continues to be an important bottle neck to be overcome if we intend to take full advantage of modern astronomical data. We describe here one of such attempts, which has been consistently focusing on anomaly detection for astronomy in the last 5 years.

This paper is an overview of the SNAD project – an international group of researchers (including astronomers, physicists, data scientists and mathematicians) which focus on developing new AL algorithms for the search for anomalous objects in large data-sets of all-sky astronomical surveys. In Sect. 2 we provide the description and main goals of the project and the team, and briefly summarise data sources and used techniques. In Sect. 3 the most prominent results of the project are presented. Section 4 describes some auxiliary products developed by the team, which are now publicly available. Section 5 gives conclusive remarks for future prospects of the project.

2   What is the SNAD Project?

2.1   Goals and Objectives

The goal of the SNAD[1] project is to develop a pipeline where human expertise and modern machine learning techniques can complement each other in the task of identifying unusual astronomical objects mostly by their photometrical features. The team concentrates on the search for unusual, rare or yet unknown objects in the sets of photometric light curves (LCs) by combining different AL AD algorithms with the additional information provided by the expert in order to label a significant fraction of the most obvious outliers and choose those which are true astronomical anomalies. Enabling reliable anomaly/outlier detection based solely on photometric observations is one of the fundamental puzzles to be solved before we can convert the full potential of large-scale surveys into scientific results. This project represents an effective strategy to guarantee we shall not overlook exciting new science hidden in the data we fought so hard to acquire.

2.2   Team Members and Expertise

The SNAD team is composed of young researchers, each offering a unique set of skills and experiences that are utilized within the project. All team members are involved in various stages of the anomaly detection process, from feature engineering to expert analysis.

In addition to their collective expertise in anomaly detection, each team member also contributes their unique knowledge from different fields: gamma-ray bursts (Alina Volnova), supernovae (Maria Pruzhinskaya), fast transients (Anastasia Lavrukhina), accretion flows in astrophysics (Konstantin Malanchev), adapting machine learning algorithms (Emille E. O. Ishida), astronomical site characterization (Matwey Kornilov), time-domain astronomy and photometric classification (Partick Aleo), astronomical image analysis and dwarf galaxy detection (Sreevarsha Sreejith), real-bogus classification (Timofey Semenikhin), aeronautical science (Vladimir Korolev), symbolic regression for light curve analysis (Etienne Russeil), observational cosmology (Emmanuel Gangler).

Over the years, several other researchers, including Florian Mondon, Anastasia Malancheva, Alexandra Novinskaya, and Anastasiya Voloshina have also contributed their skills and knowledge to the project.

SNAD activities are designed around annual meetings whose format was inspired by other initiatives which focused on boosting innovation and creativity (e.g. the silicon valley model [2] and the Cosmostatistics Initiative[2]). We also

---

[1] https://snad.space.
[2] https://cosmostatistics-initiative.org/.

invite[3] other researchers and students of different levels for collaboration in the field of astronomical anomalies detection. The resulting environment is one of the key components which play an important role in the development of products described in this work.

2.3   Data Sources and Techniques

The first data-set used for our research was extracted from the Open Supernova Catalog[4] (OSC; [18]). It contains all publicly available data on all SNe from a dozen of catalogues, including, in different cases, multi-colour LCs, spectra, redshift estimations and classification. We extracted 1999 LCs with the number of data points enough to fit the LC within some particular interval of time relative to the maximum, and presented in 3 different photometric pass-bands (gri, $g'r'i'$ or BRI). These 1999 LCs allowed us to test our main ideas by finding several peculiar and super-luminous SNe, along with a few dozens of miss-classified objects [32].

The results of the first try encouraged us to use more numerous data-set of photometric LCs, and the Zwicky Transient Facility (ZTF) survey data were chosen. The ZTF survey started on March 2018 and during its initial phase has observed around a billion objects [5]. Each object is represented with a bunch of LCs in filters g, r, i with a cadence – on average $\sim$1 day for the Galactic plane and $\sim$3 days for the Northern-equatorial sky. To minimize the affection of different cadences on the results we selected 3 different fields: 1 in the M31 galaxy, 1 in the Galactic plane, and 1 far above the Galactic disk and analysed data from the first 9.4 months of the ZTF survey, between 17 March and 31 December 2018. The total amount of objects with at least 100 data-point in the LC is 2.25 millions. The results of the search for anomalies in this set are described in Sect. 3.

Investigating all data described above we experimented with a series of different anomaly detection algorithms. Our goal was to quantify their effectiveness in identifying scientifically interesting anomalous objects within large data sets. Among others, we used Isolation Forest (IF) [26], Local Outlier Factor [9], Gaussian Mixture Models [29] and one-class Support Vector Machines [35]. These were applied to small [32] as well as large [27] data sets with encouraging results.

However, we noticed that frequently, the majority of objects with high anomaly score represent non-astrophysical artificial effects. These were most times border effects, glitches in the detector, bad CCDs, cosmic rays, unexpected telescope movements, satellites, etc. This discovery was important, and generated an additional line of investigation, described in Sect. 3.4. Nevertheless, the need of an adaptive algorithm, which could allow the user to define which type of anomaly was interesting became increasingly more evident.

We started experimenting with adaptive learning techniques by using the Active Anomaly Discovery algorithm (AAD) [13], which uses a human-in-the-loop strategy to apply a series of sequential modifications to a traditional IF. By

---

[3] https://snad.space/#contact.

[4] https://github.com/astrocatalogs.

downgrading decision paths which disagree with the expert's definition of what is an interesting anomaly the algorithm allows the construction of a personalized AD model. We showed that this strategy is effective in small [22] and large [31] data sets. The successful experiences with AAD motivated many of the results described below.

3 Key Contributions and Discoveries

3.1 Supernova Catalog

The SNAD team discovered potential, unreported supernovae—powerful and bright explosion of a star, indicating the final stage of its evolution [8]—within the ZTF DRs during a non-targeted anomaly detection search [27]. This led to the inception of a new experiment aimed at developing specialized machine learning models. The AAD algorithm (Sect. 2.3) and the SNAD Transient Miner (Sect. 4.3) have been trained and refined based on our long-term experience with supernovae. Applying these algorithms, the SNAD team identified 144 new supernova candidates hidden within the vast photometric data of the ZTF survey.

Each of these candidates has been thoroughly inspected and validated by our domain experts. Detailed information about the candidates has been made publicly available in the SNAD supernova catalog[5].

Furthermore, these supernova candidates have also been reported to the Transient Name Server[6] (TNS). The TNS is the official resource for announcing new astronomical transients, and our reports ensure that these candidates are available for further study by the international astronomical community. With this dual approach of cataloguing and reporting, we aim to foster collaboration and further our understanding of supernovae.

3.2 Superluminous Supernova Candidates

Among the supernova candidates discovered using the AAD algorithm (Sect. 2.3) we report four objects: SNAD120, SNAD121, SNAD160, SNAD187, which could belong to the superluminous supernovae class. They display an unusually broad LC when compared to the Nugent's models[7] and multiple redshift estimations indicate their high absolute brightness [31]. Among them, SNAD160 displays a particularly broad LC and further analysis indicates that it could belong to the pair instability supernovae class [33]. It is a theoretical category of thermonuclear explosions of extremely massive stars, as described by [3] [16]. There have been no reports of observational confirmation of such event yet and any good candidate contributes to enrich our understanding of the phenomena.

Additional analysis has been performed by inputting SNAD160 to a symbolic regression algorithm. It is a machine learning method based on genetic programming [25] that computes a mathematical expression with independent variables

---

[5] https://snad.space/catalog/.

[6] https://www.wis-tns.org/.

[7] https://c3.lbl.gov/nugent/nugent_templates.html.

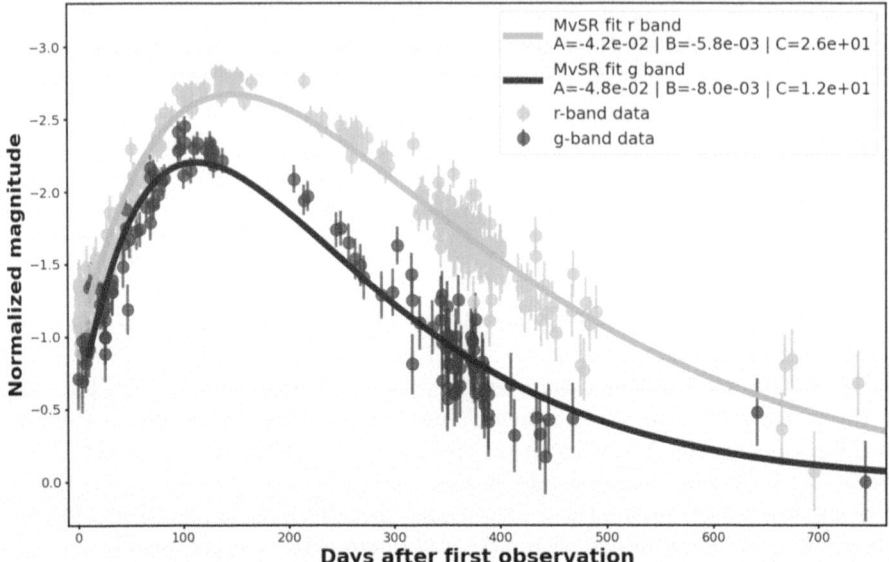

Fig. 1. Fit of SNAD160 using Eq. 1 resulting from MvSR procedure. The magnitude is normalized by subtraction of the peak magnitude.

to optimally fits some input data. We developed a more complex approach of the algorithm called Multiview symbolic regression (MvSR) [34]. It allows for the input of multiple datasets supposedly generated by the same parametric function, and directly returns the common parametric function that generated the examples. MvSR was applied to SNAD160 with LCs in passbands g and r being used as two separate datasets. The resulting parametric function describes transient like behaviors with a linear rising and an exponential decay.

$$f(t) = A(t + C) \times e^{B(t+C)} \qquad (1)$$

It is particularly well fitting SNAD160 in both pass-bands as shown in Fig. 1. It has been applied to other SLSN candidates and provides an accurate representation of the LCs. With three free parameters, this equation offers a noticeably dense description of supernova events. Using this representation for further feature extraction analysis could lead to more interesting SLSN discoveries.

3.3 Red Dwarf Flares

Within the SNAD group, we are currently working on detecting red dwarf flares in the high cadence data of the ZTF survey releases. Stellar flares are very energetic phenomena, during which the optical luminosity of a star increases by several fold over tens of seconds. The currently accepted version of stellar flares' nature is a magnetic field, generated in the convective stellar interior [17].

The detection of rapid stellar flares, such as those from red dwarfs, is of significant importance in the field of exoplanet science, particularly when studying exoplanet habitability. Additionally, creating a statistically significant sample of red dwarf flares can greatly aid in our understanding of the physical processes involved in these events.

By this time, approximately a hundred new candidates for red dwarf flares have been found [39]. Some of them were discovered using the AAD algorithm (Sect. 2.3). Others were detected by fitting the LCs with a model that accurately describes the properties and shape of red dwarf flares' LCs, followed by further analysis of the goodness of fit. The candidates found using presented methods are then subject to further analysis and approval by experts within the SNAD team, with the assistance of the SNAD ZTF Viewer (Sect. 4.1).

Examples of found red dwarf flare candidates LCs presented in Fig. 2. Found candidates vary both by a LC shape (multiple outburst or one outburst flare) and brightness, which allows exploring the complexity of flare events and forming a more comprehensive sample.

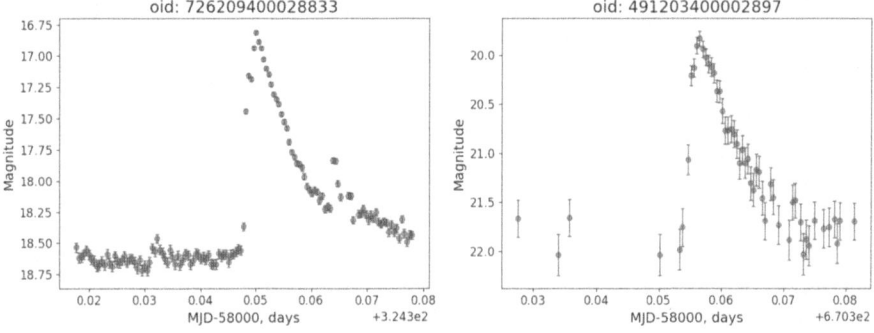

Fig. 2. Two examples of red dwarf flares candidates found by SNAD team. (Color figure online)

3.4 Catalog of Artefacts

Another interesting data product that resulted from the construction of the SNAD database is the SNAD catalog of artefacts. The term artefact is generally used as an umbrella term to denote observational, phenomenological or instrumental oddities in signal that manifest as diffraction spikes, a step-effect in brightness, colour saturation etc. to name a few. We are collating a catalog of such occurrences in the SNAD database that is aimed for both outreach and science goals. The catalog consists of the OIDs (identifying number that's referenced in ZTF), the FITS images of the respective fields and the SNAD link to the main object page denoted by the OID. Some examples of the artefacts thus identified are given in Fig. 3. While spurious in the strictest sense, several of these artefacts present as visually breathtaking making them perfect for outreach purposes.

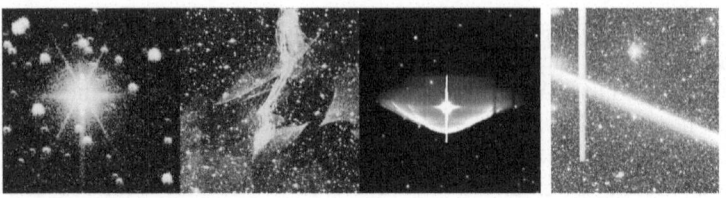

Fig. 3. Some examples from the SNAD catalog of artefacts.

The catalog of artefacts also provides a data set for training machine learning algorithms that could potentially be used for identifying and/or removing these objects from fields of interest. Using the labels of 2230 objects from SNAD knowledge database (Sect. 4.2), roughly half of which are artefacts, we developed an algorithm that predicts whether an object is an artefact or not based on the sequence of object frames from the ZTF survey. The algorithm consists of two parts: (1) a variational auto-encoder that returns the compressed representation of each frame after being trained on all the images from the sample and, (2) a recurrent neural network that takes the sequence of compressed frame representations and an object label as inputs, and returns the probability that the object is an artefact. Currently the best classification result is $ROC - AUC = 0.856 \pm 0.010$ and $Accuracy = 0.802 \pm 0.023$ [36].

4   Scientific Tools and Resources

4.1   SNAD ZTF Viewer

The SNAD Viewer[8] [28] is a web portal specifically designed for astronomers, providing a centralized view of individual objects from various data releases. It integrates data from multiple publicly available astronomical archives and sources, thereby offering a comprehensive platform for time-domain data analysis and interpretation.

In the era of big data in astronomy, the SNAD Viewer emerges as a significant tool designed to centralize and streamline the management of astronomical data. Initially conceived to facilitate expert feedback in active machine learning applications (see Sects. 4.3, 4.4), the SNAD Viewer has evolved into a valuable

---

[8] https://ztf.snad.space.

community asset. It centralizes public information and provides a multi-dimensional view of individual objects from the ZTF data releases.

The SNAD Viewer's infrastructure is characterized by its scalability and flexibility. It is designed to accommodate a wide range of data types and user needs, demonstrating its adaptability to the changing landscape of astronomical research. This adaptability extends to its ability to be personalized and used by other surveys and for various scientific goals, underscoring its broad applicability within the field.

Importantly, the SNAD Viewer is not an isolated entity but is part of a larger network of astronomical data portals. It is linked to by other significant portals such as Antares, Fink, YSE PZ, and Astro-COLIBRI, which further enhances its accessibility and utility within the astronomical community. This interconnectedness highlights the collaborative nature of modern astronomical research and the crucial role of the SNAD Viewer within this ecosystem.

The SNAD Viewer is publicly available online, emphasizing its commitment to open science and the democratization of astronomical data. It serves as a testament to the crucial roles that domain experts continue to play in the era of big data in astronomy.

4.2 SNAD Knowledge Database

In addition to the discovery of candidates in anomalies, our research also led to the creation of a knowledge database. One of the core features of the SNAD ZTF Viewer (Sect. 4.1) is the ability to assign specific tags to each ZTF object, a function our experts extensively used when reviewing potential anomalies. This tagging system incorporates a wide range of general astronomical classes, including variable stars, transients, active galactic nuclei, along with their various specific types and subtypes. We also devised custom tags for internal purposes, as well as non-astrophysical tags like artefacts and their subtypes. Multiple tags can be assigned to a single object, with the history of tag changes stored in the database.

During the expert analysis, a total of 7238^9 objects were labelled. Despite the ZTF data processing pipeline's procedure to distinguish astrophysical events from bogus ones, almost a half of them are artefacts. These assigned labels serve as a reliable source of curated data for training supervised machine learning models. Furthermore, they can help to refine the ZTF's pre-processing pipeline, ensuring more efficient identification and classification of astronomical events in the future.

4.3 SNAD Transient Miner

The SNAD Miner is an exhaustive similarity search for LC features for transient discovery, using simulated LCs as a guide to identify transients in a large dataset

---

[9] on Sep. 18, 2024

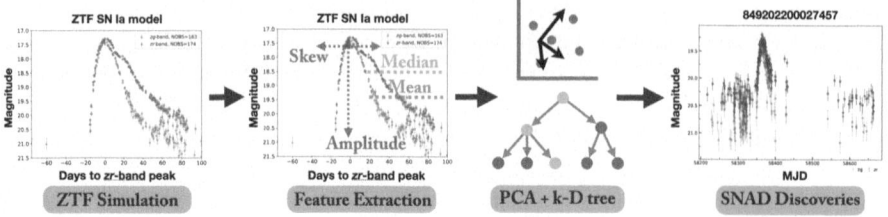

Fig. 4. SNAD Miner schematic. We use bright ZTF SNe simulations (left) and extract their LC features (left center). Then, we apply a (PCA+) k-D tree on these features, to search for real ZTF DR events nearest neighbors (right center). Some of these matched nearest neighbors were previously missed SNe (right).

mostly comprised of variable stars [11]. Specifically, it leverages extracted statistical features from simulated SNe LCs, from which a k-D tree is applied to search for similar feature values in the ZTF DR4. As a result, we discovered 11 previously missed transients, 7 SNe candidates and 4 AGN candidates.

Realistic ZTF simulations were generated using SNANA [24], with ZTF data release 3 cadence and magnitude error distribution (see [10] for details). Starting from the template models originally developed for PLAsTiCC, [20,23], we selected the seven brightest, well-sampled LCs to use as a reference (3 SLSN-I, 1 SN Ia, 1 SN II and 2 TDE with peak magnitude $\sim 17^m$).

Using these objects, we applied a k-D tree [6] to their extracted 82 non-normalized features. Subsequently, we identified the 15 nearest neighbors for each simulation (105 matches in total, resulting in 89 unique ZTF DR4 sources), and manually inspected the results. Ultimately, we discovered 11 previously unreported supernova and active galactic nucleus candidates. The remaining 94 matches (81 unique ZTF DR4 sources) were either known/already reported transients or variable stars. Given the large number of variable stars previously estimated in ZTF data releases [11], it is reasonable to expect many well sampled, high-amplitude variable stars whose coverage in parameter space significantly overlaps with the regions populated by transients (see, e.g. Figure 4 in [1]). Thus, a ratio of 18 transients (11 newly-discovered) to 44 variable stars out of 89 unique sources selected from 990,220 considered ZTF sources is a very successful result. Considering the extreme case where $\approx$3000 SNe discovered by ZTF [14] were part of the data set, the expected incidence of SNe when choosing 100 sources at random would be of $< 1(\approx 0.3)$ event.

This "SNAD Miner" process is flexible, and can use simulated or real objects as the input, any number of features, and any number of nearest neighbors to "mine" in the existing dataset. A schematic is shown in Fig. 4.

4.4 Coniferest Python Library

The coniferest library [37] is aiming to add adaptive capabilities to isolation forest algorithm which is inherently static.

The library has implementations of two adaptive learning algorithms. One of them is an already mentioned earlier AAD algorithm [13]. And the second one is the implementation of our own adaptive learning isolation forest algorithm named PineForest. It is based of tree filtering approach. After every new observation with some label given by an expert the PineForest filters out trees that do not push forward right observations.

This approach have a few remarkable advantages. First, the algorithm has not much hyperparameters, so it is easy to tune. Second, it maybe used for both – as an adaptive learning with an expert in loop and as an accumulator of prior knowledge about data. Finally, it has a very good performance characteristics making it suitable for data intensive applications.

Also, as a bonus the library includes our own implementations of classical isolation forest with much better performance in scoring compared to `scikit-learn`'s one. Refer to `coniferest` package documentation for more details.

5   Future Directions and Challenges

Large scale sky surveys have fundamentally changed the process of astronomical investigation and discovery. In the era of LSST, when millions of new transients will be detected every night, serendipitous discoveries will not happen. On the other hand, targeted searches are bound to identify objects which fall within our domain knowledge.

The SNAD team has been consistently working on the development of adaptable algorithms and tools which allow the user to probe the boundaries of their domain knowledge – thus enabling scientific discovery in large data sets. These have been proven to be effective in current state of the art catalog data.

In the near future we intend to concentrate our efforts in two important bottlenecks: ensuring that our tools are scalable to meet LSST requirements and make additional connections with expert communities to guarantee scientific impact of SNAD products. The latter will allow the SNAD team to develop increasingly more personalized tools which will fulfill the requirements of individual experts and ensure that we can fully exploit the scientific potential of modern astronomical surveys.

Acknowledgements. Authors thank the Ministry of Science and Higher Education of Russian Federation for financial support, grant 075-15-2022-1221 (2022-BRICS-8847-2335). We used the equipment funded by the Lomonosov Moscow State University Program of Development. This work made use of the Illinois Campus Cluster, a computing resource that is operated by the Illinois Campus Cluster Program (ICCP) in conjunction with the National Center for Supercomputing Applications (NCSA) and which is supported by funds from the University of Illinois at Urbana-Champaign.

References

1. Aleo, P.D., Malanchev, K.L., Pruzhinskaya, M.V., et al.: SNAD transient miner: finding missed transient events in ZTF DR4 using k-D trees. New Astron. **96**, 101846 (2022). https://doi.org/10.1016/j.newast.2022.101846
2. Annika Steiber, S.A.: The Silicon Valley Model. Springer, Cham (2016)
3. Barkat, Z., Rakavy, G., Sack, N.: Dynamics of supernova explosion resulting from pair formation. Phys. Rev. Lett. **18**, 379–381 (1967). https://doi.org/10.1103/PhysRevLett.18.379
4. Baron, D., Poznanski, D.: The weirdest SDSS galaxies: results from an outlier detection algorithm. Mon. Not. R. Astron. Soc. **465**(4), 4530–4555 (2017). https://doi.org/10.1093/mnras/stw3021
5. Bellm, E.C., Kulkarni, S.R., Graham, M.J., et al.: The Zwicky transient facility: system overview, performance, and first results. Publ. Astron. Soc. Pac. **131**(995), 018002 (2019). https://doi.org/10.1088/1538-3873/aaecbe
6. Bentley, J.L.: Multidimensional binary search trees used for associative searching. Commun. ACM **18**(9), 509–517 (1975). https://doi.org/10.1145/361002.361007
7. Blanton, M.R., Bershady, M.A., Abolfathi, B., et al.: Sloan digital sky survey IV: mapping the milky way, nearby galaxies, and the distant universe. Astron. J. **154**(1), 28 (2017). https://doi.org/10.3847/1538-3881/aa7567
8. Branch, D., Wheeler, J.C.: Supernova explosions (2017). https://doi.org/10.1007/978-3-662-55054-0
9. Breunig, M.M., Kriegel, H.P., Ng, R.T., et al.: LOF: identifying density-based local outliers. In: Proceedings of the 2000 ACM SIGMOD International Conference on Management of Data, pp. 93–104 (2000)
10. Chatterjee, D., Narayan, G., Aleo, P.D., et al.: El-CID: a filter for gravitational-wave electromagnetic counterpart identification. arXiv e-prints arXiv:2108.04166 (2021)
11. Chen, X., Wang, S., Deng, L., et al.: The Zwicky transient facility catalog of periodic variable stars. Astrophys. J. Suppl. Ser. **249**(1), 18 (2020). https://doi.org/10.3847/1538-4365/ab9cae
12. Ćiprijanović, A., Lewis, A., Pedro, K., et al.: DeepAstroUDA: semi-supervised universal domain adaptation for cross-survey galaxy morphology classification and anomaly detection. Mach. Learn.: Sci. Technol. **4**(2), 025013 (2023). https://doi.org/10.1088/2632-2153/acca5f
13. Das, S., Wong, W.K., Fern, A., et al.: Incorporating feedback into tree-based anomaly detection. In: Workshop on Interactive Data Exploration and Analytics (IDEA 2017), p. arXiv:1708.09441. KDD Workshop (2017)
14. Dhawan, S., Goobar, A., Smith, M., et al.: The Zwicky transient facility type Ia supernova survey: first data release and results. Mon. Not. Roy. Astron. Soc. **510**(2), 2228–2241 (2022). https://doi.org/10.1093/mnras/stab3093
15. Etsebeth, V., Lochner, M., Walmsley, M., et al.: Astronomaly at scale: searching for anomalies amongst 4 million galaxies. arXiv e-prints arXiv:2309.08660 (2023). https://doi.org/10.48550/arXiv.2309.08660
16. Gal-Yam, A.: The most luminous supernovae. Annu. Rev. Astron. Astrophys. **57**(1), 305–333 (2019). https://doi.org/10.1146/annurev-astro-081817-051819
17. Gershberg, R.E.: Solar-Type Activity in Main-Sequence Stars. Springer, Cham (2005). https://doi.org/10.1007/3-540-28243-2
18. Guillochon, J., Parrent, J., Kelley, L.Z., et al.: An open catalog for supernova data. Astrophys. J. **835**(1), 64 (2017). https://doi.org/10.3847/1538-4357/835/1/64

19. Hambleton, K.M., Bianco, F.B., Street, R., et al.: Rubin observatory LSST transients and variable stars roadmap. arXiv e-prints arXiv:2208.04499 (2022). https://doi.org/10.48550/arXiv.2208.04499

20. Hložek, R., Ponder, K.A., Malz, A.I., et al.: Results of the photometric LSST astronomical time-series classification challenge (PLAsTiCC). arXiv e-prints arXiv:2012.12392 (2020)

21. Ishida, E.E.O., Beck, R., González-Gaitán, S., et al.: Optimizing spectroscopic follow-up strategies for supernova photometric classification with active learning. Mon. Not. R. Astron. Soc. **483**(1), 2–18 (2019). https://doi.org/10.1093/mnras/sty3015

22. Ishida, E.E.O., Kornilov, M.V., Malanchev, K.L., et al.: Active anomaly detection for time-domain discoveries. Astron. Astrophys. **650**, A195 (2021). https://doi.org/10.1051/0004-6361/202037709

23. Kessler, R., Narayan, G., Avelino, A., et al.: Models and simulations for the photometric LSST astronomical time series classification challenge (PLAsTiCC). Publ. Astron. Soc. Pac. **131**(1003), 094501 (2019). https://doi.org/10.1088/1538-3873/ab26f1

24. Kessler, R., Bernstein, J.P., Cinabro, D., et al.: SNANA: a public software package for supernova analysis. Publ. Astron. Soc. Pac. **121**(883), 1028 (2009). https://doi.org/10.1086/605984

25. Koza, J., Koza, J.: Genetic Programming: On the Programming of Computers by Means of Natural Selection. A Bradford Book, Bradford (1992). https://books.google.com.br/books?id=Bhtxo60BV0EC

26. Liu, F.T., Ting, K.M., Zhou, Z.H.: Isolation forest. In: 2008 Eighth IEEE International Conference on Data Mining, pp. 413–422. IEEE (2008)

27. Malanchev, K.L., Pruzhinskaya, M.V., Korolev, V.S., et al.: Anomaly detection in the Zwicky transient facility DR3. Mon. Not. R. Astronon. Soc. **502**(4), 5147–5175 (2021). https://doi.org/10.1093/mnras/stab316

28. Malanchev, K., Kornilov, M.V., Pruzhinskaya, M.V., et al.: The SNAD viewer: everything you want to know about your favorite ZTF object. Publ. Astron. Soc. Pac. **135**(1044), 024503 (2023). https://doi.org/10.1088/1538-3873/acb292

29. McLachlan, G.J., Peel, D.: Finite mixture models. Wiley Series in Probability and Statistics. New York (2000)

30. Perez-Carrasco, M., Cabrera-Vives, G., Hernandez-García, L., et al.: Alert classification for the ALeRCE broker system: the anomaly detector. Astron. J. **166**(4), 151 (2023). https://doi.org/10.3847/1538-3881/ace0c1

31. Pruzhinskaya, M.V., Ishida, E.E.O., Novinskaya, A.K., et al.: Supernova search with active learning in ZTF DR3. Astron. Astrophys. **672**, A111 (2023). https://doi.org/10.1051/0004-6361/202245172

32. Pruzhinskaya, M.V., Malanchev, K.L., Kornilov, M.V., et al.: Anomaly detection in the open supernova catalog. Mon. Not. R. Astron. Soc. **489**(3), 3591–3608 (2019). https://doi.org/10.1093/mnras/stz2362

33. Pruzhinskaya, M., Volnova, A., Kornilov, M., et al.: Could SNAD160 be a pair-instability supernova? Res. Notes Am. Astron. Soc. **6**(6), 122 (2022). https://doi.org/10.3847/2515-5172/ac76cf

34. Russeil, E., et al.: Multi-view symbolic regression. arXiv e-prints arXiv:2402.04298 (2024). https://doi.org/10.48550/arXiv.2402.04298

35. Schölkopf, B., Williamson, R., Smola, A., et al.: Support vector method for novelty detection, vol. 12, pp. 582–588 (1999)

36. Semenikhin, T.A.: Neural network architecture for artifacts detection in ZTF survey. Syst. Means Inf. **34**(1), 70–79 (2024). https://doi.org/10.14357/08696527240106
37. SNAD: Coniferest python package. https://github.com/snad-space/coniferest
38. Storey-Fisher, K., Huertas-Company, M., Ramachandra, N., et al.: Anomaly detection in Hyper Suprime-Cam galaxy images with generative adversarial networks. Mon. Notes R. Astron. Soc. **508**(2), 2946–2963 (2021). https://doi.org/10.1093/mnras/stab2589
39. Voloshina, A.S., et al.: SNAD catalogue of M-dwarf flares from the Zwicky Transient Facility. Mon. Not. R. Astron. Soc. **533**(4), 4309–4323 (2024). https://doi.org/10.1093/mnras/stae2031
40. Walmsley, M., Smith, L., Lintott, C., et al.: Galaxy zoo: probabilistic morphology through Bayesian CNNs and active learning. Mon. Not. R. Astron. Soc. **491**(2), 1554–1574 (2020). https://doi.org/10.1093/mnras/stz2816

Tree-Based Machine-Learning Classifier for Stellar Flares in The Zwicky Transient Facility Survey

Anastasia Lavrukhina[(✉)]

Faculty of Space Research, Lomonosov Moscow State University, Leninsky Gori 1
bld. 52, Moscow 119234, Russia
lavrukhina.ad@gmail.com

Abstract. This work is dedicated to solving the task of detecting flares
from red dwarfs among the light curves of The Zwicky Transient Facility
(ZTF) data. The study utilizes light curves with a temporal delay of no
more than 30 min since the characteristic duration of flares ranges from
30 min to 2 h. The task is addressed using machine learning methods,
specifically a binary classifier. Two models were employed as classifiers:
random forest and gradient boosting. Both real ZTF data and synthesized flare light curves were used for model training. All models were
tested on both synthesized flares and real flares which were found in the
ZTF data previously. Based on the validation set with real flares, it was
concluded that the gradient boosting model demonstrates the best performance. The achieved model quality allows it to be used for directly
assembling a sample of red dwarf flare candidates. GitHub repository
with full pipeline and demonstrative data is available at the link: https://
github.com/snad-space/flare-classifier.

Keywords: Machine learning · Astronomy · Stellar flare

1 Introduction

Flares of red dwarf stars are incredibly energetic phenomena, spanning a wide
range of energies from $E \sim 10^{26}$ erg up to $10^{35}-10^{36}$ erg [7,8]. Their light
curves have a distinctive profile with a drastically rapid brightening and following exponential-like decline, with the entire duration lasting from tens of
minutes to several hours. Studying flares of red dwarfs is important for exoplan-
etary science because these stellar flares release a significant amount of energy
in the ultraviolet spectrum, impacting the habitability of nearby planets [2].
Furthermore, compiling a statistically significant sample size can aid in further
investigations of the population of such objects.

This work proposes solving the given problem of stellar flare identifying using
machine learning methods. In the present day, machine learning and deep learning techniques have become highly effective tools for addressing challenges in

J. Baixeries et al. (Eds.): DAMDID/RCDL 2023, CCIS 2086, pp. 209–214, 2024.
https://doi.org/10.1007/978-3-031-67826-4_16

data-intensive domains of astronomy. One of the most common task is an object classification, which is now solved efficiently based on machine learning methods [3,6]. Equally significant is the task of anomaly detection, which aids in identifying rare events or objects exhibiting unexpected physical characteristics [13,14]. We propose training a binary classifier that will help select stellar flares candidates from high-cadence data of the Zwicky Transient Facility (ZTF) [1]. Subsequently, an expert evaluation of the flare candidates obtained using the model will be conducted for further detailed analysis and potential artifact filtering.

2 Data

We used 4 249 038 968 g-, r-, and i-band light curves from the 13th release of The Zwicky Transient Facility (ZTF) as the source data. A light curve is a time series of the stellar magnitude (brightness) of an astronomical object. Light curves in the ZTF survey are based on epochal PSF-fit photometry (for more details, see [11]). We selected 420 022 light curves having duration of at least 30 min, and time intervals between consequent observation to be not less than 30 min. We also synthesized the same amount of light curves based on TESS observations of stellar flares [5]. Each light curve was pre-processed for irregular time series feature extraction, 31 features in total (see [9,10] for light-curve[1] package description).

An example of red dwarf flare light curve found in the ZTF data is presented in Fig. 1.

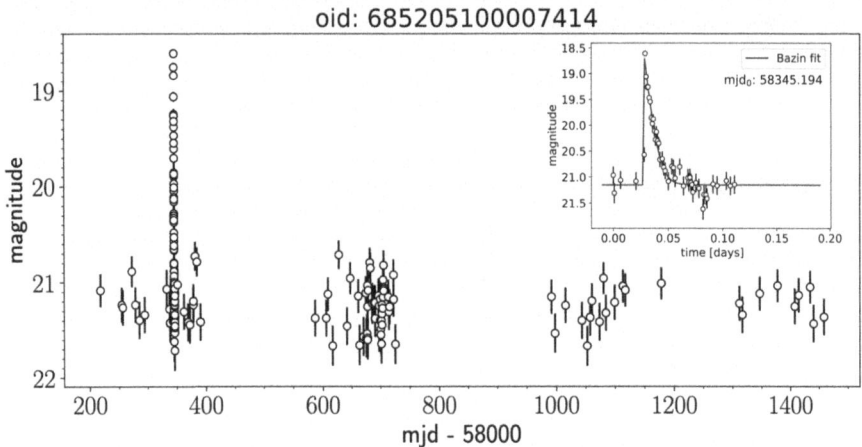

Fig. 1. Red dwarf flare light curve found in ZTF data. (Color figure online)

---

[1] https://github.com/light-curve.

3   Methods

3.1   Random Forest

Random Forest is a machine learning model based on an ensemble of decision trees. The ensemble refers to training multiple models on bootstrap samples and averaging their responses to obtain predictions. We used the implementation of random forest from `scikit-learn` [12] package with the default hyperparameters and 100 trees as a model.

3.2   CatBoost

Unlike ensemble methods, boosting builds the base algorithms sequentially. Each subsequent base model is constructed to reduce the error of the current model. Boosting that uses decision trees as base algorithms is called gradient boosting with decision trees.

In this study, we used the `CatBoost` [4] implementation of gradient boosting. The model's hyperparameters were set as follows: learning rate of 0.001, depth of 5 and the logistic loss function. The model was trained for 10 000 iterations.

4   Models Evaluation

4.1   Validation on Test Dataset

The following metrics were used to evaluate the performance of the models on the test dataset: recall, precision, accuracy and F_β-score. F_β-score use a factor β to reweight an importance of recall metrics in comparison to precision:

$$F_\beta = (1 + \beta^2) \cdot \frac{precision \cdot recall}{(\beta^2 \cdot precision) + recall}$$

For the F_β-score metric, a value of β equals to 0.3 was chosen, as precision is more significant in our task. All candidates identified by the classifier will be intended to undergo a detailed expert analysis to exclude possible artifacts (observable phenomena with non-astrophysical nature). Therefore, it is necessary to optimize the number of objects that will be further analyzed by the expert.

Prior to evaluating the performance for each model, a threshold optimization procedure was conducted. The threshold for each model was selected based on the validation dataset in order to maximize the value of the F_β-score metric (β = 0.3).

The metrics for all described models, along with the optimal threshold, are presented in Table 1. All metrics were obtained from the same test dataset.

Table 1. The metric results on the test dataset and the optimal threshold for the two trained models: random forest and gradient boosting (CatBoost).

	Precision	Recall	Accuracy	F_β-score	Threshold
Random forest	0.983	0.791	0.889	0.964	0.83
CatBoost	0.978	0.777	0.880	0.958	0.86

4.2   Validation on Real Flares

Since the positive sample of flares used for training was generated synthetically, while the negative sample was taken from real light curves in the ZTF data, it was necessary to verify the considered models indeed learned to classify objects of the specified type rather than distinguishing synthesized light curves from real ones. For this reason, all models were tested on a dataset consisting of 104 real red dwarf flares previously identified in the ZTF data using other methods (Voloshina et al., in prep.) and 1 000 random negative objects taken from the test dataset. The values of the precision, recall, accuracy and F_β-score metrics were calculated for each model on this dataset (see Table 2). ROC-curve and ROC AUC values for each model presented in Fig. 2.

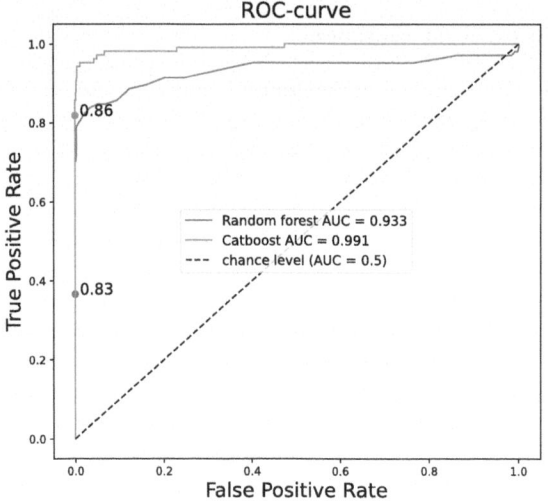

Fig. 2. ROC curves and ROC AUC values for random forest (blue curve) and CatBoost (orange curve) models. The bold points define a position of the optimal thresholds on ROC curve. (Color figure online)

Based on the metrics obtained on real data, the gradient boosting-based classifier demonstrates the best performance so far.

Table 2. Results of the metrics on the dataset with real flares for the two trained models: random forest and gradient boosting (CatBoost).

	Precision	Recall	Accuracy	F_β-score
Random forest	1.0	0.356	0.940	0.870
CatBoost	1.0	0.827	0.984	0.983

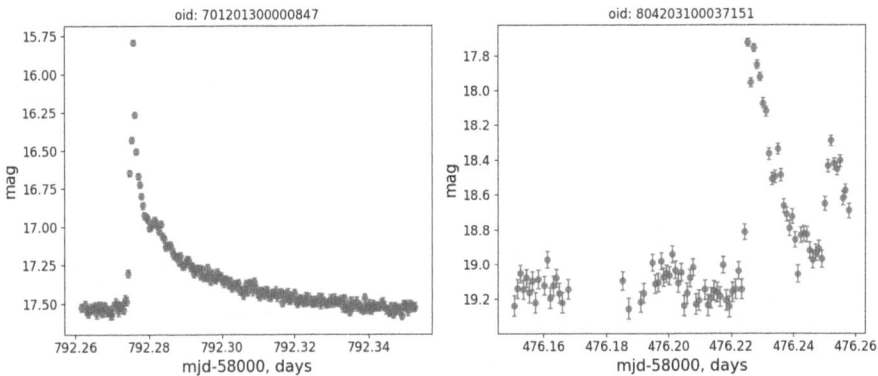

Fig. 3. Examples of red dwarf flares light curves found in ZTF data using CatBoost classifier. (Color figure online)

5 Conclusion

Within this work, methods for analyzing a large volume of photometric data were developed to solve the object classification task using machine learning methods. A subsample with high observation cadence was extracted from the ZTF photometric data. Based on this subsample and synthesized light curves, two models were trained for classification: random forest and gradient boosting. Comparing the metrics on the test dataset and the dataset with real flares revealed that the gradient boosting model achieved the best performance. Initially, the model was applied to 2% of the target dataset, resulting in the detection of 25 new candidates for flaring events (see Fig. 3). This outcome provides hope that applying the classifier to the entire target dataset will lead to the discovery of approximately 1 000 new candidates with various galactic declinations, which significantly surpasses previously published datasets of this kind based on all-sky surveys.

Acknowledgements. This work is supported by Nonprofit Foundation for the Development of Science and Education "Intellect".

References

1. Bellm, E.C., et al.: The Zwicky transient facility: system overview, performance, and first results. Publ. Astron. Soc. Pac. **131**(995), 018002 (2019). https://doi.org/10.1088/1538-3873/aaecbe

2. Bogner, M., Stelzer, B., Raetz, S.: Effects of flares on the habitable zones of m dwarfs accessible to TESS planet detections. Astronom. Nachrichten **343**(4) (2021). https://doi.org/10.1002/asna.20210079

3. Donoso-Oliva, C., Becker, I., Protopapas, P., Cabrera-Vives, G., Vishnu, M., Vardhan, H.: ASTROMER. A transformer-based embedding for the representation of light curves. Astron. Astrophys. **670**, A54 (2023). https://doi.org/10.1051/0004-6361/202243928

4. Dorogush, A.V., Ershov, V., Gulin, A.: CatBoost: gradient boosting with categorical features support. CoRR abs/1810.11363 (2018). http://arxiv.org/abs/1810.11363

5. Günther, M.N., et al.: Stellar flares from the first TESS data release: exploring a new sample of M dwarfs. Astronom. J. **159**(2), 60 (2020). https://doi.org/10.3847/1538-3881/ab5d3a

6. Kim, D.W., et al.: The epoch project. Astron. Astrophys. **566**, A43 (2014). https://doi.org/10.1051/0004-6361/201323252

7. Kowalski, A.F., Hawley, S.L., Holtzman, J.A., Wisniewski, J.P., Hilton, E.J.: A white light megaflare on the dM4.5e star YZ CMi. Astronom. J. Lett. **714**(1), L98–L102 (2010). https://doi.org/10.1088/2041-8205/714/1/L98

8. Lacy, C.H., Moffett, T.J., Evans, D.S.: UV Ceti stars: statistical analysis of observational data. Astrophys. J. **30**, 85–96 (1976). https://doi.org/10.1086/190358

9. Lavrukhina, A., Malanchev, K.: Performant feature extraction for photometric time series (2023)

10. Malanchev, K.L., et al.: Anomaly detection in the Zwicky transient facility DR3. Mon. Not. R. Astron. Soc. **502**(4), 5147–5175 (2021). https://doi.org/10.1093/mnras/stab316

11. Masci, F.J., et al.: The Zwicky transient facility: data processing, products, and archive. Publ. Astron. Soc. Pac. **131**(995), 018003 (2019). https://doi.org/10.1088/1538-3873/aae8ac

12. Pedregosa, F., et al.: Scikit-learn: machine learning in Python. J. Mach. Learn. Res. **12**, 2825–2830 (2011)

13. Pruzhinskaya, M.V., et al.: Supernova search with active learning in ZTF DR3. Astron. Astrophys. **672**, A111 (2023). https://doi.org/10.1051/0004-6361/202245172

14. Villar, V.A., et al.: A deep-learning approach for live anomaly detection of extragalactic transients. Astrophys. J. Suppl. Ser. **255**(2), 24 (2021). https://doi.org/10.3847/1538-4365/ac0893

Classification of Long Gamma-Ray Transients from INTEGRAL Data Using Machine Learning Approach

Georgiy Mozgunov[1]([✉])(ID), Alexei Pozanenko[1,2](ID), Pavel Minaev[1](ID),
Ivan Chelovekov[1](ID), and Sergei Grebenev[1](ID)

[1] Space Research Institute, Russian Academy of Sciences, Profsoyuznaya ul. 84/32,
Moscow 117997, Russia
`georgiy99@bk.ru`
[2] National Research University "Higher School of Economics", Myasnitskaya ul. 20,
Moscow 101000, Russia

Abstract. In this paper we use 19 years of data from INTEGRAL detectors to train classification model for gamma-ray bursts. We present algorithms for automated processing of the light curve of gamma-ray bursts, consisting of a background approximation, distinguishing an event from background, and duration calculation. Candidates are crossmatched with several catalogues of transient events. This provided us labels for supervised machine learning. Gradient boosting classifier is employed for training to find Solar flares and gamma-ray bursts. Estimated accuracy is $\sim$91% for latter events. Similar machine learning approach can be applied to other types of transients.

Keywords: Machine learning · Crossmatching · Gamma-ray bursts · INTEGRAL

1 Introduction

Gamma-ray bursts (GRB) where discovered in 1973 as a sudden and rapid increase in gamma-ray flux [21]. Since then GRB trigger algorithms rely on search for excess of signal above background on short timescale; for example, the trigger scale in IBAS varies from 0.05 to 5 s [30]. This works well for usual short (duration less than 2 s) and long (greater that 2 s) gamma-ray bursts, but can be problematic in case of events with large signal rise time and/or duration. An example of such an event is ultralong gamma-ray bursts [15]: their duration varies from hundreds up to tens thousands of seconds [5] and may consist of several episodes. Also, it was reported [37] that duration distribution of long bursts is skewed, possibly, because of not found long events with duration $\gtrsim$100 seconds.

There is a way to fill this gap in long events - "offline" blind search, carried out on archival data. It is named blind, because search procedure does not

J. Baixeries et al. (Eds.): DAMDID/RCDL 2023, CCIS 2086, pp. 215–224, 2024.
https://doi.org/10.1007/978-3-031-67826-4_17

have information about presence of known transient at any given time. One of them is Bayesian blocks [35], it is well suited for observatories with stable background, such as Konus-WIND (KW) [7]. A more conventional approach when working with more variable background is using running averages and search for excesses on different timescales [34]. More precise background approximation can be achieved by modeling based on physical properties of observatory and environment. It can be conducted by either physical approach [9] or by training a Neural Network [13]. After that search algorithm can be applied: simple significance threshold or, for example, FOCuS-Poisson [40]. A step further would be using Recurrent Neural Networks, perfectly suited for sequential data, for both background approximation and anomaly detection [28,33].

Determining origin of detected transients is the main problem, especially for detectors without imaging and spectral capabilities. Often, transients are distinguished by a set of different parameters in several dimensions, such as comparison of flux in different detectors or shape of the light curve. It can be carried out by traditional machine learning approaches (for example, random forest [14,25,41]) or by deep learning [33]. Classification requires labels, which are usually derived from catalogues of transients via crossmatching procedure.

In this paper we present our approach to problem of classification long transients using INTEGRAL data. We employ new method, that combines statistical approach to background modeling and data processing with Gradient Boosting classifier.

2   Instruments and Data Processing

2.1   INTEGRAL

INTEGRAL was launched in 2002 on highly elliptical orbit, with apogee about 150000 km [18]. We use data from almost every detector onboard observatory, namely SPectrometer of INTEGRAL (SPI), SPI AntiCoincidence Shield (SPI-ACS), INTEGRAL Radiation Environment Monitor (IREM), JEM-X, Imager on-Board the INTEGRAL Satellite (IBIS) consisting of Integral Soft Gamma-Ray Imager (ISGRI) and Pixellated Caesium-Iodide Telescope (PICsIT). SPI-ACS is a scintillator, which main purpose is to protect SPI from background photons outside its field of view (FOV), but it has been proven as an effective separate gamma-ray detector [32,39]. Its main disadvantage is lack of imaging and spectral capabilities. IREM is a semiconductor detector, aimed at monitoring radiation environment and prevent damage to other instruments [17]. It can also be used as a full-fledged detector of charged particles. Among 15 channels we use only TC3. It is sensitive to electrons from 0.8 MeV and protons from 12 MeV and has the widest energy range and sensitivity. JEM-X [26], SPI [38] and IBIS [23,24,29] are imaging detectors, that differ in operating energy range: from low energy X-ray in JEM-X to gamma-ray in SPI. Information from all these detectors can give a broad insight about transients spectra, temporal structure in high energy range (from 3 keV and above), and its origin (either radiation or particles).

2.2 Dataset

As a dataset we use sample of 4364 transient events found in SPI-ACS data with more than 3σ significance between 2003 and 2021. One fraction of these events are random fluctuations, while another are real Solar and astrophysical events. Distribution by detection year is presented in Fig. 1. The number of detections is correlated with the Solar activity cycle. This is expected due to increasing instability of particle environment and number of flares in this period.

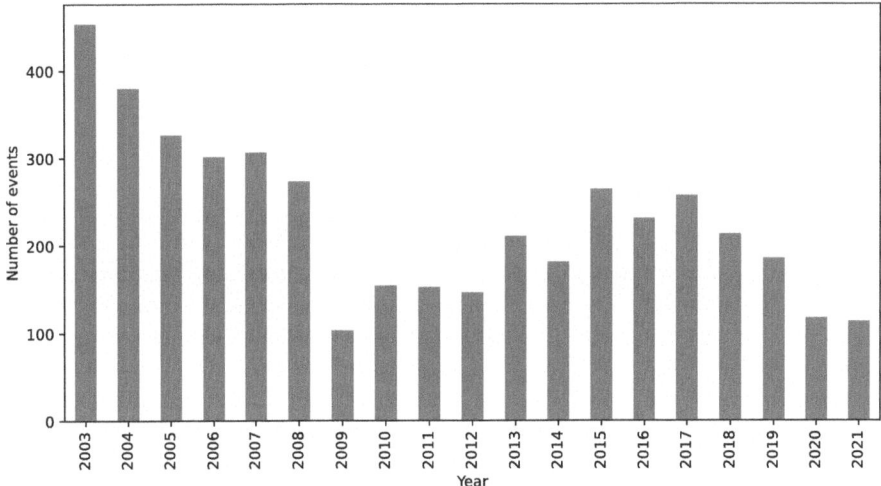

Fig. 1. Distribution of triggered events by year.

2.3 Data Processing

Preprocessing procedure is needed to determine duration and flux for each detector, which will be the main features of the model. These parameters can be found in almost every existing catalog of gamma-ray transients, which allows the model to be applied to them without additional processing.

The duration is calculated using the SPI-ACS light curve according to the procedure whose block schema is presented in Fig. 2.

Algorithm requires initial timescale, that varies from 20 to 200 s and is selected individually for each event. Firstly, the light curve is binned into current (initial) timescale t_{curr}. Then follows the main part, background subtraction and event extraction. In this section an array of $\{(t_i, c_i)\}$ is referred to as event. On the first iteration this array is empty. Algorithm consists of 3 steps:

1. Approximate background with 3rd degree polynomial using all bins in light curve, except event. Resulting model is subtracted from original data.
2. Determine point (t_{max}, c_{max}) with maximum statistical significance. Statistical significance above background is determined as $\frac{F}{\sqrt{D}}$, where F is number of counts in a bin after background subtraction and D is the dispersion, calculated on the background intervals from previous stage (1).

3. Calculate $T_{1\sigma}$. In this paper $T_{1\sigma}$ is number of seconds between first and second bin with significance more than 1σ without gaps. We compared 1σ values to manually calculated duration parameters T_{50}, T_{90}, T_{100} of sample of bursts with extended emission [27]. In majority of cases ($\sim$ 65%) value of $T_{1\sigma}$ is between T_{50} and T_{100}. The remaining events belong to the case when the burst consists of several episodes.

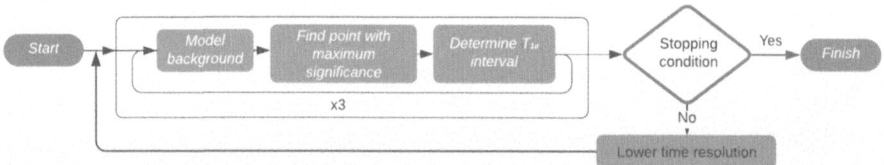

Fig. 2. Procedure of processing SPI-ACS light curve.

Steps (1)–(3) are repeated 3 times. On second and third step bins corresponding to event (calculated from previous steps) are taken into account (excluded) in the process of background modeling (1), which increases quality if approximation.

After that, event is checked on two stopping conditions: limit value of t_{curr} is reached or event has 10 or more bins. Latter condition is introduced to save computation time in case of long transients. If neither of this conditions are met, then algorithm is repeated from the beginning with $t_{curr} = \frac{t_{curr}}{2}$. Also on the first step $T_{1\sigma}$ interval is excluded from background approximation.

Finally, background subtracted integral flux over $T_{1\sigma}$ interval is calculated. We also want to add light curve shape to feature space. Therefore we bin light curve into 10 bins. Their duration varies for different events, because it depends on $T_{1\sigma}$. After binning counts are normalized to be in range [0, 1] and added as separate features.

For other detectors procedure is much simpler. Background model is a 1st degree polynomial, calculated on light curve except for the $T_{1\sigma}$ interval. Then, we calculate background subtracted integral flux in this interval. All code is written in Python and can be found at Github. Final dataset is composed of 21 feature: duration $T_{1\sigma}$, mean distance to Earth in $T_{1\sigma}$ interval, 10 bins for light curve shape and 9 integral fluxes of INTEGRAL detectors: SPI-ACS, IBIS veto [29], IREM TC3, ISGRI (20–100 keV), JMX 1 and 2 (3–20 keV), PICsIT (event mode and spectrum mode, 175–500 keV) and SPI (20–500 keV).

3   Crossmatching and Labeling

To train a classifier our dataset must be labeled. This is done by crossmatching our events with events from different catalogues of confirmed events. In this paper we consider 3 classes of transients: gamma-ray bursts, Solar flares and background fluctuation, is the class which represents either absence of an event or presence of an unknown background activity. For gamma-ray bursts we use K. Hurleys "masterlist" [3], compilation of confirmed GRB from different catalogues up to 2021. The reason we chose Konus-WIND lies in its position. Location in L2 point guarantees stable background, which is important for detecting long events with duration $\gtrsim 100$ s. Difficulties in detecting long and ultra-long events with near-Earth observatories are described for example in [9,13]. For Solar flares we use GOES [1] and RHESSI [4] catalogues as well as Konus-WIND SF catalog [2]. It has close to SPI-ACS energy range (unlike former ones working in X-ray) so we eliminate coincide triggers. Events crossmatched only with GOES and RHESSI catalogues are marked as potential Solar flares and are not used in training process. There are several types of events, that has non-radiative nature: SEPEs [31], electron clusters in magnetosphere tail [22], crossing of van Allens belts [17] etc. These events has unstable radiation environment and background model could not describe it well. Therefore we classify all events with background $\chi^2/d.o.f. > 3$ as background events.

For crossmatching we use trigger times and duration $T_1\sigma$ of the event. An event is crossmatched with catalog if time interval of candidate intersects with time interval of catalog event. If duration or time interval is not provided, we check if catalog time belongs to $T_{1\sigma}$ interval. One event can belong to multiple classes: this may happen by accident, when gamma-ray burst coincide with Solar flare, or confirmed event may be surrounded by an unstable background. This is physically impossible and show imperfection of our crossmatching algorithm. In this case we consider one event belonging to several classes, which can be represented as duplicates in a dataset with different labels.

After the crossmatching step, 2420 events are labeled, remaining 1944 events are not found in the used catalogues and have stable background. The duration distribution is shown in Fig. 3. As expected the Solar flares generally longer than the gamma-ray bursts [16,36]. Also, number of background (1909) events is bigger then GRB and, especially, Solar events. Such imbalance needs to be taken into account during training.

4   Model

In this paper we use gradient boosting as classifier. It is one of universal algorithms which has proven to be one of the best for tabular data [20]. It requires less data than NN, which is important due to our relatively small dataset (only 3198 events). Hyperparameter optimization is fulfilled via optuna [6] package with objective to maximize precision of predicting gamma-ray bursts. Therefore we use $F_\beta = (1 + \beta^2) \cdot \frac{precision \cdot recall}{(\beta^2 \cdot precision) + recall}$ score [8] with $\beta = 0.5$. Two commonly used values for β are 2, which weighs recall higher than precision, and 0.5, which weighs recall lower than precision. We perform 500 iteration, each time generating new train and test splits in ratio of 80 to 20. Optimal parameters are then used during training. Training is executed on 80% of data, the rest is used to control accuracy metrics. In Fig. 4 we present a ROC (Receiver Operating Characteristic), which binds the True positive rate to the False positive rate and can be used to evaluate the quality of prediction as follows: the closer the area under ROC curve (ROC AUC) to 1 the better. ROC AUC for gamma-ray bursts is 0.97.

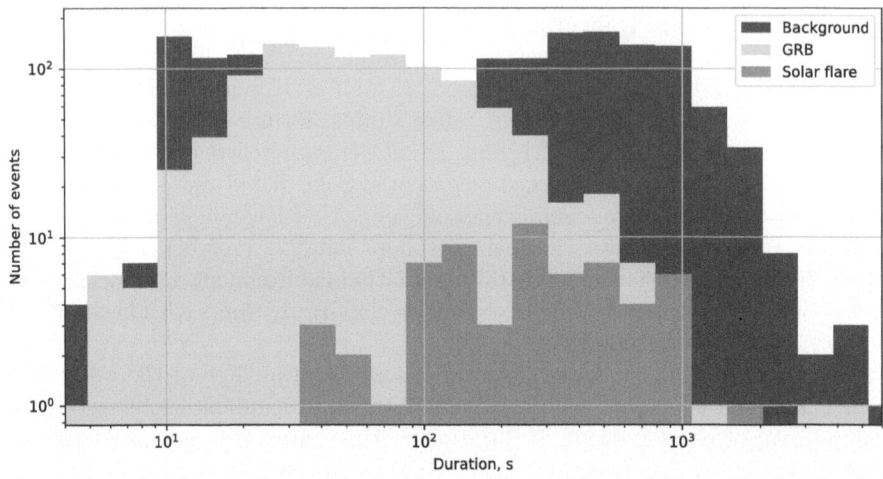

Fig. 3. Duration distribution of labeled events.

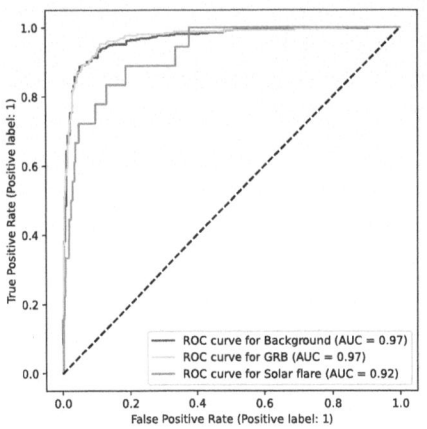

Fig. 4. ROC curve for classificator.

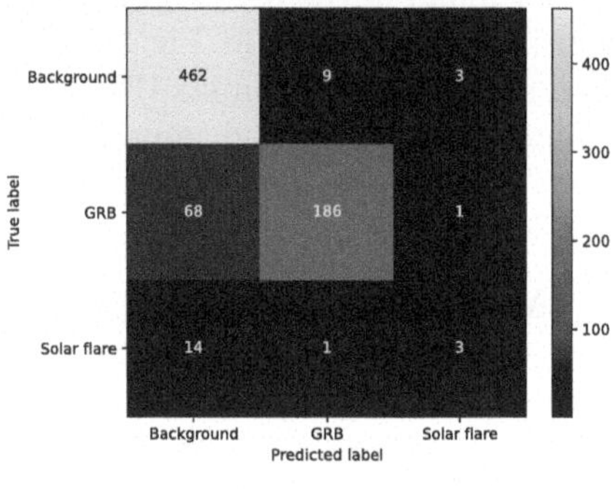

Fig. 5. Confusion matrix for the classification on test set.

We calculate custom probability thresholds for GRB and Solar flares maximizing F_β score. The optimal probability thresholds are 0.828 for GRB and 0.5 for Solar flares.

We applied similar training and evaluation procedures for other models: Logistic Regression [10] and Random Forest [12]. The results are presented in the

Table 1. Gradient boosting outperforms both of the models and is much faster than Random Forest classifier (possibly due to different implementations).

Table 1. Comparison of different ML models. sklearn implementation of Logistic Regression and Random Forest is used.

Model	f_β	Balanced Accuracy
Logistic Regression	0.51	0.38
Random Forest	0.87	0.60
Gradient Boosting	0.90	0.67

We also checked whether the model is overfitting. One model is trained as described above and another one with early stopping: 10% of training data is allocated for validation set, which is used to stop training when metrics stop improving for 100 rounds. Then the results are compared. No significant difference is found between this models, all metrcis are withing margin of errors.

5   Results

We analyzed feature importances of our model, i.e., the relative values, describing impact of each feature on the final prediction. It is calculated via LightGBM [20] embedded functions based on Gini impurity [11]. Top 4 important features are total fluence in SPI-ACS, duration, ISGRI and IREM fluence. It is explained by strong distinction of events by duration and the fact, that background fluctuations tend to be more faint than real GRB (for the same duration). Moreover, importance of IREM fluence indicate that background events, unlike GRB, may consist of charged particles.

Confusion matrix for test set is presented in Fig. 5. The number in each square in the confusion matrices represents the overall classification for each event type compared to the actual classification obtained through crossmatching procedure. To evaluate the accuracy of our classifier, we use the method of cross-fold validation on 10 folds. One fold consist of 10 sets, nine are used for training, last one uses the created model to classify the remaining sample set. This procedure is repeated for each fold and metrics are averaged. Total accuracy is $91 \pm 4\%$ for gamma-ray bursts. Such high accuracy is achieved at the cost of recall, its value is $\sim 73\%$. This fact does not contradict the goals of our work, to make reliable classification model. For Solar flares accuracy is $35 \pm 32\%$. Such low value can be a result of low amount of samples, only 61 event for both training and test samples. Also, it can be low due to high similarity between GRB and Solar flares with gamma-ray emission. INTEGRAL detectors could lack sensitivity to pick out differences in the spectrum. The last reason might be the crossmatching procedure. We compare events based only on time intervals, which does not negate possible coincidences.

We checked the hypothesis that hyperparameter selection created a bias towards GRB and lowered precision for Solar flares. We repeated hyperparameter optimization for 2 more cases with objectives of maximizing F_β for Solar flares and maximizing sum of F_β scores for SF and GRB. Neither of this approaches show significantly better results, all metrics were within margin of error.

6   Conclusion

Our model shows consistent result in classification of gamma-ray bursts with accuracy ∼91% on test set. This makes it possible to classify events in the absence of data from other observatories. Accuracy can be improved by several ways. Firstly, we can expand our training dataset; it can include more known confirmed gamma-ray bursts and Solar flares. Secondly, we can enrich feature space with new parameters. We can expand usage of the shape of transient on all events and apply Furie Transform with dimensionality reduction to use them as features [19]. Thirdly, use more complex algorithm, for example ensemble of multiple machine learning models or Neural Network. Also we can make use of imaging detectors: if transient is seen inside FOV we can use Convolutional Neural Network to help with classification.

The method, proposed in this work can be extended to any observatory, having a sufficient set of detectors. The more detectors with non-overlapping energy ranges it has, the better should be the accuracy for different classes of events.

Acknowledgments. Georgiy M. and Sergei G. acknowledge the support from the Foundation for the Advancement of Theoretical Physics and Mathematics "BASIS" (project 22-1-1-57-1). Alexei P. and Pavel M. acknowledge the support from the Russian Science Foundation (project 23-12-00198) in part of the data comparison with GRB catalogues.

References

1. GOES flare list. ftp://ftp.swpc.noaa.gov/pub/warehouse/. Accessed 07 May 2023
2. Konus-WIND Solar Flares. http://www.ioffe.ru/LEA/Solar/index.html. Accessed 07 May 2023
3. "Masterlist", compilation of confirmed gamma-ray bursts. www.ssl.berkeley.edu/ipn3/masterli.txt. Accessed 07 Aug 2022
4. RHESSI flare list. https://hesperia.gsfc.nasa.gov/hessidata/dbase/. Accessed 07 May 2023
5. Ultra-long gamma-ray burst candidates. http://www.ioffe.ru/LEA/kw/wm/ulong/index.html. Accessed 07 May 2023
6. Akiba, T., Sano, S., Yanase, T., Ohta, T., Koyama, M.: Optuna: a next-generation hyperparameter optimization framework (2019)
7. Aptekar, R.L., et al.: Konus-WIND experiment for cosmic gamma-ray bursts: Observational capabilities. In: Rothschild, R.E., Lingenfelter, R.E. (eds.) High Velocity Neutron Stars. American Institute of Physics Conference Series, vol. 366, pp. 158–163 (1996). https://doi.org/10.1063/1.50233

8. Baeza-Yates, R., et al.: Modern information retrieval (1999)
9. Biltzinger, B., Kunzweiler, F., Greiner, J., Toelge, K., Burgess, J.M.: A physical background model for the iFermi/i gamma-ray burst monitor. Astron. Astrophys. **640**, A8 (2020). https://doi.org/10.1051/0004-6361/201937347
10. Bishop, C.M.: Pattern Recognition and Machine Learning (Information Science and Statistics). Springer, Heidelberg (2006)
11. Breiman, L., Friedman, J., Stone, C., Olshen, R.: Classification and Regression Trees. Taylor & Francis (1984). https://books.google.com.ua/books?id=JwQx-WOmSyQC
12. Breiman, L.: Random forests. Mach. Learn. **45**, 5–32 (2001). https://doi.org/10.1023/A:1010933404324
13. Crupi, R., Dilillo, G., Bissaldi, E., Fiore, F., Vacchi, A.: Searching for long faint astronomical high energy transients: a data driven approach (2023)
14. Farrell, S.A., Murphy, T., Lo, K.K.: Autoclassification of the variable 3XMM sources using the random forest machine learning algorithm. **813**(1), 28 (2015). https://doi.org/10.1088/0004-637X/813/1/28
15. Gendre, B.: Ultralong GRBs as proxy of Population III stars. In: 40th COSPAR Scientific Assembly, vol. 40, pp. E1.17–10–14 (2014)
16. Grieder, P.K.: Chapter 6 - heliospheric phenomena. In: Grieder, P.K. (ed.) Cosmic Rays at Earth, pp. 893–974. Elsevier, Amsterdam (2001). https://doi.org/10.1016/B978-044450710-5/50008-7, https://www.sciencedirect.com/science/article/pii/B9780444507105500087
17. Hajdas, W., Bühler, P., Eggel, C., Favre, P., Mchedlishvili, A., Zehnder, A.: Radiation environment along the INTEGRAL/i orbit measured with the IREM monitor. Astron. Astrophys. **411**(1), L43–L47 (2003). https://doi.org/10.1051/0004-6361:20031251
18. Jensen, P.L., et al.: The INTEGRAL spacecraft - in-orbit performance. **411**, L7–L17 (2003). https://doi.org/10.1051/0004-6361:20031173
19. Jespersen, C.K., et al.: An unambiguous separation of gamma-ray bursts into two classes from prompt emission alone. **896**(2), L20 (2020). https://doi.org/10.3847/2041-8213/ab964d
20. Ke, G., et al.: LightGBM: a highly efficient gradient boosting decision tree. In: Guyon, I., et al. (eds.) Advances in Neural Information Processing Systems, vol. 30. Curran Associates, Inc. (2017). https://proceedings.neurips.cc/paper_files/paper/2017/file/6449f44a102fde848669bdd9eb6b76fa-Paper.pdf
21. Klebesadel, R.W., Strong, I.B., Olson, R.A.: Observations of Gamma-Ray Bursts of Cosmic Origin. In: Bulletin of the American Astronomical Society, vol. 5, p. 322 (1973)
22. Konradi, A.: Electron and proton fluxes in the tail of the magnetosphere. **71**(9), 2317–2325 (1966). https://doi.org/10.1029/JZ071i009p02317
23. Labanti, C., et al.: The Ibis-Picsit detector onboard Integral. **411**, L149–L152 (2003). https://doi.org/10.1051/0004-6361:20031356
24. Lebrun, F., et al.: ISGRI: the INTEGRAL soft gamma-ray imager. **411**, L141–L148 (2003). https://doi.org/10.1051/0004-6361:20031367
25. Lo, K.K., Farrell, S., Murphy, T., Gaensler, B.M.: Automatic classification of time-variable X-Ray. Sources **786**(1), 20 (2014). https://doi.org/10.1088/0004-637X/786/1/20
26. Lund, N., Westergaard, N.J., Budtz-Jørgensen, C.: JEM-X: joint European X-ray monitor. **70**, 1303 (1999)

27. Mozgunov, G.Y., Minaev, P.Y., Pozanenko, A.S.: Extended emission of cosmic gamma-ray bursts detected in the SPI-ACS/INTEGRAL experiment. Astron. Lett. **47**(3), 150–162 (2021). https://doi.org/10.1134/S1063773721030038

28. Parmiggiani, N., et al.: A deep-learning anomaly-detection method to identify gamma-ray bursts in the ratemeters of the AGILE anticoincidence system. **945**(2), 106 (2023). https://doi.org/10.3847/1538-4357/acba0a

29. Quadrini, E.M., et al.: IBIS veto system. Background rejection, instrument dead time and zoning performance. **411**, L153–L157 (2003). https://doi.org/10.1051/0004-6361:20031259

30. Rau, A., Kienlin, A.V., Hurley, K., Lichti, G.G.: The 1st INTEGRAL SPI-ACS gamma-ray burst catalogue. **438**(3), 1175–1183 (2005). https://doi.org/10.1051/0004-6361:20053159

31. Reames, D.V.: The two sources of solar energetic particles. Space Sci. Rev. **175**(1–4), 53–92 (2013). https://doi.org/10.1007/s11214-013-9958-9

32. Rodríguez-Gasén, R., Kiener, J., Tatischeff, V., Vilmer, N., Hamadache, C., Klein, K.L.: Exploring the capabilities of the anti-coincidence shield of the international gamma-ray astrophysics laboratory (INTEGRAL) spectrometer to study solar flares. **289**(5), 1625–1641 (2014). https://doi.org/10.1007/s11207-013-0418-1

33. Sadeh, I.: Deep learning detection of transients (ICRC-2019). arXiv e-prints arXiv:1908.01615 (2019). https://doi.org/10.48550/arXiv.1908.01615

34. Savchenko, V., Neronov, A., Courvoisier, T.J.L.: Timing properties of gamma-ray bursts detected by SPI-ACS detector onboard INTEGRAL. **541**, A122 (2012). https://doi.org/10.1051/0004-6361/201218877

35. Scargle, J.D., Norris, J.P., Jackson, B., Chiang, J.: Studies in astronomical time series analysis. VI. Bayesian block representations. **764**(2), 167 (2013). https://doi.org/10.1088/0004-637X/764/2/167

36. Tarnopolski, M.: Analysis of Fermi gamma-ray burst duration distribution. **581**, A29 (2015). https://doi.org/10.1051/0004-6361/201526415

37. Tarnopolski, M.: Analysis of gamma-ray burst duration distribution using mixtures of skewed distributions. Mon. Not. R. Astron. Soc. **458**(2), 2024–2031 (2016). https://doi.org/10.1093/mnras/stw429

38. Vedrenne, G., et al.: SPI: the spectrometer aboard INTEGRAL. **411**, L63–L70 (2003). https://doi.org/10.1051/0004-6361:20031482

39. von Kienlin, A., et al.: INTEGRAL spectrometer SPI's GRB detection capabilities. GRBs detected inside SPI's FoV and with the anticoincidence system ACS. **411**, L299–L305 (2003). https://doi.org/10.1051/0004-6361:20031231

40. Ward, K., Dilillo, G., Eckley, I., Fearnhead, P.: Poisson-FOCuS: an efficient online method for detecting count bursts with application to gamma ray burst detection. arXiv e-prints arXiv:2208.01494 (2022). https://doi.org/10.48550/arXiv.2208.01494

41. Yang, H., Hare, J., Kargaltsev, O., Volkov, I.: Machine learning classification of variable galactic X-ray sources from Chandra source catalog. In: AAS/High Energy Astrophysics Division. AAS/High Energy Astrophysics Division, vol. 54, p. 110.03 (2022)

AWARE: Alert Watcher and Astronomical Rapid Explorer

Nicolai Pankov[1,2(✉)] [ID], Artem Prokhorenko[3], Eugene Schekotihin[1],
Alexei Pozanenko[1,2], Pavel Minaev[2,4], Sergei Belkin[1,2] [ID], and Alina Volnova[2]

[1] National Research University Higher School of Economics, 21/4 Staraya,
Basmannaya, Ulitsa 105066, Russian Federation
npankov@hse.ru
[2] Space Research Institute of the Russian Academy of Sciences,
Profsoyuznaya ul. 84/32, Moscow 117997, Russian Federation
[3] Institute of Physics and Technology, Institutskiy Pereulok, 9,
Dolgoprudny 141701, Russian Federation
[4] P.N. Lebedev Physical Institute of the Russian Academy of Sciences, 53 Leninsky
Avenue, 119991 Moscow, Russian Federation

Abstract. We present the software AWARE for automatic scheduling
the observations with the ground based network of optical telescopes
GRB-IKI-FuN (extendable on other facilities). It was designed to provide
the optimal coverage of large uncertainty contours of LIGO/Virgo/KA-
GRA (LVK) gravitational wave events, and gamma-ray bursts (GRBs),
detected by space missions, which is crucial for fast look-up for their optical counterparts. In addition, AWARE provides capabilities for notifying
users on new alert messages received from General Coordinate Network
(GCN), and sharing final products generated by the planner via Telegram bot. The implementation of the software allows both scheduler,
and Telegram bot to run concurrently in near real-time. In this paper,
we discuss main follow-up tactics of LVK events, and GRBs. We overview
the automatic planning software AWARE in terms of an architecture and
algorithms. The example products made by AWARE since the beginning
of the LVK O4 run on 25 May 2023 are given. The plans for future work
are briefly discussed.

Keywords: observations · gamma-ray bursts · LIGO/Virgo/
KAGRA · planner

Introduction

The observation and following study of the electromagnetic counterparts to
gravitational-wave (GW) events registered by the LIGO [24], Virgo [5] and
KAGRA [22] as the LVK collaboration [1,2] is a crucial objective in the modern
multi-messenger era of astronomy. Gravitational waves are emitted from coa-
lescing binary systems (CBS) such as binary neutron star mergers BNS, neutron

J. Baixeries et al. (Eds.): DAMDID/RCDL 2023, CCIS 2086, pp. 225–245, 2024.
https://doi.org/10.1007/978-3-031-67826-4_18

star – black hole mergers NS-BH, and binary black hole (BBH) mergers. In the case of a BNS merger, electromagnetic transients could be generated: a short GRB, and a KN (less probable for an NS-BH merger). Up to the moment, there is a single LIGO/Virgo event GW170817 [3,4] that is known to be accompanied by both these electromagnetic counterparts: a short GRB 170817A [15,30,35] and kilonova AT2017gfo [11]. Although, there is a short GRB 190425A, which is most probably connected with GW 190425 [31]. Continuous multi-messenger observations will allow to shed more light on processes in forming GRBs/KNe from CBSs [13,29,39]. The major problem that astronomers are currently facing is the sky localization uncertainty of the GW events. The entire localization area estimated only by the LVK detectors is up to a thousand square degrees (in order of magnitude). However, in case of electromagnetic counterpart detection by space gamma- and X-ray telescopes, the refined error box *may be* reduced to a few arcminutes and arcseconds squared, correspondingly. Still, observations of such localization regions remain difficult, but they represent a significant goal for astronomers.

The attempts to create a generic observation planner were made, for example, in [9,18,20,28]. The SALT scheduling system [9] is a queue-scheduled system and requires the proposal to start observations of the specified targets. PIPT [20] proposal tool for the SALT provides exposure calculator based on the simulated spectra, visibility, and slew utilities. However, this approach is designed for a single telescope, but not the network. The solutions proposed by [19,34] for the LCOGT network, and SPHERE [18] offer algorithms for optimal observation scheduling of targets across several nights with multiple telescopes. In contrast, [32] designed the scheduler GWEMOPT[1] specifically for optimal observing the LVK sky error boxes with a distributed network of narrow-field telescopes. [32] demonstrate multiple algorithms for planning the observations with GWEMOPT, but the main idea is to re-grid uncertainty contour in a set of sky fields, combined in the observation blocks. Algorithms [32] optimize field traverse in terms of airmass, and cummulative probability enclosed in the uncertainty contour. Let's highlight separately ASTROPLAN [28]. It is positioned as a basement for a custom observation planners, providing such features as setting the observation constraints (e.g. on Sun altitude, the Moon separation, telescope rotation angles, etc.), grouping targets in blocks, and performing ranging by airmass (lowest targets have greater score). Also ASTROPLAN supports the calculation of the transition time between targets (accounts for filter changes, telescope slews, and CCD read-out time).

AWARE[2] is a lightweight Python double-threaded application for scheduling observations of LVK sky maps, and astrophysical transient sources (e.g. GRBs) in optimal order. AWARE also serves as a message broker between the GCN[3] and Telegram[4] subscribers. Additionally, it provides uploading of observation plots

---

[1] https://github.com/skyportal/gwemopt.

[2] https://github.com/mickolaua/aware-repo.

[3] https://gcn.nasa.gov.

[4] https://telegram.org.

and plans, generated by the scheduler thread. In this paper, we discuss methods of observations for both narrow-field and wide-field telescopes in Sect. 1. We provide an overview of AWARE implementation, data system, operation, and algorithms used in observational planning (Sect. 2). In the last section (Sect. 3), we summarize our results, make conclusions, and suggest ideas on the future enhancement of AWARE.

1 Observation Tactics

Localization areas of astrophysical events spans orders of magnitudes: from a few square arcseconds for gamma-ray bursts (in optical or soft X-ray bands) up to several hundred square degrees for gravitational wave events and gamma-ray bursts detected in hard X-rays or gamma-rays. It depends on the detector construction and sensitivity in the corresponding band. In the first case, when the uncertainty region does not exceed a few square degrees, it is observed, as a rule, even with a telescope with a narrow field of view in a series of observations [40,41,43]. However, in order to cover the gravitational-wave sky map error box, a network of optical telescopes is involved. In this scenario, the observation planning should account for each telescope local target visibility and the distribution of targets or sky fields between the telescopes for better efficiency. Depending on the field of view (FOV) of telescopes being used, two main observation tactics are utilized: target observations of galaxies (probable hosts of optical transients) inside the error box, and scanning the error box with tiles of the FOV size. The third tactic of observation can be defined as a combination of the previous two.

1.1 Target Observations

For BNS and NSBH events it is possible to perform target observations of the sample of hosts from the specially prepared catalog GLADE+[5] [12] to find optical transients (e.g. [14,33]). The mean sky density of the GLADE+ galaxies can be calculated as $\rho = N/A_{sky} \approx 561$ deg^{-2}, where $N \sim 25 \times 10^6$, and $A_{sky} \approx 41309$ deg^2 are the total number of GLADE+ sources, and the whole sky area, respectively. Although the sky density per square degree is large, it can be reduced to a few galaxies per FOV on average, when taking into account a typical FOV of a telescope, constraint on the gravitational event distance, and catalog completeness. The first optical counterpart confirmed to be related to the BNS merger occurred on the outskirts of the lenticular galaxy NGC 4993 [13]. However, it has been well localized to an elliptical contour with a 90% area of ~ 30 deg^2. It was possible due to quasi-synchronous detection of GW event by LIGO/Virgo, and GRB 170817A by *INTEGRAL*, and Fermi observatories. Let us estimate a maximal performance of target observations in the O4 run, based on 1.5-meter AZT-33IK telescope of Sayan Observatory (Mondy), which has the FOV = 0.015 deg^2. Here, we assuming the initial sky error box of LVK

---

[5] https://glade.elte.hu.

S230518h [25] with a 90% area $A \approx 1002$ deg^2, a galaxy density $\rho = 13$ deg^{-2} (at a luminosity distance of $D_L = 276 \pm 79$ Mpc), a single exposure time of $\Delta t = 120$ s, and a slew rate $\Delta T = 3$ min. Hence, the sky error box would be observed for $\tau \sim A \, \rho \, (\Delta T + \Delta t) \sim 2$ days with only one telescope. Usage of a network of 5 telescopes dramatically reduces the observation time to about 9.6 h. Thus, single telescope target observations have low efficiency, several instruments are required. Hence, an automatic software becomes necessary to coordinate them.

1.2   Mosaic Scanning

The other method is the scanning of LVK error boxes in a series of sky tiles with size, equivalent to the telescope's FOV [27]. The approach is adopted by many wide-field surveys, for instance *ZTF* [10,17], *BOOTES* [21], *Pan-STARRS* [38], *GOTO* [16], *MASTER* [26]. We estimate the performance of mosaic scanning to be $\tau \sim 30$ hr for the 0.7-meter AS-32 ($FOV = 0.5$ deg^2) of Abastumani Observatory (AbAO). Here, we assume parameters of the LVK event the same as in the previous subsection. However, instruments with large FOV may perform much faster. For example, ZTF's Schmidt telescope with $FOV = 47.7$ deg^2 could observe the whole area in under an hour (lower limit). As can be seen, a single wide-field telescope has a performance compared with 5 narrow-field instruments. Despite being slow, a typical narrow-field telescope could have greater sensitivity due to a large aperture, so can detect fainter transients. Therefore, combination of both kinds of telescopes may cancel out each other disadvantages.

1.3   Combined Observations

In a more realistic scenario for a large collaboration or network, both narrow-field and wide-field telescopes could be available. In this scenario the method simultaneously schedule mosaic observations, and target observations of GLADE+ galaxies. The GRB-IKI-FuN collaboration [41] collaboration among other groups have access to small aperture (less than 1 m) wide-field and large aperture narrow-field telescopes. We adopt this method in the LVK O4 run, rather than just target observations that we carried out in the previous runs.

In this section we reviewed primary techniques to observe GW skymaps for look-up for transients. Figure 1 illustrates all these methods in a schematic way. It has been shown that tile observations with wide-field provide significantly better performance than target observations in terms of traversing the sky error contour. Target observations in its turn allow reach deeper magnitudes, which is important for faint and distant transient detection. We pointed out that a network of telescope boosts both mosaic scanning and target observations. However, operation of multiple telescopes requires centralized software for the automatic observation scheduling. Although, there are several approaches presented for this problem, we believe that they are not the best suitable for a distributed and heterogeneous survey such as GRB-IKI-FuN. The next section describes our planning application designed to solve this problem.

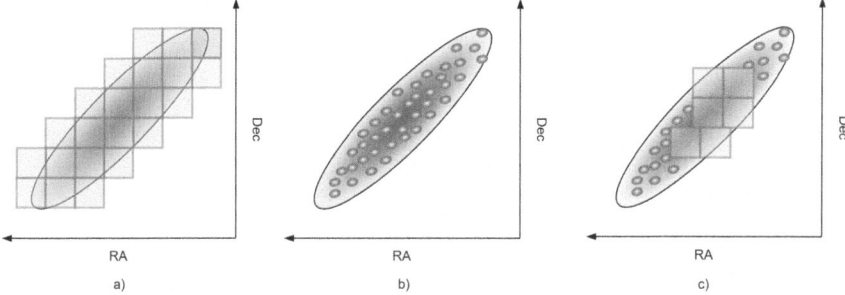

Fig. 1. The schematic big picture of observation tactics: a) mosaic scanning, b) target observations, and c) combined observations. The vertical axis is R.A. and the horizontal one is Dec (both in arbitrary units). The highlighted quasi-elliptical contour represents a localization region: the redder the color, the higher the probability of finding an LVK event there. The purple squares in the a) and c) cases indicate sky tiles to be observed with a wide-field telescope. The small blue ellipses in the b) and c) represent GLADE+ galaxies. (Color figure online)

2 Application Design

AWARE is a double-threaded asynchronous Python application: the alert parser and scheduler are run in the first thread, while the second thread is a Telegram bot that in concurrent mode sends alert messages in a human-readable form, observing plots and plans to subscribers. For a faster development process and easier deployment, we decided to make a monolithic application, i.e. solution, which is self-contained and independent from other programs. The application processing flow is presented in the Fig. 2.

The application configuration is stored in the form of a YAML file. Specifically for AWARE, we designed `CfgOption` class, which allows the setting of a proper option name, default value, and type, based on their location in the module tree. By default, the `aware.yaml` will be loaded if placed in the working folder. Otherwise, a person should set the environmental variable `AWARE_CONFIG_FILE` to the path of the actual configuration file. However, the configuration file is not required to utilize AWARE.

2.1 Alert Processing

The AWARE connects to the GCN Kafka Cluster for alert stream via Confluent client for Python (packages GCN_KAFKA[6] and CONFLUENT_KAFKA[7]). Note, it is required to obtain GCN credentials to be able to receive messages. For security reasons credentials are passed via environment variables `GCN_KAFKA_CLIENT_ID` and `GCN_KAFKA_CLIENT_SECRET`. Although the GCN continues to support socket

---

[6] https://github.com/nasa-gcn/gcn-kafka-python.
[7] https://github.com/confluentinc/confluent-kafka-python.

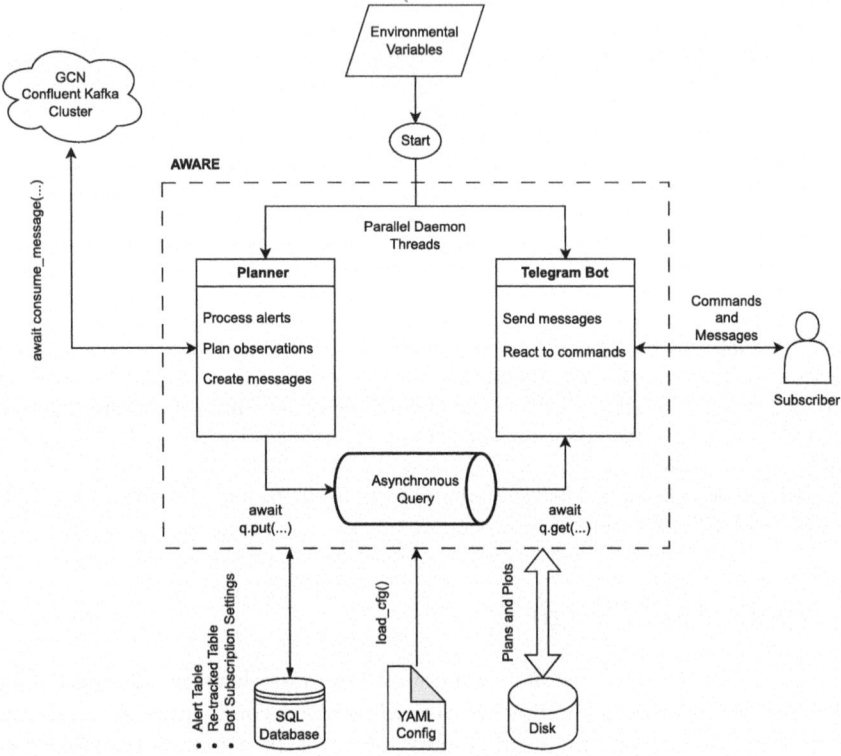

Fig. 2. The processing flow of the monolithic application AWARE. On startup, the planner and the bot daemon threads executed. The planner thread communicates with GCN via TCP and with the bot thread via the asynchronous query. This thread dumps data to the database and on disk. The bot thread reads data from the query and sends human-readable messages, observational plots, and plans (read from disk) to subscribers.

connection for receiving alert messages, this practice is considered outdated and not secure. This is the reason, why PYGCN[8] client was not chosen as a candidate solution for consuming alerts. The AWARE provides a special object `ConsumerLoop` for consuming alert messages asynchronously, which is important for not blocking the execution threads. In comparison, HOP CLIENT[9] as a part of SCiMMA [7,8], provides only a synchronous interface. A raw stream of alert messages should be parsed and processed to extract necessary information events. The GCN provides three types of data formats: text messages, binary messages, and VOEvents. There is a fourth type that seems to be only supported by LVK at the moment is JSON structures, which are validated against specified schemes. Due to the fact, VOEvents are most supported among astronomers we

---

[8] https://github.com/nasa-gcn/pygcn.
[9] https://github.com/scimma/hop-client.

decided to support only it. Migration to more convenient JSON in the future could be possible by only rewriting some parts of the parsing functions. Thus, AWARE receives alert messages in the form of VOEvents [36,42] (XML files with a special header), and then parses them. When the alert message is received, the application tries to find a suitable `AlertParser` from the registry that matches a special attribute, id, of the parser with the alert topic, for example, `gcn.classic.voevent.LVC_PRELIMINARY`. The Python package LXML[10] (v4.9.2) used in AWARE provides a convenient interface and great performance for XML tree parsing. We temporarily replace VOEvent headers before parsing, but preserve the original headers. This is implemented with a code from VOEVENT-PARSE[11] package, which provides the convenient interface for VOEvent trees representation as objects with attributes, using `lxml.objectify` internally. After parsing, the parameters of the event, a name, a trigger date and time, skymap, coordinates of the localization barycenter, and its uncertainties are retrieved and assigned to a `TargetInfo` object. A specific set of event parameters assigned to the `TargetInfo` is then dumped to the database (more details in Sect. 2.4). LVK skymaps are already represented as HEALPix multi-ordered coverage maps (MOCs) [28], but for some other instruments, for instance, *Swift* X-Ray Telescope (XRT), conic MOCs are created based on the localization center coordinates and error radius with MOCPY[12] [6]. It is valid because they have Gaussian-like uncertainty contours.

Event Crossmatching. While receiving new alert messages, it is important to establish a relationship with events already stored in the AWARE database. It allows us to reveal if the event contained in a new message is actually coincident with an already stored event under a different name. We do not want to create and send observational data for such events if they provide a larger localization error box. To solve this problem, the event from a new message is cross-matched against the SQL database. The crossmatching is performed with a nearest-neighbor algorithm in both time and sky coordinates. A pair of events is matched in the coordinate space using their MOCs: if the intersection of MOCs is not empty, then the events are matched. The events with a trigger date difference below several minutes are considered matched in time. Mathematically speaking, the crossmatching conditions are expressed by the inequalities (1):

$$matching = \begin{cases} M_1 \cap M_2 \neq \{0\} \\ |T_1 - T_2| < \Delta T \end{cases} \tag{1}$$

where, M_1, M_2 are MOCs of the events, T_1, T_2 are alert trigger dates, and ΔT is the trigger date matching uncertainty. Typically, ΔT should be at least a minute because different instruments could trigger a different phase of the event due to various energy range sensitivities of their detectors, especially if the event

---

[10] https://lxml.de.

[11] https://github.com/timstaley/voevent-parse.

[12] https://github.com/cds-astro/mocpy/.

spectrum is softening/hardening over time. We estimate the performance of the cross-matching to be a few seconds for a database containing about 500 rows.

2.2   Planning Observations

AWARE provides a special class `aware.site.Site` that defines a telescope inherited from `astroplan.observer.Observer`, but with additional attributes such as default exposure time and number of frames, field of view, observation plan file format, default limiting exposure, aperture size, short and full name. This information is used to create observation programs and plots. We implemented a special factory function `create_observation_program` in the module `aware.planning.program` to generate observational plans in given formats. At the moment of writing (29 Jan 2024), there are two formats available: a simple .txt ASCII table, and .list for automatic schedulers KDS, FORTE or HOUSE [23]. The mentioned file formats can be described by expressions 2 (first row corresponds to .txt, and second for .list formats, respectively):

$$hh:mm:ss.s \pm dd:mm:ss.s \ T.TxN \ F$$
$$Jhhmmss.ss \pm ddmmss.ss \ = \ 0 \ hhmmss.ss \pm ddmmss.ss \ m.mm \ NxT.T[*F]$$
$$(2)$$

where, `hh:mm:ss.s` is the RA in hours, `dd:mm:ss.s` is the Dec in degrees, `T.T` is the single exposure time (s), `N` is the number of exposures, `F` is the filter name. Additionally, there are fields unique for .list format: `Jhhmmss.ss±ddmmss.ss` is the target name, `0` is the target type (fixed or moving), `m.mm` is the estimate magnitude of a target, and square brackets indicate an optional field (filter can be omitted).

The planning process begins as soon as the processed alert message is sent to Telegram bot subscribers. Let us consider the planning process, which differs slightly for GRBs and LVK events below.

Planning of GRBs. Alert messages on GRB detections sent by observatories such as *Fermi, Swift, INTEGRAL*, etc., provide an approximately circular sky error region in the sky according to the Gaussian 2D probability distribution to locate events inside the region. For these events, the planner creates the visibility plot for the center of the localization region, if the target could be observable by a telescope at the nearest observational window[13] within 24 h from the current date. The observational window is calculated using ASTROPLAN. In addition, in the current implementation of AWARE we assume that the telescope must observe

---

[13] The nearest observational window is defined by two times when the Sun's altitude is below a certain value, specified for each telescope. Commonly, these correspond to the sunrise, and the sunset at the telescope location. However, for some locations, such as the Mondy observatory, the Sun may never cross the horizon for a few months in a year. This is the reason for setting Sun altitude constraints for an instrument separately.

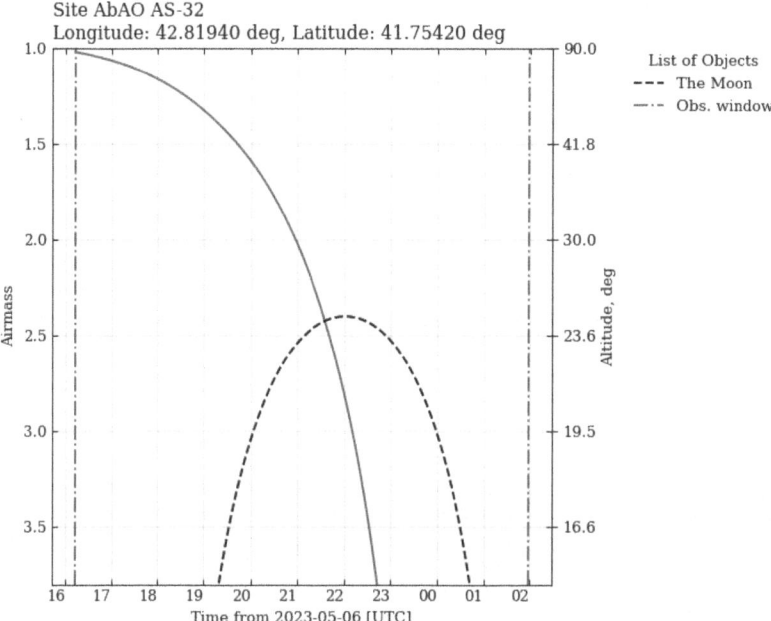

Fig. 3. Visibility plot for *Swift* BAT Trigger 1167288 (GRB 230506C) calculated for AS-32 telescope of Abastumani Astrophysical Observatory (AbAO). The blue solid line is the airmass of the target, the black dash line is the Moon, and the blue dash-dot lines denote the observational window borders. The left and right vertical axes are airmass and corresponding altitude (deg), respectively. The horizontal axis is UTC. (Color figure online)

GRB localization in a single scan, i.e. localization area must be less than 90% of the FOV[14]. Finally, the constraint is placed on airmass $X = \sec Z$, where Z is the zenith angle, $X \geq 3.5$ or $Alt = 90° - Z \geq 16.6°$ for a target/sky field to be considered observable. Thus, the observation plan file will contain only one target. An example of planning products, the visibility plot, and the target lists (in two formats) for GRB 230506C generated for Abastumani Observatory (AbAO) are presented in the Fig. 3, and Listing 1.1, respectively.

Planning of LVK Events. In comparison with GRBs, uncertainty contours of LVK GW-events typically have non-regular shapes (e.g. banana-like). They tend to span larger areas, thus it is more difficult to observe and find its electromagnetic counterparts. The more advanced scheduling technique is required for their observation. The planning procedure is step-wise. In the first step, the 90% (by default) cumulative probability contours are calculated using Python package LIGO.SKYMAP. Secondly, for each telescope in the list of specified telescopes, planner applies different observational tactics depending on the FOV, as

---

[14] 10% is accounted for possible edge artifacts or aberrations.

Listing 1.1. The examples of planning files for GRB 230506C for different formats.

```
# Output of .txt file
ra dec exp filter
10:13:14.9 +48:12:13  120.0x3  R

# Output of .list file
J101314.90+481213.00 = F  101314.90  +481213.00  0.00  3x120.0*R
```

was stated in the Sect. 1. If a telescope has the $FOV \leq 0.25$ deg^2 a target observation plan is computed, otherwise, for a wide-field telescope, a mosaic scanning plan is created.

Target Planning. The target planning is performed as following. The reduced GLADE+ catalog is loaded, containing only necessary fields RA (Right ascension in degrees), Dec (Declination in degrees), B_{mag} (Apparent magnitude in the B-filter), and d_L (Luminosity distance in Mpc). The catalog is cross-matched in coordinate space against a 90% cumulative probability volume, the depth of which is defined by the distance constraint (Eq. 3):

$$d_{L,GLADE+} \in [\hat{d}_L - \sigma_{d_L}, \hat{d}_L + 2\sigma_{d_L}] \tag{3}$$

where, $d_{L,GLADE+}$ are photometric distances of GLADE+ galaxies, $\hat{d}_L$ is the mean photometric distance (estimation) to the event, and σ_{d_L} is the standard deviation error of d_L. These assymetric borders were found to be more relevant than $\pm 1\sigma$ by considering the GLADE+ completeness, which drops after 50 Mpc (see Fig. 2 in [12]). Then, the targets are sorted according to the following Algorithm 1.

The algorithm for sorting the targets in optimal observational order, implemented in AWARE, represents a slightly modified nearest-neighbor method. The first target to observe is the one with the highest airmass (or lowest altitude) on average during the night. This is achieved using the `astroplan.Observer.altaz` object for construction of the airmass-time $X_i(T_{obs})$ curves, where $i = 1 \ldots n -$ is a target index, and $n -$ is the number of targets. We use default precision of the curve 150 points per night duration. Secondly, for each subsequent target, the closest neighbor is selected within a circle of a radius of several arcminutes (defined in config-file). The radius is chosen to minimize sharp transitions from one target to another because a regular mounting cannot rotate the telescope fast and stabilize it simultaneously. Hereafter, the algorithm repeats step #2 while the list of sorted targets is being updated. The estimated time complexity of the algorithm is $O(N^3)$, where N is the number of targets to sort. It may be reduced to $O(N^2) + O(N \log N) = O(N^2)$ if the distance matrix is computed before sorting targets. Here, we assume that the distance matrix is presented in the form of a k-d tree.

Algorithm 1. Target sorting algorithm

Require: $\exists X_i \geq 3.5$

1:   $R \leftarrow 5$ arcmin   ▷ Initial look-up radius for nearest neighbors
2:   $Step \leftarrow 5$ arcmin   ▷ Radius increment step
3:   $N \leftarrow \frac{\Delta t}{\Delta T \times n}$ ▷ Maximum observable targets during the night
4:   $t_0 \leftarrow Targets[\arg\min X]$ ▷ First target
5:   $SortedTargets \leftarrow t_0$
6:   **for** $i = 0$ to $i = N$ **do**
7:      $j \leftarrow i + 1$
8:      $D \leftarrow \{0\}$
9:      $d \leftarrow R$
10:     **while** $D = \{0\}$ **do**
11:        $D \leftarrow \{s \leftarrow dist(t_i, t_k), k = j..N, k \neq i, t_k \notin SortedTargets, s \leq d\}$ ▷
     Distances to all other targets from previous one
12:        $d \leftarrow d + Step$ ▷ Increase radius with this step, if no neighbors found
13:     **end while**
14:     $j \leftarrow \arg\min D$
15:     $t_j \leftarrow Targets[j]$ ▷ Next target (closest)
16:     $SortedTargets.$append(t_j)
17: **end for**

Scheduling Tile Observations. The tile observation planning method implies re-projection of the LVK skymap in the form of `HEALPix` into Cartesian grid with the pixel size of the telescope FOV. This process is performed using HEALPY package for Python. The scanning order is defined by the probability gradient: from the most probable to least probable pixel. Such an approach also allows to avoid large re-directions between tiles, when exceed a few FOV in size. The resulting curve of traversing the sky fields has a spiral shape.

Skymap Planner. To incorporate both observation routines in a convenient manner, we implemented a special class `SkymapPlanner`. It performs the mosaic planning of the skymap first, then scheduling the GLADE+ galaxies. `SkymapPlanner` has two modes of operation, unique distribution, and intersection. In the unique distribution mode, it prevents planning of the same area or GLADE+ galaxies by painting the already scheduled skymap pixels with zeros. The other mode (enabled by default) does not take it to account, therefore different telescope plans may contain the same tiles/fields. A planned skymap is written on disk (as a copy of the original skymap, that is modified), as well as, a CSV-list of galaxies. As a benefit of this, planning observations for the next epoch may be continued.

Final Products. As the result of the processing, the AWARE alert client/planner thread creates human-readable messages (including ones containing a payload) using `TargetInfo` objects as information sources and sends them via an asynchronous query to the Telegram bot. There are two types of messages available, `TelegramAlertMessage` and `TelegramDataPackage`. The first one con-

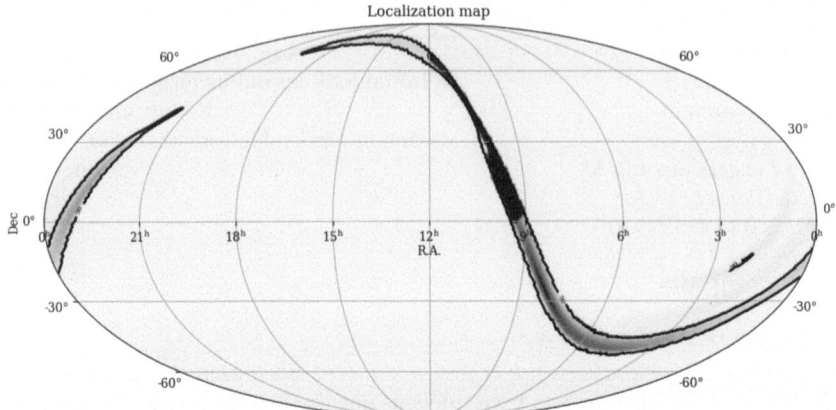

Fig. 4. The sky error box coverage map for LVK S230528A, drawn for AbAO AS-32 telescope. The black solid line traces 90% probability contours on the sphere. The contours are filled with yellow and red color: the more reddish, the higher the probability. The black highlighted area is the telescope coverage. (Color figure online)

tains only text information about the event name, trigger date, origin, and first mention. Additionally, `TelegramAlertMessage` includes other experiment-based parameters of the event. Such messages are prioritized in the query, ensuring the Telegram subscribers are notified on the event as soon as possible. `TelegramDataPackage` messages, stores the telescope parameters, generates visibility and skymap plots, and observational plans in telescope-specific format. They are placed in the message query only if the observational plots and plans can be created. The example localization coverage plot is shown in Fig. 4.

2.3   User Interaction

A user (or subscriber) receives alert messages and final products from observation planner by interacting with the bot. We have chosen AIOGRAM[15] package for an asynchronous Telegram client, which offers a wide spectrum of features to create bots, including message handlers, finite state machine (FST), different text rendering, keyboards, middleware support, etc. In addition, we used the package AIOGRAM_DIALOG[16] for a faster way of a dialog creation. In order to subscribe to the bot, a user must send the `/sub` command. Altough, we do not use it directly, AIOGRAM_DIALOG builds the FST internally. The corresponding transition diagram of the dialog's FST is illustrated in the Fig. 5.

The dialog is started from a state `content_types`, where a user has to choose which kind of content to receive. Available options are alert messages, observational data, or both. The next state is where the user selects the types (topics

---

[15] https://aiogram.dev.
[16] https://github.com/Tishka17/aiogram_dialog.

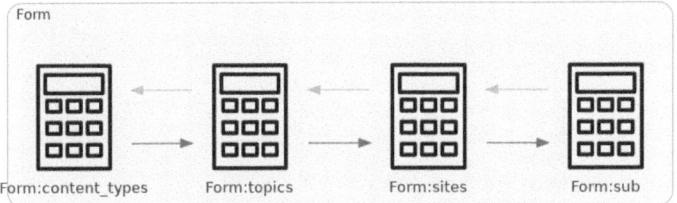

Fig. 5. The Telegram subscription dialog transition diagram. The states are denoted with windows having buttons. Bright blue arrows (points east) show the forward direction of the dialog (from start to finish). Faint gray arrows (points west) represent a backward direction (when a user wants to return to the previous form). The transition diagram has been drawn using `aiogram_dialog` (Color figure online).

in Kafka terminology) of alert messages to receive. At further state, it is possible to choose the telescope names for which one wants to receive observational materials (visibility plots, lists of targets/sky fields) or just skip this form for not obtaining the observational planning. Finally, the user should either approve the decision to subscribe or cancel the dialog. Note that at each step, excluding the first one, the person can switch back to the previous form, e.g., when some of the alert types need to be checked or unchecked (Fig. 5).

Despite, the `/sub` command, the Telegram bot has a set of auxiliary commands that a user may find helpful (see Table 1).

Table 1. The available bot commands

Command	Description
/findchart	Display $15' \times 15'$ finding chart from DSS2 survey (as shown in the Fig. 6)
/mysettings	Display current subscription plan
/start	Initiate dialog with the bot
/sub	Subscribe to the bot
/status	Display the bot status
/telescopes	Display the list of available telescopes
/unsub	Unsubscribe from the bot

Currently, the bot is running only in long-polling mode. Communication between the planner and the bot is established via an asynchronous query from the Python standard library ASYNCIO. Note that the environmental variable `AWARE_TG_API_TOKEN` that stores the Telegram bot token granted from *BotFather*[17] must be set before running AWARE. Additionally, the application can be

---

[17] https://t.me/botfather.

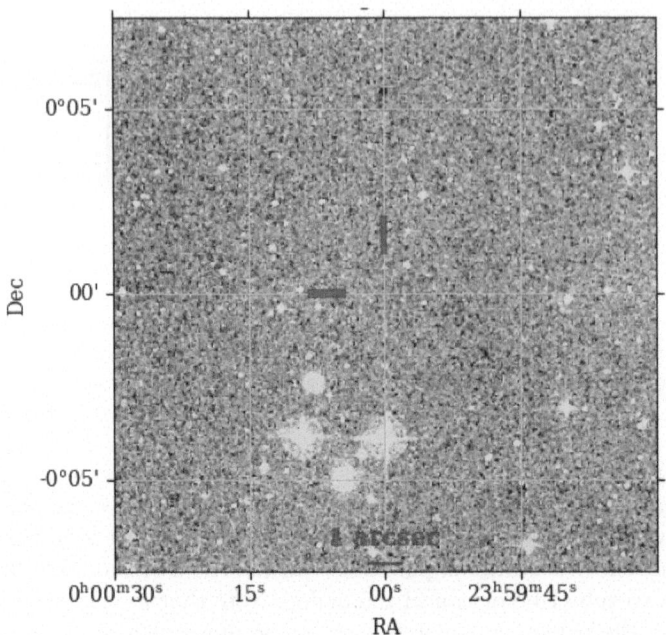

Fig. 6. The DSS2 finding chart plot of the artificial target at $RA = 00^h\ 00^m\ 00^s$, $Dec = 00^d\ 00^m\ 00^s$. The red crosshair is pointed at the target coordinates. (Color figure online)

executed with a disabled Telegram bot in the case someone wants only planning features or prefers to connect the planner with another service or GUI. A service should then listen for the database updates, and files generated on disk device.

2.4   Data Storage

The AWARE generates not so much data; only observational plots and plans are created and stored inside the working directory, specified in the configuration file. Typically, for a single event, there are up to several megabytes of disk volume consumed. Thus, it does not require special disk partitioning or large cloud storage to store data. Given this, we deployed the AWARE instance on a regular PC server[18].

On startup, AWARE creates two sub-folders inside the current working directory: products and cache. The products sub-folder is a root directory for further tree that stores observation products per telescope. Each downloaded skymap file is located inside the cache folder. The complete directory tree is shown in the Fig. 7.

The application also creates a SQL database on startup (if it is not exist yet), which is necessary for its operation. All incoming alert messages are stored

---

[18] We recommend to have at-least dual-core CPU, and 16 GB of RAM to be installed.

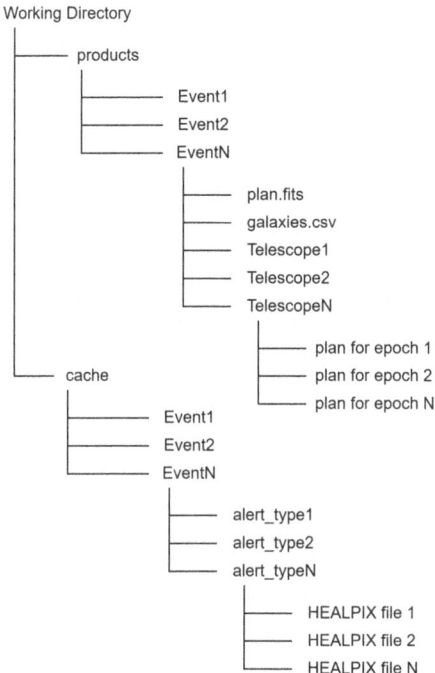

Fig. 7. The working directory tree. The event data is placed in separate sub-folders in the **products** directory. Also, the event sub-folder holds a separate folder for observational data per each telescope used. The downloaded HEALPix localization files are stored in the **cache** folder.

in the database. The main reasons for that are the possibility to cross-match alert events for localization updates and to store user subscription settings. The SQL database contains two tables. The first one is **alert** designed for storing alert messages, and event information. The second table, **settings**, stores user subscription plans. The table schemes are depicted in the Fig. 8.

We rely on the SQLALCHEMY package for querying the database in an objected-oriented style. SQLALCHEMY provides API for most used database engines such as PostgreSQL, MySQL, or SQLite. Since SQLite does not support login sessions, the SQLCIPHER3 package is applied to get access to the previously AES-512 encrypted database. The keyphrase is stored in the environment variable **AWARE_SQL_KEYPHRASE**. The reason behind it is to achieve secure storage of sensitive data, in our case, identifiers of the Telegram bot subscribers and their subscriptions. We recommend that users choose more advanced and robust options such as PostgreSQL or MySQL instead of using SQLite in production.

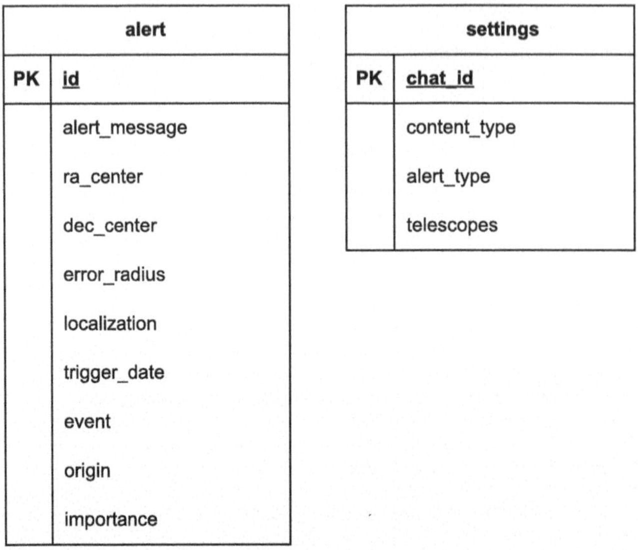

alert	
PK	**id**
	alert_message
	ra_center
	dec_center
	error_radius
	localization
	trigger_date
	event
	origin
	importance

settings	
PK	**chat_id**
	content_type
	alert_type
	telescopes

Fig. 8. The tables of the AWARE SQL database. On the left side, there is **alert** table schema. The main purpose of this table is to store alert messages and target information (event name, origin, trigger date, importance) alongside with localization skymap. The other table, shown on the right, is **settings**, which serves as a storage of user subscriptions.

3   Conclusions

3.1   Current Status and Results

Thus, in this paper, we provided an overview of our own lightweight open-source project, AWARE, designed as an LVK event and GRB alert stream client, and observation scheduler. Compared with other free software, AWARE offers an automatic three-in-one solution for processing alert messages, scheduling optical observations of their sky error boxes, and notifying users via Telegram in near real-time. The planner performs optimal sorting of GLADE+ galaxies for target observations, and efficient mosaic scanning of skymap tiles. The asynchronous implementation helps to perform these tasks together. The deployment process of AWARE is straightforward. We installed it on the laboratory server, configured and set in production mode. During operation, AWARE responded to several *Swift* and *Fermi* GBM gamma-ray bursts, and two *LVK* S230518h and S230528a in the O4 run (only events classified by *LVK* pipelines as NS-BH or BNS mergers). As an example, the Table 2 contains list of GRBs, which optical afterglow have been detected by *GRB-IKI-FuN* telescopes after the AWARE signal.

Table 2. Optical afterglow candidates to GRBs found in observations of the telescopes from *GRB-IKI-FuN* collaboration planned with AWARE.

Name	T_0^a UTC	Origin^b	Afterglow^c	t-T_0^d days	Magnitude^e mag	Filter^f	GCN^g
GRB 230414B^h	2023-04-14 16:14:21	Swift-XRT	Mondy AZT-33IK	0.01859	19.2 ± 0.2	R	33615
GRB 230506C	2023-05-06 17:09:19	Swift-BAT	TShAO Zeiss-1000	0.02914	20.1 ± 0.1	R	33734
GRB 230512A	2023-05-12 06:27:45.41	Fermi-LAT	Kitab RC-36	0.57624	19.43 ± 0.25	Clear	33796
GRB 231017A	2023-10-17 08:05:03.30	Swift-XRT	Mondy AZT-33IK	3.43519	22.98 ± 0.20	R	34867

^{a} Time of the GRB/LVK detection

^{b} Instrument that detected the event, and which localization uncertainty was observed.

^{c} Instrument that found the optical afterglow

^{d} Time since detection (in days)

^{e} Apparent magnitude of the afterglow, taken in the filter ^{f}, and submitted to the GCN circular no.^{g}.

^{h} Early version of AWARE v0.3a used.

AWARE is a growing project, and of course, we experienced some problems and bugs. First, we found two memory leakages, one in MATPLOTLIB.PYPLOT library, and other caused by `crossmatch` function of LIGO.SKYMAP package. Forcing the Python garbage collector `gc.collect` seems to solve the problem. Also we investigated large memory consumption by alert crossmatching functionality, particularly, in SQL queries. The limit was set on the number of rows selected in a single query, 100 rows found to be optimal count. We experienced unexpected runs of the scheduler on some of *Swift XRT* alerts, because they do not provide metadata on failures in the star-tracker lock. However, in general, *Swift BAT* alerts precede them, and indicate star-tracker lost of lock events. In that cases, the scheduler is not triggered (as it should be).

3.2   Future Plans

We plan to continue the development and further support the AWARE project in the future. There are many features we want to implement or improve. Custom telescope plugins provided as JSON or Python files would be convenient for extensibility to large telescope network without the source code modification. The accounting for the Moon separation angle and phase will enhance the visibility determination. Also, visibility plots will contain the Moon separation angle drawn every two hours in future releases. We found it convenient to use observation constraints and blocks provided by ASTROPLAN package API in the next versions of AWARE. Probably, the GCN will force everyone to migrate from VOEvents to JSON-messages in the near future, and we should adapt to this change. The web-UI/Telegram dashboards may be great for controlling the scheduling query, a manual user management, and the application status display. Therefore, an asynchronous query is needed to be replaced by a message broker (e.g. RABBITMQ[19]). Popular frameworks such as DJANGO[20], FLASK[21], or FASTAPI[22] could be used for dashboard implementation. The cache database

---

[19] https://rabbitmq.com.

[20] https://www.djangoproject.com.

[21] https://github.com/pallets/flask.

[22] https://github.com/tiangolo/fastapi.

(e.g. REDIS[23]) could serve as a suitable storage for recent requests (e.g. application database queries or HTTP-requests to SKYSERVER). It may be more convenient to store HEALPix skymaps in the database in the form of range data types, which are implemented in HEALPIX ALCHEMY package for Python, and supported by POSTGRESSQL. HEALPIX ALCHEMY[24] [37] allows, for instance, to query for GLADE+ galaxies in the skymap at a given probability threshold. On the database side, it is also planned to add several new tables for faster event cross-matching and retraction. These tables are need to be linked with `alert` table. Finally, it is considered to decoupling AWARE into micro-services for better reliability and flexibility. This will also provide more capabilities to integrate AWARE with other services.

We encourage developers who are interested in observing and studying optical transients to contribute to the project. Originally, AWARE was designed for *GRB-IKI-FuN* with its variety of telescopes. With a minor effort it, may be extended to another telescope network.

Funding Information. NP, AP, PM, SB, and AV acknowledge the Ministry of Science and Higher Education of the Russian Federation for financial support, grant 075-15-2022-1221 (2022-BRICS-8847-2335).

Data Availibility Statement. The AWARE is a free software (licensed under GPLv2), the source code is available here: https://github.com/mickolaua/aware-repo. We are glad to ask researchers to cite this article and insert the URL of the code repository in the footnote if your work has made use of AWARE.

References

1. Abbott, B.P., et al.: Prospects for observing and localizing gravitational-wave transients with Advanced LIGO, Advanced Virgo and KAGRA. Living Rev. Relativ. **21**(1), 3 (2018). https://doi.org/10.1007/s41114-018-0012-9
2. Abbott, B.P., et al.: Prospects for observing and localizing gravitational-wave transients with Advanced LIGO, Advanced Virgo and KAGRA. Living Rev. Relativ. **23**(1), 3 (2020). https://doi.org/10.1007/s41114-020-00026-9
3. Abbott, B.P., et al.: Gravitational waves and gamma-rays from a binary neutron star merger: GW170817 and GRB 170817A. ApJ **848**(2), L13 (2017). https://doi.org/10.3847/2041-8213/aa920c
4. Abbott, B.P., et al.: Multi-messenger observations of a binary neutron star merger. ApJ **848**(2), L12 (2017). https://doi.org/10.3847/2041-8213/aa91c9
5. Acernese, F., et al.: Advanced Virgo: a second-generation interferometric gravitational wave detector. Class. Quantum Gravity **32**(2), 024001 (2015). https://doi.org/10.1088/0264-9381/32/2/024001
6. Boch, T.: MOCPy, a python library to manipulate spatial coverage maps. In: Molinaro, M., Shortridge, K., Pasian, F. (eds.) Astronomical Data Analysis Software and Systems XXVI. Astronomical Society of the Pacific Conference Series, vol. 521, p. 487 (2019)

---

[23] https://redis.io.
[24] https://github.com/skyportal/healpix-alchemy.

7. Brazier, A., et al.: SCiMMA HopSkotch: coordinating multi-messenger astrophysical observatories during LVK run O4. In: American Astronomical Society Meeting Abstracts. American Astronomical Society Meeting Abstracts, vol. 55, p. 268.05 (2023)
8. Brazier, A., et al.: SCiMMA Hopskotch: performant, robust, and secure messaging cyberinfrastucture for multi-messenger astrophysics. In: American Astronomical Society Meeting Abstracts. American Astronomical Society Meeting Abstracts, vol. 54, p. 348.10 (2022)
9. Brink, J., Charles, A., Hettlage, C., Husser, T.-O., Koeslag, A., Romero-Colmenero, E.: The SALT observation control system. In: Bridger, A., Radziwill, N.M. (eds.) Advanced Software and Control for Astronomy II. Society of Photo-Optical Instrumentation Engineers (SPIE) Conference Series, vol. 7019, p. 70190N (2008). https://doi.org/10.1117/12.787632
10. Coughlin, M.W., et al.: GROWTH on S190425z: searching thousands of square degrees to identify an optical or infrared counterpart to a binary neutron star merger with the zwicky transient facility and Palomar Gattini-IR. ApJ **885**(1), L19 (2019). https://doi.org/10.3847/2041-8213/ab4ad8
11. Coulter, D.A., et al.: Swope supernova survey 2017a (SSS17a), the optical counterpart to a gravitational wave source. Science **358**(6370), 1556–1558 (2017). https://doi.org/10.1126/science.aap9811
12. Dálya, G., et al.: GLADE+: an extended galaxy catalogue for multimessenger searches with advanced gravitational-wave detectors. MNRAS **514**(1), 1403–1411 (2022). https://doi.org/10.1093/mnras/stac1443
13. Drout, M.R., et al.: Light curves of the neutron star merger GW170817/SSS17a: implications for R-process nucleosynthesis. Science **358**(6370), 1570–1574 (2017). https://doi.org/10.1126/science.aaq0049
14. Gehrels, N., Cannizzo, J.K., Kanner, J., Kasliwal, M.M., Nissanke, S., Singer, L.P.: Galaxy strategy for LIGO-Virgo gravitational wave counterpart searches. ApJ **820**(2), 136 (2016). https://doi.org/10.3847/0004-637X/820/2/136
15. Goldstein, A., et al.: An ordinary short gamma-ray burst with extraordinary implications: fermi-GBM detection of GRB 170817A. ApJ **848**(2), L14 (2017). https://doi.org/10.3847/2041-8213/aa8f41
16. Gompertz, B.P., et al.: Searching for electromagnetic counterparts to gravitational-wave merger events with the prototype Gravitational-Wave Optical Transient Observer (GOTO-4). MNRAS **497**(1), 726–738 (2020). https://doi.org/10.1093/mnras/staa1845
17. Graham, M.J., et al.: A light in the dark: searching for electromagnetic counterparts to black hole-black hole mergers in LIGO/Virgo O3 with the Zwicky transient facility. ApJ **942**(2), 99 (2023). https://doi.org/10.3847/1538-4357/aca480
18. Gross, A., Meunier, J.C., Surace, C., Le Coroller, H., Agneray, F.: SPHERE TDB: catalogue management, observation planning and detection visualization. In: Lorente, N.P.F., Shortridge, K., Wayth, R. (eds.) Astronomical Data Analysis Software and Systems XXV. Astronomical Society of the Pacific Conference Series, vol. 512, p. 527 (2017)
19. Hawkins, E., et al.: Scheduling observations on the LCOGT network. In: Silva, D.R., Peck, A.B., Soifer, B.T. (eds.) Observatory Operations: Strategies, Processes, and Systems III. Society of Photo-Optical Instrumentation Engineers (SPIE) Conference Series, vol. 7737, p. 77370P (2010). https://doi.org/10.1117/12.857756
20. Hettlage, C., et al.: Handling observation proposals for SALT. In: Silva, D.R., Peck, A.B., Soifer, B.T. (eds.) Observatory Operations: Strategies, Processes, and Sys-

tems III. Society of Photo-Optical Instrumentation Engineers (SPIE) Conference Series, vol. 7737, p. 77371B (2010). https://doi.org/10.1117/12.858025

21. Hu, Y.D., et al.: The burst observer and optical transient exploring system in the multi-messenger astronomy era. Front. Astron. Space Sci. **10**, 952887 (2023). https://doi.org/10.3389/fspas.2023.952887

22. Kagra Collaboration, et al.: KAGRA: 2.5 generation interferometric gravitational wave detector. Nat. Astron. **3**, 35–40 (2019). https://doi.org/10.1038/s41550-018-0658-y

23. Kouprianov: 6th European Conference on Space Debris, Proceedings of the conferencje held 22-25 April 2013, in Darmstadt, Germnay. Edited by L. Ouwehand. ESA SP-723 (2013). ISBN 978-92-9221-287-2, 2013, id.21. The ADS link is as follows: https://ui.adsabs.harvard.edu/abs/2013ESASP.723E..21K/abstract

24. LIGO Scientific Collaboration, et al.: Advanced LIGO. Class. Quantum Gravity **32**(7), 074001 (2015). https://doi.org/10.1088/0264-9381/32/7/074001

25. Ligo Scientific Collaboration, VIRGO Collaboration, and Kagra Collaboration: LIGO/Virgo/KAGRA S230518h: Identification of a GW compact binary merger candidate. GRB Coordinates Network **33813**, 1 (2023)

26. Lipunov, V., et al.: Strategy and results of MASTER network follow-up observations of LIGO and Virgo gravitational wave events within the observational sets O1, O2, and O3. Astron. Rep. **66**(12), 1118–1253 (2022). https://doi.org/10.1134/S1063772922110129

27. Lundquist, M.J., et al.: Searches after gravitational waves using ARizona observatories (SAGUARO): system overview and first results from advanced LIGO/Virgo's third observing run. ApJ **881**(2), L26 (2019). https://doi.org/10.3847/2041-8213/ab32f2

28. Morris, B.M., et al.: Astroplan: an open source observation planning package in python. ApJ **155**(3), 128 (2018). https://doi.org/10.3847/1538-3881/aaa47e

29. Pian, E., et al.: Spectroscopic identification of R-process nucleosynthesis in a double neutron-star merger. Nature **551**(7678), 67–70 (2017). https://doi.org/10.1038/nature24298

30. Pozanenko, A.S., et al.: GRB 170817A associated with GW170817: multi-frequency observations and modeling of prompt gamma-ray emission. ApJ **852**(2), L30 (2018). https://doi.org/10.3847/2041-8213/aaa2f6

31. Pozanenko, A.S., Minaev, P.Y., Grebenev, S.A., Chelovekov, I.V.: Observation of the second LIGO/Virgo event connected with a binary neutron star merger S190425z in the gamma-ray range. Astron. Lett. **45**(11), 710–727 (2020). https://doi.org/10.1134/S1063773719110057

32. Rana, J., Singhal, A., Gadre, B., Bhalerao, V., Bose, S.: An enhanced method for scheduling observations of large sky error regions for finding optical counterparts to transients. ApJ **838**(2), 108 (2017). https://doi.org/10.3847/1538-4357/838/2/108

33. Salmon, L., Hanlon, L., Jeffrey, R.M., Martin-Carrillo, A.: Galaxy-targeted robotic telescope follow-up of gravitational wave events. In: Revista Mexicana de Astronomia y Astrofisica Conference Series. Revista Mexicana de Astronomia y Astrofisica Conference Series, vol. 53, pp. 67–74 (2021). https://doi.org/10.22201/ia.14052059p.2021.53.17

34. Saunders, E., Lampoudi, S.: Multi-telescope observing: the LCOGT network scheduler. In: Wozniak, P.R., Graham, M.J., Mahabal, A.A., Seaman, R. (eds.) The Third Hot-Wiring the Transient Universe Workshop, pp. 117–123 (2014)

35. Savchenko, V., et al.: *INTEGRAL* detection of the first prompt gamma-ray signal coincident with the gravitational-wave event GW170817. ApJ **848**(2), L15 (2017). https://doi.org/10.3847/2041-8213/aa8f94

36. Seaman, R., et al.: Sky Event Reporting Metadata Version 2.0. IVOA Recommendation 11 July 2011 (2011). https://doi.org/10.5479/ADS/bib/2011ivoa.spec.0711S

37. Singer, L.P., et al.: HEALPix alchemy: fast all-sky geometry and image arithmetic in a relational database for multimessenger astronomy brokers. AJ **163**(5), 209 (2022). https://doi.org/10.3847/1538-3881/ac5ab8

38. Smartt, S.J., et al.: Pan-STARRS and PESSTO search for an optical counterpart to the LIGO gravitational-wave source GW150914. MNRAS **462**(4), 4094–4116 (2016). https://doi.org/10.1093/mnras/stw1893

39. Tanvir, N.R., et al.: The emergence of a lanthanide-rich kilonova following the merger of two neutron stars. ApJ **848**(2), L27 (2017). https://doi.org/10.3847/2041-8213/aa90b6

40. Vlasyuk, V.V., Sokolov, V.V.: Multi-messenger astronomy: the alert observations of gamma-ray bursts afterglows, supernovae and search for optical counterparts to neutrino events and gravitational waves. In: The Multi-Messenger Astronomy: Gamma-Ray Bursts, Search for Electromagnetic Counterparts to Neutrino Events and Gravitational Waves, pp. 243–254 (2019). https://doi.org/10.26119/SAO.2019.1.35556

41. Volnova, A., et al.: IKI GRB-FuN: observations of GRBs with small-aperture telescopes. Anais da Academia Brasileira de Ciências **93**(1) (2021). https://doi.org/10.1590/0001-3765202120200883

42. Williams, R.D., Seaman, R.: VOEvent: information infrastructure for real-time astronomy. In: Gabriel, C., Arviset, C., Ponz, D., Enrique, S. (eds.) Astronomical Data Analysis Software and Systems XV. Astronomical Society of the Pacific Conference Series, vol. 351, p. 637 (2006)

43. Xin, L., et al.: The observations of GRB afterglows and the plan to search for optical counterparts of gravitational wave events. In: The Multi-Messenger Astronomy: Gamma-Ray Bursts, Search for Electromagnetic Counterparts to Neutrino Events and Gravitational Waves, pp. 260–267 (2019). https://doi.org/10.26119/SAO.2019.1.35558

Information Extraction from Text

Cross-Domain Robustness of Transformer-Based Keyphrase Generation

Anna Glazkova[1,3]([✉]) [iD] and Dmitry Morozov[2,3] [iD]

[1] University of Tyumen, Tyumen, Russia
a.v.glazkova@utmn.ru
[2] Novosibirsk State University, Novosibirsk, Russia
[3] The Institute for Information Transmission Problems (Kharkevich Institute),
Moscow, Russia

Abstract. Modern models for text generation show state-of-the-art results in many natural language processing tasks. In this work, we explore the effectiveness of abstractive text summarization models for keyphrase selection. A list of keyphrases is an important element of a text in databases and repositories of electronic documents. In our experiments, abstractive text summarization models fine-tuned for keyphrase generation show quite high results for a target text corpus. However, in most cases, the zero-shot performance on other corpora and domains is significantly lower. We investigate cross-domain limitations of abstractive text summarization models for keyphrase generation. We present an evaluation of the fine-tuned BART models for the keyphrase selection task across six benchmark corpora for keyphrase extraction including scientific texts from two domains and news texts. We explore the role of transfer learning between different domains to improve the BART model performance on small text corpora. Our experiments show that preliminary fine-tuning on out-of-domain corpora can be effective under conditions of a limited number of samples.

Keywords: Keyphrase extraction · BART · Transfer learning · Scholarly document · Text summarization

1 Introduction

The task of keyphrase generation aims at predicting a set of keyphrases summarizing the content of the source text. Keyphrases are often indexed in databases to improve the performance of information retrieval tools. Researchers select keyphrases for their papers to increase their visibility in the scientific community. Automatic selection of keyphrases for scholarly documents helps to analyze the current research trends, recommend papers, and identify potential peer reviewers [39].

Supported by the grant of the President of the Russian Federation no. MK-3118.2022.4.

J. Baixeries et al. (Eds.): DAMDID/RCDL 2023, CCIS 2086, pp. 249–265, 2024.
https://doi.org/10.1007/978-3-031-67826-4_19

Abstract. The study considers robust estimation of linear regression parameters by the regularization method, the pseudoinverse method, and the Bayesian method allowing for correlations and errors in the data. Regularizing algorithms are constructed and their relationship with pseudoinversion, the Bayesian approach, and BLUE is investigated

Keyphrases: linear regression problems regularization, robust estimation, linear regression parameters, pseudoinverse method, Bayesian method, pseudoinversion, Bayesian approach, BLUE, Bayes methods, estimation theory, probability, statistical analysis

Fig. 1. An example of a source text with the corresponding list of from the Inspec corpus [22]. The keyphrases that appear in the text are underlined.

Figure 1 demonstrates an example of a source text and its keyphrases. Some keyphrases are present in the source text while others are absent. Most unsupervised approaches for keyphrase selection have the purpose of keyphrase extraction, in other words, the ranking and selection of phrases that appear in the text. Recent generative approaches produce both keyphrases present in the text and those absent from it. These approaches utilize deep learning methods using the encoder-decoder architecture [8,32,48] and various training techniques, such as incorporating a copying mechanism [44], reinforcement learning [10], hierarchical decoding [11], and multitask learning [26]. Currently, the models for automatic text generation achieve high results in various natural language processing tasks. Since a list of keyphrases is some type of summary of a scientific text, pre-trained models for abstractive summarization appear to be effective for generating keyphrases as a sequence. In our previous work [18,19], we explored the performance of some of these models for keyphrase generation. It was shown that BART [27] fine-tuned for generating lists of keyphrases on texts from the target domain showed competitive results as compared to several baselines. However, it can show lower performance on the texts from other corpora and domains similar to other fine-tuned models. Our goal is to evaluate whether we can transfer knowledge from the BART model that was fine-tuned to generate keyphrases for one domain to another ones. We seek to answer the following research questions:

- **RQ1.** How effective is a text summarization model fine-tuned on one corpus or one domain for generating keyphrases from texts of other corpora or domains in zero-shot settings?
- **RQ2.** Can we improve the model performance by adding training examples from other corpora and domains?
- **RQ3.** With a small number of training examples, can the model perform as effectively as a model fine-tuned on larger corpora?
- **RQ4.** Can transfer learning improve the model performance using a varying size of training data?

The paper is organized as follows. Section 2 presents related works in the field. In Sect. 3, we describe the corpora. Section 4 contains a brief description of the models we used. Section 5 presents the experimental setup. The results are reported and discussed in Sect. 6. Section 7 concludes this paper.

2 Related Work

2.1 Abstractive Text Summarization Using Pre-trained Transformers

Pre-trained language models show impressive results in many natural language processing (NLP) tasks. A pre-trained model is a saved network that is previously trained on a large dataset. This is a common and highly effective approach to deep learning on small datasets [15]. Automatic text summarization is a relevant trend in NLP. A summary can be generated through extractive, as well as abstractive, methods. Abstractive methods are difficult to implement because they need a lot of natural language processing. However, abstractive models, such as BART [27], PEGASUS [49], and many others, allow us to generate novel samples by either rephrasing or using new words, instead of simply extracting the important sentences [21,41].

Neural abstractive summarization based on pre-trained language models has been studied by many researchers and showed a high performance with the aid of large text corpora. In particular, abstractive summarization models were applied for generating summaries in news [5,13,20,52], scientific [6,35,45], sport [31], and financial domains [42,51]. One of the main challenges related to neural abstractive summarization is that domain-shifting problems and overfitting could occur with a small number of samples for the target corpora [12]. The use of additional texts for other corpora is not always successful since different corpora contain texts of different writing styles and forms. The annotation for abstractive summarization is costly. Therefore, exploring approaches to low-resource abstractive summarization is very relevant and attracts the attention of scientists.

2.2 Keyword Selection

Keyword selection approaches can be roughly divided into three categories: i) actual keyword extraction, ii) keyword assignment, and iii) keyphrase generation [1,8]. The actual keyword extraction involves extracting words directly presented in the text. In keyword assignment, keywords are chosen from a predefined set of terms, while documents are classified into thematic categories according to their topics. Keyphrase generation aims to produce a set or string of keywords using recent advances in sequence-to-sequence applications of neural networks. In this work we focus on keyphrase generation but use some keyword extraction approaches as baselines. Keyphrase generation allows us to generate broad terms and keyphrases that are not presented in the source text in an explicit form.

Table 1. A summary statistics for the corpora. The average number of tokens was obtained using NLTK [2]. The "±" sign is utilized to indicate a standard deviation. The abbreviations in this table are CS—computer science, BM—biomedical, A—abstract, and T—text (body).

Characteristic	Krapivin-A	Krapivin-T	Inspec	PubMed	NamedKeys	DUC-2001	KPTimes
Size	2,294	2,293	2,000	1,320	3,049	308	20,000
Domain	scientific, CS			scientific, BM		news	
Type of texts	A	T	A	T	A		
Avg. number of tokens	169.06 ± 68.58	8597.63 ± 2411.77	127.35 ± 65.03	5270.97 ± 2690.67	274.67 ± 99.88	848.22 ± 563.41	733.78 ± 477.49
Avg. number of sentences	6.64 ± 2.69	343.95 ± 107.3	5.3 ± 2.73	206.81 ± 127.11	10.52 ± 3.66	34.74 ± 23.33	26.65 ± 22.19
Avg. keyphrases per text	5.34± 2.77		14.11 ± 6.41	5.4 ± 2.17	14.15 ± 5.2	8.08 ± 1.87	5.03 ± 1.88
Absent keyphrases, %	51.3 ± 25.99	18.04 ± 19.69	43.8 ± 17.83	13.52 ± 19.7	1.04 ± 5.83	2.45 ± 7.94	35.72 ± 29.22
Number of unique keyphrases	8,703		19,066	5,580	20,804	1,850	21,126

To date, some scholars have examined neural models to generate multiple keyphrases as a sequence [8,39]. Chowdhury et al. [14] demonstrated that fine-tuned BART shows competitive results in keyphrase generation compared with the existing extractive neural models. In [23,46], the authors experimented with controllable text generation for producing keyphrases. The authors of [26] proposed KeyBART, a new pre-training setup for the BART model that learns to generate keyphrases in their original order in the source document. Shen and Le [37] investigated the advantages of title attention and sequence code representing phrase order in a keyphrase sequence in improving Transformer-based keyphrase generation.

The authors of [38] provided a comprehensive survey on recent advances in keyphrase selection from pre-trained language models. They emphasize that most existing keyphrase extraction datasets and studies are based on a few of the most common topics and lack datasets and research related to other domains. Therefore, transferring knowledge from one domain to another to build domain-specific keyphrase extraction models is one of the major challenges for keyphrase generation.

3   Data

The experiments are carried out on six corpora for keyphrase selection:

- Krapivin [25] and Inspec [22] containing scientific texts from the computer science domain;
- PubMed [36] and NamedKeys [17], which include scientific texts from the biomedical domain;
- DUC-2001 [43] and KPTimes [16] consisting of news texts.

Table 2. Top-10 common keyphrases for the corpora.

Corpus	Keyphrases (keyphrase—number of occurrences)
Krapivin	scheduling – 36, performance evaluation – 25, data mining – 24, computational complexity – 24, parallel algorithms – 22, fault tolerance – 22, approximation algorithms – 22, model checking – 21, distributed systems – 21, preconditioning – 20
Inspec	internet – 199, information resources – 97, probability – 70, computational complexity – 69, optimisation – 60, gender issues – 49, matrix algebra – 47, psychology – 46, human factors – 46, academic libraries – 45
PubMed	children – 24, breast cancer – 21, epidemiology – 20, internet – 19, quality of life – 19, preconception care – 16, pregnancy – 16, apoptosis – 15, cancer – 13, magnetic resonance imaging – 13
NamedKeys	CI – 245, OR – 144, reactive oxygen species – 142, ROS – 134, confidence interval – 131, nitric oxide – 112, NO – 111, HR – 95, ER – 84, oxidative stress – 83
Duc-2001	police brutality – 12, mad cow disease – 11, illegal aliens – 10, Census Bureau – 10, Ben Johnson – 10, Clarence Thomas – 10, investigation – 9, firefighters – 9, welfare reform – 8, crash – 8
KPTimes	U.S. – 1,472, Donald Trump – 1,274, China – 1,122, terrorism – 525, baseball – 510, Russia – 474, elections – 435, Shinzo Abe – 386, football – 364, North Korea – 350

Table 3. A comparison of corpora content proximity, evaluated as in [24]. The value of 1 indicates identical corpora. The higher the score, the greater the difference between corpora.

	Krapivin-A	Krapivin-T	Inspec	PubMed	NamedKeys	DUC-2001	KPTimes
Krapivin-A	1.00	46.24	51.63	141.92	214.31	295.17	267.85
Krapivin-T	46.24	1.00	61.44	125.14	190.75	277.01	238.56
Inspec	51.63	61.44	1.00	98.11	146.79	216.29	202.35
PubMed	141.92	125.14	98.11	1.00	68.72	194.62	121.91
NamedKeys	214.31	190.75	146.79	68.72	1.00	309.29	285.13
DUC-2001	295.17	277.01	216.29	194.62	309.29	1.00	162.96
KPTimes	267.85	238.56	202.35	121.91	285.13	162.96	1.00

The Krapivin corpus contains full papers divided into titles, abstracts, and bodies. In this work, we separately utilized the abstract and body of the paper to select keyphrases (Krapivin-A and Krapivin-T respectively). The original KPTimes corpus is composed of 279,923 article-keyphrase pairs. Here, we used only a test set of the original corpus containing 20,000 samples. Summary statistics for the corpora are presented in Table 1. The most popular keyphrases are shown in Table 2. A comparison of the contents of the corpora is given in Table 3.

4   Models

For keyphrase generation, we utilized BART-base [27], a transformer-based denoising autoencoder for pre-training a seq2seq model. The model has 12 layers, 768 hidden units per layer, and a total of 139M parameters. BART was pre-trained by corrupting documents and then optimizing a reconstruction loss-the cross-entropy between the decoder's output and the original document. We fine-tuned BART-base for six epochs with a maximum sequence length of 256 tokens. We utilized a standard cross-entropy loss and the AdamW optimizer [30]. We used the source text as an input of the model and a list of keyphrases in a string format as an output. Keyphrases included in lists of keyphrases were separated with commas.

As baselines, we used the implementations of TopicRank [4] and YAKE! [7] from the PKE library [3] and KeyBART [26] that represents pre-trained BART-based architecture to produce a sequence of keyphrases pre-trained on the OAGKX dataset [9], which consists of 23 million scientific documents across multiple domains.

5   Experimental Setup

We randomly split each corpus into a 70% training set and a 30% test set. For BART, we performed three runs for each model and then calculated the average results. Since TopicRank and YAKE! are unsupervised methods and they require a pre-defined number of keyphrases to select, we extracted 5, 10, and 15 keyphrases for each corpus and chose the best value for each metric. KeyBART was used in zero-shot settings. The models were evaluated in terms of the full-match F1-score (F1), ROUGE-1 (R1), ROUGE-L (RL) [28], and BERTScore (BS) [50].

The full-match F1-score evaluates the number of exact matches between the original and generated sets of keyphrases. It is calculated as a harmonic mean of precision and recall.

The ROUGE-1 score calculates the number of matching unigrams between the model-generated text and the reference. The ROUGE-L score works in a similar way but measures the longest common subsequence. To measure ROUGE-1 and ROUGE-L, the keyphrases for each text were combined into a string with a comma as a separator.

BERTScore utilizes the pre-trained contextual embeddings from BERT-based models and matches words in the source and generated texts using cosine similarity. It has been shown that human judgment correlates with this metric on sentence-level and system-level evaluation. To calculate BERTScore, we use contextual embeddings from RoBERTa-large [29], a modification of BERT that is pre-trained using dynamic masking.

6   Results and Discussion

To answer **RQ1** and **RQ2**, we fine-tuned BART on one corpus and applied it to the other corpora in zero-shot settings. Then we fine-tuned BART on mixed data. For this purpose, we evaluated four strategies:

1. $Domain_{eq}$, fine-tuning the model on the texts of all corpora from one domain (for example, CS domain includes Krapivin-a, Krapivin-T, and Inspec), then testing on each corpus separately. In this strategy, we use an equal number of texts for each corpus. For example, if the size of training sets for Krapivin-A, Krapivin-T, and Inspec are 1,606, 1,605, and 1,400 respectively, we utilize 1,400 random texts from Krapivin-A and Krapivin-T and all texts from Inspec. The overall size of training data is 4,200. The texts from different corpora are mixed in random order.
2. $Domain_{all}$, the strategy is similar to the previous one but we use all texts from each corpus. In this case, the overall size of training data for the example above is 4,611, i.e. 1,606+1,605+1,400.
3. Mix_{eq}, fine-tuning the model on the texts of all corpora using an equal number of texts for each corpus, then testing it on each corpus separately. The texts from different corpora are mixed in random order.
4. Mix_{all}, the strategy is similar to Mix_{eq} but we use all texts from each corpus.

Table 4 shows the performance of baselines on test sets. The best baseline results are underlined. The performance of different methods varies depending on the corpus. For example, KeyBART performs worse on the news domain since this model was pre-trained on scientific texts.

The BART results are presented in Table 5. The results obtained for models fine-tuned on data containing the target corpus are highlighted in gray. Training data are italicized. The scores outperforming baselines are underlined. For mixed training data, we indicate the overall number of training examples in brackets and highlighted in bold the scores that exceed the results of the BART fine-tuned only on the target corpus. The best results among all models (Tables 4 and 5) are marked with an asterisk (*). Table 6 in Appendix A shows a standard deviation for three runs of BART.

Table 4. Baseline results, %.

Target	F1	R1	RL	BS	F1	R1	RL	BS	F1	R1	RL	BS
	KeyBART				*YAKE!*				*TopicRank*			
Krapivin-A	8.58	23.34	19.81	88.26	8.14	20.75	17.58	86.35	6.89	17.68	14.86	87.94
Krapivin-T	5.42	16.61	14.67	87.18	7.09	16.43	14.13	86.14	5.89	15.17	12.48	87.44
Inspec	10.66	29.09	23.55	86.72	13.84	33.80	27.00	86.45	16.32	35.68	25.01	87.38
PubMed	5.70	13.41	12.27	85.56	13.35	20.00	17.28	85.43	11.11	18.61	15.41	86.83
NamedKeys	9.01	21.11	18.30	82.96	20.80	30.62*	22.06*	84.80	19.40	30.55	22.00	84.80
DUC-2001	5.57	11.54	10.37	86.12	13.58	26.93	22.16	85.63	20.88*	30.91*	23.59*	88.51*
KPTimes	4.50	8.52	7.87	83.95	10.05	18.92	16.18	84.83	10.40	14.44	12.74	86.24

The BART fine-tuned on a target corpus outperforms baselines in many cases (Krapivin-A, Inspec, and KPTimes – all metrics; Krapivin-T – F1, R1, and RL; PubMed – R1, RL, and BS; NamedKeys – RL and BS). For DUC-2001, the results of BART are lower than the ones of unsupervised methods, which is probably due to the smaller size of this corpus. The out-of-corpus results are generally lower than the in-corpus ones. For example, when fine-tuning on Inspec (CS domain), the performance in terms of F1 is reduced by 37% and 30% for Krapivin-A and Krapivin-T respectively (both – CS), by 51% and 56% for PubMed and NamedKeys (BM), and by 34% and 78% for DUC-2001 and KPTimes (news). The only exception is the model fine-tuned on Krapivin-A. For Krapivin-T, its results are higher than the in-corpus scores. Thus, fine-tuning on abstracts demonstrated higher scores than the fine-tuning on texts of the papers for the same corpus. The lengths of abstracts and texts were limited to the first 256 tokens due to restrictions on the length of the input sequence and resource limits.

Figure 2 illustrates the effect of adding training examples from other corpora and domains in terms of F1. In our experiments, the effectiveness of the use of additional data varies depending on the characteristics of the corpus. For DUC-2001, which contains few training examples, the use of training examples from other corpora and domains increased the results for all strategies. In contrast, the highest result for KPTimes, which is the largest corpus in our experiments, is obtained using the only target training set. The use of the $Domain_{eq}$ and Mix_{eq} strategies led to a sharp decrease in the size of the training set and the number of targeted examples and negatively affected the model performance. In general, the Mix_{eq} strategy reduces the scores for all corpora except DUC-2001 due to a strong reduction in the amount of training data from the target corpus. Mix_{all} generally improves the performance or at least does not lead to a strong degradation of results[1]. This strategy showed the best results among all models for Krapivin-A (in terms of F1 and BS), Krapivin-T (BS), Inspec (R1, RL, and BS), PubMed (F1, R1, RL), and NamedKeys (RL, BS). Reducing the size of the dataset naturally leads to a decrease in training time. For instance, the training time is 53 min 59 s for Mix_{all} (21,885 training examples) and 3 min 59 s for Mix_{eq} (1,512 training examples). In this case, the training time decreases by about 20 times using the NVIDIA Tesla T4 GPU.

---

[1] This model is available at: https://huggingface.co/beogradjanka/bart_finetuned_keyphrase_extraction.

Table 5. BART results, %.

Target	Training data											
	F1	R1	RL	BS	F1	R1	RL	BS	F1	R1	RL	BS
	Krapivin-A				*Krapivin-T*				*Inspec*			
Krapivin-A	11.77	24.88*	21.46*	88.37	8.38	20.87	17.98	86.86	7.44	21.61	16.94	85.04
Krapivin-T	8.25	18.05	16.05	87.39	7.17	17.25	15.65	87.02	5.05	15.51	13.31	84.30
Inspec	9.85	24.01	19.37	86.63	6.66	19.27	15.97	85.76	20.18*	42.11	35.09	88.42
PubMed	7.97	15.67	13.67	85.74	4.44	11.25	10.12	83.76	6.61	16.98	14.01	83.62
NamedKeys	9.34	16.05	13.63	83.44	5.56	11.87	10.22	81.85	8.89	22.30	16.68	82.58
DUC-2001	6.01	12.84	11.85	86.00	3.19	8.40	7.80	84.34	7.81	17.56	15.48	85.27
KPTimes	3.84	7.17	6.63	83.64	2.46	5.35	4.99	82.30	6.67	11.71	10.25	83.52
	PubMed				*NamedKeys*				*DUC-2001*			
Krapivin-A	6.92	15.67	13.88	86.81	5.22	11.88	10.54	84.81	4.50	14.85	13.05	85.34
Krapivin-T	4.19	11.35	10.33	85.89	2.97	7.77	7.21	83.38	2.65	9.93	9.16	84.14
Inspec	6.08	16.33	13.92	85.46	4.48	11.04	9.51	82.97	4.69	16.22	14.07	83.77
PubMed	13.35	20.96	18.69	86.89*	10.62	17.92	15.43	84.55	6.00	16.14	14.26	85.17
NamedKeys	12.47	20.31	17.13	84.16	20.11	27.04	22.53	85.31	6.42	15.64	13.25	82.75
DUC-2001	5.23	11.64	10.81	85.79	6.27	12.65	11.32	84.61	11.78	24.14	20.65	87.45
KPTimes	6.38	9.89	9.06	84.46	8.94	12.24	10.97	84.22	2.80	8.47	7.78	83.44
	KPTimes				CS_{eq} (4,200)				CS_{all} (4,611)			
Krapivin-A	3.73	7.58	7.20	84.81	12.04	24.35	21.14	88.24	**12.08**	24.49	21.02	88.30
Krapivin-T	2.35	6.27	6.00	84.55	**8.32**	**18.41**	**16.38**	**87.42**	**8.36***	**18.54***	**16.50***	**87.48**
Inspec	3.79	7.80	7.16	83.07	20.04	42.00	34.86	**88.47**	20.01	42.09	34.88	**88.47**
PubMed	8.22	10.77	10.02	85.02	8.11	16.66	14.45	85.57	8.20	16.84	14.55	85.68
NamedKeys	6.31	7.75	7.16	81.76	7.99	16.60	13.75	83.06	7.82	16.52	13.62	83.01
DUC-2001	6.86	14.64	13.15	86.12	4.54	11.36	10.58	85.76	5.16	12.09	11.30	85.87
KPTimes	30.97*	33.98*	28.92*	88.12*	7.11	10.81	9.79	84.51	7.43	10.98	10.05	84.58
	BM_{eq} (1,848)				BM_{all} (3,058)				$News_{eq}$ (432)			
Krapivin-A	6.63	14.89	13.13	86.50	6.30	14.52	12.70	86.11	5.43	13.61	12.05	85.77
Krapivin-T	4.63	11.41	10.52	85.66	4.41	11.20	10.30	85.47	3.34	9.69	8.96	84.89
Inspec	6.11	15.26	13.10	85.10	5.78	14.64	12.36	84.64	5.91	13.74	12.09	83.70
PubMed	13.29	**21.19**	18.48	86.33	13.24	**21.39**	18.46	85.88	7.39	15.80	14.34	85.78
NamedKeys	18.44	25.38	21.21	85.10	**20.69**	**27.77**	**23.13**	**85.42**	7.60	15.03	12.83	83.18
DUC-2001	6.37	13.88	12.48	85.80	6.48	13.53	12.32	85.53	**13.76**	**24.64**	**21.15**	**87.83**
KPTimes	9.43	12.50	11.26	84.83	8.85	12.47	11.21	84.60	4.76	9.34	8.60	84.56
	$News_{all}$ (14,216)				Mix_{eq} (1,512)				Mix_{all} (21,885)			
Krapivin-A	4.68	10.38	9.50	85.61	9.57	21.67	18.61	87.75	**12.52***	24.82	21.41	**88.41***
Krapivin-T	3.41	9.61	8.99	85.58	6.10	15.91	14.27	86.68	**8.24**	**18.09**	**16.19**	**87.50***
Inspec	6.95	15.77	13.81	85.00	13.47	33.34	26.65	87.24	20.00	**42.25***	**35.10***	**88.51***
PubMed	9.71	13.77	12.46	85.60	13.19	20.45	17.85	86.77	**13.71***	**21.89***	**18.94***	86.25
NamedKeys	8.37	11.75	10.43	82.72	13.40	20.36	17.20	84.33	20.79	27.93	**23.26***	**85.53***
DUC-2001	**12.76**	23.62	20.45	**87.93**	**13.88**	**24.15**	**21.29**	**87.96**	13.31	25.63	22.60	88.02
KPTimes	30.53	33.78	28.79	88.10	5.56	9.96	9.18	84.80	30.22	33.49	28.54	88.07

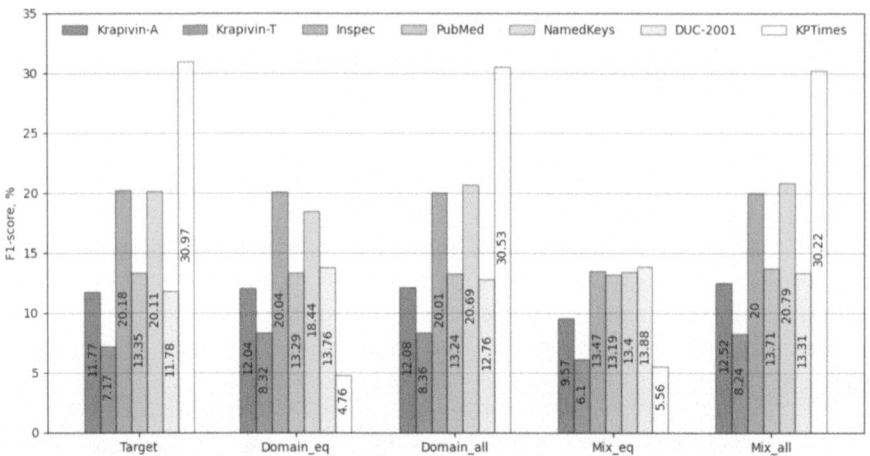

Fig. 2. Adding training examples from other corpora and domains.

To answer **RQ3**, we fine-tuned BART on a smaller number of training examples and evaluated the performance by increasing the size of the training data. Similarly to [33], we used the following few-shot transfer procedure. We randomly sampled 50 texts from a target training set, fine-tuned the pre-trained model on this subset, and then tested it on a target test set. Next, we increased the sample size by 50 texts of the target training set and repeated the described procedure, doing so up to 1,000 texts or the end of the training set. We compared the results with the scores obtained using the full target corpus and the out-of-domain corpora mixed in equal proportions. For instance, for Krapivin-A, the mix of out-of-domain corpora includes PubMed, NamedKeys, DUC-2001, and KPTimes. To answer **RQ4**, we first fine-tuned BART on a mixture of out-of-domain corpora, and then fine-tuned the same model on the texts from the target corpus using the above strategy. We evaluated two options for two-stage fine-tuning. In the first case, we fine-tuned the model on out-of-domain data during half of the epochs (three epochs out of six) and then continued fine-tuning on the target data during the remaining three epochs. In the second case, we doubled the number of epochs and fine-tuned the model within six epochs on both out-of-domain and target data.

The results in terms of F1 are presented in Fig. 3. The figure uses the following conventions. *Best baseline* – the best baseline result for the dataset. *Full target (6 ep)* – fine-tuning on the full target corpus. *Not target_eq (6 ep)* – fine-tuning on out-of-domain data. *Target (6 ep)* – fine-tuning on a part of the target corpus. *Not target_eq (3 ep) → Target (3 ep)* – fine-tuning on out-of-domain data for three epochs, then fine-tuning on a part of a target corpus for three epochs. *Not target_eq (6 ep) → Target (6 ep)* – fine-tuning on mixed out-of-domain data for six epochs, then fine-tuning on a part of the target corpus for six epochs. For PubMed, the best baseline result coincides with the line of Full target (6 ep).

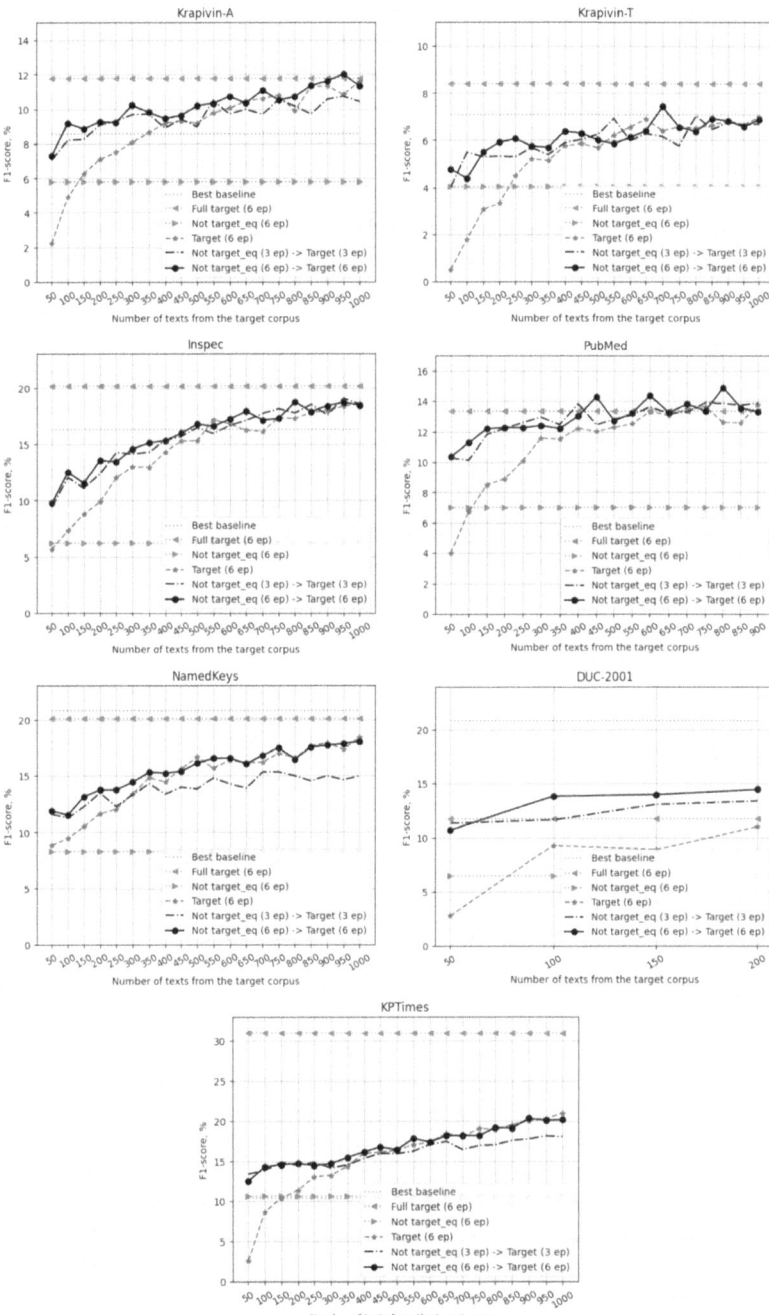

Fig. 3. Performance of BART with and without preliminary fine-tuning on out-of-domain corpora using a varying size of training data.

The models with two-stage fine-tuning outperform the ones fine-tuned only on a target corpus on a small target sample size ($\approx$ up to 300 texts). Therefore, the use of out-of-domain corpora allows the use of fewer target data. For some corpora (Krapivin-A, PubMed, and DUC-2001), the models fine-tuned in a two-stage manner outperformed the ones fine-tuned on full target corpora. For Krapivin-A, the F1 outperforming the full target score was obtained using 59% of the target training set. For PubMed and DUC-2001, it took us 43% and 46% respectively. For other corpora with a large full target size, we did not observe the results exceeding the F1 on a full target corpus during this experiment.

7   Conclusion

We explored the robustness of the abstractive text summarization models fine-tuned for the task of keyphrase generation. Our experiments are based on BART, a transformer-based denoising autoencoder for pre-training a seq2seq model. We studied the cross-domain limitations of the BART fine-tuned for keyphrase generation across six corpora from three different domains. We also investigated the impact of preliminary out-of-domain fine-tuning to improve the performance of the models under conditions of a small amount of training data.

We found that preliminary fine-tuning on out-of-domain data improves the performance of the keyphrase generation in few-shot settings and allows the use of fewer target data. Our findings add to a series of results concerning the effectiveness of a two-stage fine-tuning procedure where the transformer-based model is first fine-tuned on the source domain dataset before fine-tuning with the target domain dataset. For instance, similar studies conducted for text classification [34,47] and named entity recognition [33,40] have shown that the two-step training procedure can outperform the baseline models fine-tuned only on the target corpus. Our future research will focus on transfer learning from a high-resource language, for example, English, to other languages and to Russian in particular.

A   Appendix

Table 6. The values of standard deviation for the BART results.

Target corpus	F1	R1	RL	BS	F1	R1	RL	BS	F1	R1	RL	BS
	Krapivin-A				Krapivin-T				Inspec			
Krapivin-A	0.36	0.34	0.25	0.05	0.27	0.54	0.43	0.05	0.40	0.32	0.48	0.11
Krapivin-T	0.09	0.16	0.23	0.08	0.18	0.42	0.35	0.05	0.17	0.16	0.03	0.11
Inspec	0.14	0.26	0.18	0.13	0.25	0.41	0.24	0.07	0.21	0.29	0.35	0.07
PubMed	0.27	0.10	0.27	0.08	0.30	0.52	0.38	0.17	0.32	0.30	0.25	0.09
NamedKeys	0.16	0.75	0.55	0.11	0.40	0.80	0.57	0.21	0.13	0.26	0.12	0.07
DUC-2001	0.92	1.36	1.21	0.17	0.26	0.47	0.35	0.08	4.15	7.49	6.35	0.84
KPTimes	0.17	0.23	0.21	0.08	0.25	0.50	0.46	0.16	3.43	5.21	4.29	0.98

(continued)

Table 6. (*continued*)

Target corpus	F1	R1	RL	BS	F1	R1	RL	BS	F1	R1	RL	BS
	Krapivin-A				*Krapivin-T*				*Inspec*			
	PubMed				*NamedKeys*				*DUC-2001*			
Krapivin-A	0.38	0.48	0.43	0.18	0.12	0.05	0.26	0.03	0.23	0.22	0.33	0.08
Krapivin-T	0.11	0.17	0.18	0.20	0.27	0.29	0.27	0.20	0.06	0.15	0.15	0.06
Inspec	0.31	0.37	0.20	0.22	0.21	0.06	0.16	0.11	0.48	0.88	0.56	0.22
PubMed	0.25	0.43	0.51	0.06	0.39	0.57	0.54	0.22	0.77	0.91	0.64	0.16
NamedKeys	0.32	0.55	0.40	0.07	0.84	1.03	0.77	0.23	0.15	0.16	0.04	0.05
DUC-2001	0.24	0.80	0.75	0.41	1.55	1.44	0.84	0.11	0.52	0.99	0.48	0.16
KPTimes	0.10	0.13	0.10	0.02	0.25	0.28	0.28	0.14	0.14	0.05	0.07	0.12
	KPTimes				*CS_{eq}*				*CS_{all}*			
Krapivin-A	0.18	0.28	0.25	0.14	0.27	0.80	0.60	0.06	0.04	0.14	0.13	0.05
Krapivin-T	0.21	0.40	0.35	0.10	0.28	0.43	0.31	0.02	0.22	0.12	0.24	0.11
Inspec	0.28	0.30	0.26	0.16	0.60	0.46	0.46	0.08	0.44	0.41	0.45	0.02
PubMed	0.59	0.71	0.70	0.17	0.48	1.00	0.99	0.24	0.27	0.26	0.45	0.09
NamedKeys	0.12	0.28	0.26	0.09	0.23	0.36	0.28	0.09	0.12	0.54	0.39	0.10
DUC-2001	0.96	0.65	0.70	0.26	0.65	1.32	0.96	0.07	0.43	1.33	0.98	0.28
KPTimes	0.22	0.19	0.21	0.04	0.56	0.70	0.54	0.12	0.25	0.09	0.02	0.05
	BM_{eq}				*BM_{all}*				*News_{eq}*			
Krapivin-A	0.27	0.30	0.28	0.18	0.14	0.31	0.32	0.13	0.28	0.63	0.55	0.13
Krapivin-T	0.11	0.06	0.23	0.22	0.08	0.16	0.28	0.17	0.08	0.26	0.28	0.09
Inspec	0.09	0.69	0.53	0.14	0.25	0.49	0.46	0.23	0.43	1.24	0.93	0.46
PubMed	0.83	0.38	0.41	0.10	0.48	0.06	0.33	0.10	0.43	0.92	0.75	0.07
NamedKeys	0.52	0.59	0.37	0.13	0.14	0.01	0.22	0.05	0.22	0.45	0.39	0.07
DUC-2001	0.69	2.30	2.30	0.52	0.07	0.62	0.52	0.17	0.41	0.51	1.42	0.19
KPTimes	0.28	0.48	0.35	0.01	0.29	0.11	0.15	0.06	0.05	0.11	0.14	0.05
	News_{all}				*Mix_{eq}*				*Mix_{all}*			
Krapivin-A	0.56	0.90	0.72	0.25	0.50	0.51	0.68	0.09	0.50	0.52	0.23	0.07
Krapivin-T	0.12	0.58	0.47	0.19	0.36	0.34	0.28	0.21	0.02	0.57	0.42	0.10
Inspec	0.53	1.85	1.28	0.33	0.30	0.20	0.31	0.10	0.33	0.22	0.46	0.02
PubMed	0.15	0.22	0.22	0.10	0.84	0.20	0.11	0.02	0.62	0.35	0.42	0.07
NamedKeys	0.55	0.73	0.52	0.21	0.23	0.38	0.33	0.06	0.21	0.10	0.02	0.08
DUC-2001	0.74	0.27	0.42	0.19	0.54	0.31	0.20	0.08	0.24	0.73	1.09	0.09
KPTimes	0.12	0.13	0.17	0.01	0.16	0.13	0.11	0.01	0.28	0.28	0.17	0.02

References

1. Beliga, S.: Keyword extraction: a review of methods and approaches. University of Rijeka, Department of Informatics, Rijeka, vol. 1, no. 9 (2014)
2. Bird, S.: NLTK: the natural language toolkit. In: Proceedings of the COLING/ACL 2006 Interactive Presentation Sessions, pp. 69–72 (2006). https://doi.org/10.3115/1225403.1225421

3. Boudin, F.: PKE: an open source python-based keyphrase extraction toolkit. In: Proceedings of COLING 2016, the 26th International Conference on Computational Linguistics: System Demonstrations, pp. 69–73 (2016)

4. Bougouin, A., Boudin, F., Daille, B.: TopicRank: graph-based topic ranking for keyphrase extraction. In: International Joint Conference on Natural Language Processing (IJCNLP), pp. 543–551 (2013)

5. Bukhtiyarov, A., Gusev, I.: Advances of transformer-based models for news headline generation. In: Filchenkov, A., Kauttonen, J., Pivovarova, L. (eds.) AINL 2020. CCIS, vol. 1292, pp. 54–61. Springer, Cham (2020). https://doi.org/10.1007/978-3-030-59082-6_4

6. Cachola, I., Lo, K., Cohan, A., Weld, D.S.: TLDR: extreme summarization of scientific documents. In: Findings of the Association for Computational Linguistics: EMNLP 2020, pp. 4766–4777 (2020). https://doi.org/10.18653/v1/2020.findings-emnlp.428

7. Campos, R., Mangaravite, V., Pasquali, A., Jorge, A., Nunes, C., Jatowt, A.: YAKE! keyword extraction from single documents using multiple local features. Inf. Sci. **509**, 257–289 (2020). https://doi.org/10.1016/j.ins.2019.09.013

8. Çano, E., Bojar, O.: Keyphrase generation: a text summarization struggle. In: Proceedings of the 2019 Conference of the North American Chapter of the Association for Computational Linguistics: Human Language Technologies, Volume 1 (Long and Short Papers), pp. 666–672 (2019). https://doi.org/10.18653/v1/n19-1070

9. Çano, E., Bojar, O.: Two huge title and keyword generation corpora of research articles. In: Proceedings of the 12th Language Resources and Evaluation Conference, pp. 6663–6671 (2020)

10. Chan, H.P., Chen, W., Wang, L., King, I.: Neural keyphrase generation via reinforcement learning with adaptive rewards. In: Proceedings of the 57th Annual Meeting of the Association for Computational Linguistics, pp. 2163–2174 (2019). https://doi.org/10.18653/v1/p19-1208

11. Chen, W., Chan, H.P., Li, P., King, I.: Exclusive hierarchical decoding for deep keyphrase generation. In: Proceedings of the 58th Annual Meeting of the, Association for Computational Linguistics, pp. 1095–1105 (2020). https://doi.org/10.18653/v1/2020.acl-main.103

12. Chen, Y.S., Shuai, H.H.: Meta-transfer learning for low-resource abstractive summarization. In: Proceedings of the AAAI Conference on Artificial Intelligence, vol. 35, pp. 12692–12700 (2021). https://doi.org/10.1609/aaai.v35i14.17503

13. Chen, Y., Song, Q.: News text summarization method based on BART-TextRank model. In: 2021 IEEE 5th Advanced Information Technology, Electronic and Automation Control Conference (IAEAC), pp. 2005–2010. IEEE (2021). https://doi.org/10.1109/iaeac50856.2021.9390683

14. Chowdhury, M.F.M., Rossiello, G., Glass, M., Mihindukulasooriya, N., Gliozzo, A.: Applying a generic sequence-to-sequence model for simple and effective keyphrase generation. arXiv preprint arXiv:2201.05302 (2022). https://doi.org/10.48550/arXiv.2201.05302

15. Dung, C.V., et al.: Autonomous concrete crack detection using deep fully convolutional neural network. Autom. Constr. **99**, 52–58 (2019). https://doi.org/10.1016/j.autcon.2018.11.028

16. Gallina, Y., Boudin, F., Daille, B.: KPTimes: a large-scale dataset for keyphrase generation on news documents. In: Proceedings of the 12th International Conference on Natural Language Generation, pp. 130–135 (2019). https://doi.org/10.18653/v1/w19-8617

17. Gero, Z., Ho, J.C.: NamedKeys: unsupervised keyphrase extraction for biomedical documents. In: Proceedings of the 10th ACM International Conference on Bioinformatics, Computational Biology and Health Informatics, pp. 328–337 (2019). https://doi.org/10.1145/3307339.3342147

18. Glazkova, A., Morozov, D.: Multi-task fine-tuning for generating keyphrases in a scientific domain. In: 2023 IX International Conference on Information Technology and Nanotechnology (ITNT), pp. 1–5. IEEE (2023). https://doi.org/10.1109/ITNT57377.2023.10139061

19. Glazkova, A., Morozov, D.: Applying transformer-based text summarization for keyphrase generation. Lobachevskii J. Math. **44**(1), 123–136 (2023). https://doi.org/10.1134/S1995080223010134

20. Goloviznina, V., Kotelnikov, E.: Automatic summarization of Russian texts: comparison of extractive and abstractive methods. In: Computational Linguistics and Intellectual Technologies: Proceedings of the International Conference "Dialogue 2022", pp. 223–235 (2022). https://doi.org/10.28995/2075-7182-2022-21-223-235

21. Gupta, S., Gupta, S.K.: Abstractive summarization: an overview of the state of the art. Expert Syst. Appl. **121**, 49–65 (2019). https://doi.org/10.1016/j.eswa.2018.12.011

22. Hulth, A.: Improved automatic keyword extraction given more linguistic knowledge. In: Proceedings of the 2003 Conference on Empirical Methods in Natural Language Processing, pp. 216–223 (2003). https://doi.org/10.3115/1119355.1119383

23. Jiang, Y., Meng, R., Huang, Y., Lu, W., Liu, J.: Generating keyphrases for readers: a controllable keyphrase generation framework. J. Am. Soc. Inf. Sci. (2023). https://doi.org/10.1002/asi.24749

24. Kilgarriff, A.: Comparing corpora. Int. J. Corpus Linguist. **6** (2001). https://doi.org/10.1075/ijcl.6.1.05kil

25. Krapivin, M., Autaeu, A., Marchese, M.: Large dataset for keyphrases extraction (2009)

26. Kulkarni, M., Mahata, D., Arora, R., Bhowmik, R.: Learning rich representation of keyphrases from text. In: Findings of the Association for Computational Linguistics: NAACL 2022, pp. 891–906 (2022). https://doi.org/10.18653/v1/2022.findings-naacl.67

27. Lewis, M., et al.: BART: denoising sequence-to-sequence pre-training for natural language generation, translation, and comprehension. In: Proceedings of the 58th Annual Meeting of the Association for Computational Linguistics, pp. 7871–7880 (2020). https://doi.org/10.18653/v1/2020.acl-main.703

28. Lin, C.Y.: ROUGE: a package for automatic evaluation of summaries. In: Text Summarization Branches Out, pp. 74–81 (2004)

29. Liu, Y., et al.: RoBERTa: a robustly optimized BERT pretraining approach. arXiv preprint arXiv:1907.11692 (2019). https://doi.org/10.48550/arXiv.1907.11692

30. Loshchilov, I., Hutter, F.: Decoupled weight decay regularization. In: International Conference on Learning Representations (2018)

31. Malykh, V., Porplenko, D., Tutubalina, E.: Generating sport summaries: a case study for Russian. In: van der Aalst, W.M.P., et al. (eds.) AIST 2020. LNCS, vol. 12602, pp. 149–161. Springer, Cham (2021). https://doi.org/10.1007/978-3-030-72610-2_11

32. Meng, R., Zhao, S., Han, S., He, D., Brusilovsky, P., Chi, Y.: Deep keyphrase generation. In: ACL 2017-55th Annual Meeting of the Association for Computational Linguistics, Proceedings of the Conference (Long Papers), pp. 582–592 (2017). https://doi.org/10.18653/v1/P17-1054

33. Miftahutdinov, Z., Alimova, I., Tutubalina, E.: On biomedical named entity recognition: experiments in interlingual transfer for clinical and social media texts. In: Jose, J.M., et al. (eds.) ECIR 2020. LNCS, vol. 12036, pp. 281–288. Springer, Cham (2020). https://doi.org/10.1007/978-3-030-45442-5_35

34. Rietzler, A., Stabinger, S., Opitz, P., Engl, S.: Adapt or get left behind: domain adaptation through BERT language model finetuning for aspect-target sentiment classification. In: Proceedings of the 12th Language Resources and Evaluation Conference, pp. 4933–4941 (2020)

35. Rubio, A., Martínez, P.: HULAT-UC3M at SimpleText@ CLEF-2022: scientific text simplification using BART. In: Proceedings of the Working Notes of CLEF (2022)

36. Schutz, A.T.: Keyphrase extraction from single documents in the open domain exploiting linguistic and statistical methods (2008)

37. Shen, L., Le, X.: An enhanced method on transformer-based model for one2seq keyphrase generation. Electronics 12(13), 2968 (2023). https://doi.org/10.3390/electronics12132968

38. Song, M., Feng, Y., Jing, L.: A survey on recent advances in keyphrase extraction from pre-trained language models. In: Findings of the Association for Computational Linguistics: EACL 2023, pp. 2108–2119 (2023). https://doi.org/10.18653/v1/2023.findings-eacl.161

39. Swaminathan, A., Zhang, H., Mahata, D., Gosangi, R., Shah, R., Stent, A.: A preliminary exploration of GANs for keyphrase generation. In: Proceedings of the 2020 Conference on Empirical Methods in Natural Language Processing (EMNLP), pp. 8021–8030 (2020). https://doi.org/10.18653/v1/2020.emnlp-main.645

40. Syed, M.H., Chung, S.T.: MenuNER: domain-adapted BERT based NER approach for a domain with limited dataset and its application to food menu domain. Appl. Sci. 11(13), 6007 (2021). https://doi.org/10.3390/app11136007

41. Tank, M., Thakkar, P.: Text summarization approaches under transfer learning and domain adaptation settings-a survey. In: Buyya, R., Hernandez, S.M., Kovvur, R.M.R., Sarma, T.H. (eds.) ICCIDA 2022, pp. 73–88. Springer, Cham (2022). https://doi.org/10.1007/978-981-19-3391-2_5

42. Vaca, A., Segurado, A., Betancur, D., Jiménez, Á.B.: Extractive and abstractive summarization methods for financial narrative summarization in English, Spanish and Greek. In: Proceedings of the 4th Financial Narrative Processing Workshop@ LREC2022, pp. 59–64 (2022)

43. Wan, X., Xiao, J.: Single document keyphrase extraction using neighborhood knowledge. In: AAAI, vol. 8, pp. 855–860 (2008)

44. Wang, S., Jiang, J., Huang, Y., Wang, Y.: Automatic keyphrase generation by incorporating dual copy mechanisms in sequence-to-sequence learning. In: Proceedings of the 29th International Conference on Computational Linguistics, pp. 2328–2338 (2022)

45. Wright, D., et al.: Generating scientific claims for zero-shot scientific fact checking. In: Proceedings of the 60th Annual Meeting of the Association for Computational Linguistics (Volume 1: Long Papers), pp. 2448–2460 (2022). https://doi.org/10.18653/v1/2022.acl-long.175

46. Wu, D., Ahmad, W.U., Chang, K.W.: Pre-trained language models for keyphrase generation: a thorough empirical study. arXiv preprint arXiv:2212.10233 (2022). https://doi.org/10.48550/arXiv.2212.10233

47. Yadav, A., Milde, B.: forumBERT: topic adaptation and classification of contextualized forum comments in German. In: Proceedings of the 17th Conference on Natural Language Processing (KONVENS 2021), pp. 193–202 (2021)

48. Ye, H., Wang, L.: Semi-supervised learning for neural keyphrase generation. In: Proceedings of the 2018 Conference on Empirical Methods in Natural Language Processing, pp. 4142–4153 (2018). https://doi.org/10.18653/v1/D18-1447
49. Zhang, J., Zhao, Y., Saleh, M., Liu, P.J.: PEGASUS: pre-training with extracted gap-sentences for abstractive summarization. In: Proceedings of the 37th International Conference on Machine Learning, pp. 11328–11339 (2020)
50. Zhang, T., Kishore, V., Wu, F., Weinberger, K.Q., Artzi, Y.: BERTScore: evaluating text generation with BERT. In: International Conference on Learning Representations
51. Zmandar, N., El-Haj, M., Rayson, P.: A comparative study of evaluation metrics for long-document financial narrative summarization with transformers. In: Métais, E., Meziane, F., Sugumaran, V., Manning, W., Reiff-Marganiec, S. (eds.) NLDB 2023. LNCS, vol. 13913, pp. 391–403. Springer, Cham (2023). https://doi.org/10.1007/978-3-031-35320-8_28
52. Zolotareva, E., Tashu, T.M., Horváth, T.: Abstractive text summarization using transfer learning. In: ITAT, pp. 75–80 (2020)

MOROCCO: Model Resource Comparison Framework

Valentin Malykh[1](✉), Alexander Kukushkin[2], Maria Tikhonova[3], and Tatiana Shavrina[3]

[1] MISiS University, Moscow, Russia
`valentin.malykh@phystech.edu`
[2] Alexander Kukushkin Data Science Laboratory, Moscow, Russia
`alex@alexkuk.ru`
[3] HSE University, Moscow, Russia

Abstract. A new generation of pre-trained transformer language models has established new state-of-the-art results on many tasks, even exceeding the human level in standard NLU benchmarks. Despite the rapid progress, the benchmark-based evaluation has generally relied on the downstream task performance as a primary metric, limiting the scope of model comparison in their practical use, which is also limited by the resources required by the models to run. This paper presents **MO**del **R**es**O**ur**C**e **CO**mparison (MOROCCO), a publicly available framework (https://github.com/RussianNLP/MOROCCO/) that allows assessing models concerning their downstream quality, combined with two computational efficiency metrics such as memory consumption and throughput during the inference stage. The framework allows flexible integration with popular leaderboards compatible with `jiant` environment, e.g. SuperGLUE. We demonstrate the MOROCCO applicability by evaluating ten transformer models on two multi-task GLUE-style benchmarks in English and Russian and provide the model analysis.

1 Introduction

The field of NLP has been centered around the "pre-train & fine-tune" paradigm, which involves pre-training a language model (LM) on an extensive text corpus and its further fine-tuning for a downstream task in a supervised fashion. Many transformer LMs Vaswani et al. (2017) fall under this paradigm which has established new state-of-the-art results for the majority of NLP tasks such as text classification Sun et al. (2019), part-of-speech tagging Tsai et al. (2019), machine translation Zhu et al. (2019), and many others. The models have demonstrated various capabilities, ranging from cross-lingual zero-shot transfer Pires et al. (2019) to generating texts that are hard to distinguish from the human-written ones Zellers et al. (2020), and have even outperformed human solvers in standard NLU benchmarks He et al. (2021).

However, the rich diversity of LMs that differ in the number of parameters and the architecture design Liu et al. (2020) has been mainly assessed by

J. Baixeries et al. (Eds.): DAMDID/RCDL 2023, CCIS 2086, pp. 266–281, 2024.
https://doi.org/10.1007/978-3-031-67826-4_20

the downstream performance as a primary metric on many standard benchmarks such as GLUE Wang et al. (2018), XGLUE Liang et al. (2020), SuperGLUE Wang et al. (2019) and XTREME Hu et al. (2020). Although the benchmarks provide a standard for direct model comparison, the performance-oriented approach limits the scope of the evaluation methods Ethayarajh and Jurafsky (2020). Motivated by the need of expanding the methodology, various benchmarks and contests have been proposed targeting computational and technical aspects of the models (see Sect. 2), highlighting the problem of model scaling Rogers (2019).

In line with these works, we introduce **MO**del **R**es**O**ur**C**e **CO**mparison (MOROCCO), a publicly available framework for model evaluation in terms of their practical use. The contributions of this paper are framed as follows. First, we present a methodology to measure the downstream performance and computational efficiency of the models in a fixed environment. Second, we present a software framework that is adopted to `jiant` environment Pruksachatkun et al. (2020) that supports over 50 downstream tasks[1], including GLUE-style ones. We demonstrate the MOROCCO applicability by evaluating ten transformer models on two SuperGLUE benchmarks for English and Russian and provide the model analysis. This way of model evaluation provides the researcher with the opportunity of the model comparison from different perspectives, specifically those that meet the user's needs.

2   Related Work

NLP Benchmarks. The trend for model-agnostic evaluation has been recently set by canonical multi-task NLU benchmarks such as GLUE Wang et al. (2018) and SuperGLUE Wang et al. (2019). Such evaluation method does not consider any computational and technical aspects of the models that differ significantly by the number of parameters and architecture design choices, such as the number of transformer blocks, attention mechanism, pre-training objectives, etc. Besides, the benchmarks do not support the interaction with the user models, limiting the leaderboard results' reproducibility Rogers (2019); Ethayarajh and Jurafsky (2020).

Efficient NLP. The trade-off between model performance and computational efficiency has been explored in multiple shared tasks and competitions. The series of Efficient Neural Machine Translation challenges Birch et al. (2018); Hayashi et al. (2019); Heafield et al. (2020) jointly measured the model downstream performance on the task of machine translation and computational efficiency parameters, ranging from memory consumption to size of a Docker image. The organizers selected the Pareto-optimal solutions Aleskerov et al. (2007), i.e. those that require less computational resources when delivering a prominent downstream performance.

---

[1] https://github.com/nyu-mll/jiant/blob/master/guides/tasks/supported_tasks.md.

The EfficientQA competition Min et al. (2021) aims at creating effective NLP systems for open-domain question answering (ODQA). The submissions are limited by many performance and technical requirements, which stimulate the community to develop optimal ODQA systems that can achieve superior performance while satisfying the technical needs and operating on an optimal amount of retrieval corpora.

The SustaiNLP challenge Wang and Wolf (2020) targets developing efficient yet accurate models. The efficiency is estimated as the power consumed throughout the inference time calculated utilizing experiment impact tracker Henderson et al. (2020). The submitted systems improve total energy consumption over the BERT-base as much as 20×, but the results are around two points lower on average. Although using the same testbed, the MOROCCO framework was developed with the opposite goal in mind. It provides adequate estimates of how many resources consume the models that reach human-level performance. As MOROCCO supports Docker images, it can be easily integrated into any benchmark or probing task, built upon `jiant` framework.

Dynaboard Ma et al. (2021) is a cloud-based platform on which a submitted model is evaluated according to five different criteria, including task performance, throughput, memory consumption, fairness, and robustness scores. The aggregating Dynascore is designed according to multi-criteria optimization theory to reflect user preferences. Supported tasks include several NLI, QA, sentiment classification, and hate speech detection. The authors use a few pre-existed datasets and several newly collected ones. This forbids their results to be a direct extension to the SuperGLUE-like leaderboards. Another important point is that their methodology is somewhat questionable in fairness and robustness, since it is subjective to the augmentations they use.

3   Evaluation Framework

MOROCCO can be used to rank the benchmark leaderboard models by computational metrics (see Sect. 3.1). To demonstrate that MOROCCO is compatible with GLUE-style benchmarks, we perform experiments using SuperGLUE tasks for English and Russian (see Sect. 3.2) over popular transformer-based models (see Sect. 3.3), which are publicly released as a part of the HuggingFace library Wolf et al. (2019).

Submission Details. To evaluate a model's performance in MOROCCO on the [Russian]SuperGLUE tasks, a person should prepare their submission as a Docker container and send it to the testbed. The testbed platform runs the submitted Docker container with limited memory, CPU/GPU, and running time. The container is expected to read the texts from the standard input channel and output the answers to the standard output. During the inference, the running time is recorded for the submission scoring. We perform several runs and compute the median values to eliminate the running time and memory footprint dispersion caused by technical reasons. Next, the output from the container is evaluated with task-specific metrics. The results compute the final evaluation score for the whole submission. To ensure the comparability of the collected metrics, we fix the computation hardware. We use Yandex.Cloud[2] virtual instances, where the following hardware is guaranteed: 1× Intel Broadwell CPU, 1× NVIDIA Tesla V100 GPU. The Docker containers are equipped with Ubuntu 20.04. Following the SuperGLUE infrastructure, our framework is designed to comprise with `jiant` framework, alongside simple requirements for the evaluation containers built upon other frameworks. It also can be run locally using the released codebase. The details on the running process are provided in Appendix B.

3.1   Metrics

We report the computational efficiency of the tested model utilizing the memory footprint and inference speed.

Memory footprint implicitly allows accounting for the model's size and the number of weights, as there is strong dependency. To measure model GPU RAM usage M, we run a container with a single record as input, calculate the maximum GPU RAM consumption, repeat the procedure five times, and compute a median value.

Inference speed measures directly how much time the model consumes on specific hardware, implicitly estimating its complexity. To measure the inference speed T_p we first compute T_N, which we run a container with N records as input with batch size 32^3 for. We also estimate initialization time T_{init} with running a container with an input of size 1. Inference speed Tp is computed as follows: $Tp = \frac{N}{T_N - T_{\text{init}}}$. In our experiments, we use $N = 2000$, which the user can adjust. We repeat the procedure five times to compute a median value.

Quality Q is an aggregated value computed from task-specific metric values by averaging. The averaging itself is somewhat questionable aggregation method due to quality measure semantic difference as shown in Shavrina and Malykh (2021), nevertheless we strictly follow the methodology presented in Super-GLUE Wang et al. (2019) and RussianSuperGLUE Shavrina et al. (2020) to ease the integration of our framework with the mentioned ones.

Fitness. Overall, our evaluation procedure utilizes three different scores, namely the aggregated performance score Q, the inference speed Tp, and the memory footprint M. We propose to take into account these three characteristics of a model and make an integral measure of its "**fitness**" F that combines quality and computational metrics. There is a plethora of ways to calculate fitness, but we can propose several assumptions based on common sense. Namely, we suppose that memory consumption should be lowered, while the achieved quality and processing speed should be increased. To achieve this we propose a following formula, which follows an idea from van Rijsbergen (1979):

---

[2] https://cloud.yandex.com/.
[3] The batch size of 32 is chosen empirically and utilizes the GPU almost at 100% on the experiment tasks. Note that it can be adjusted to meet the user's needs.

$$F = f_Q(Q) \times \frac{f_{Tp}(Tp)}{f_M(M)}, \tag{1}$$

where Q is the aggregated metric-based score for the specific tasks, M is measured in bytes, Tp is measured in records per second (RPS). The f_* stand for some function which is applied to specified parameter.

Basing on our experiments, we propose to set f_Q and f_{Tp} to identity function I, while f_M should be set to logarithmic function. We ground that on the exponential model size increase as shown in Sanh et al. (2019). We result with the following formula used in the next experiments:

$$F = Q \times \frac{Tp}{\log(M)}. \tag{2}$$

The main idea of the proposed metric is to provide a practitioner with simple to use single metric or "rule-of-thumb" in their search for the best model in mythic general case. We understand that there is no general case and also provide all the underlying evaluation details to make their choice more informed. We provide details on the other possible formulations and their limitations in Sect. 5.

3.2 Tasks

We use all 9 tasks from both SuperGLUE and RussianSuperGLUE. For the tasks' description please refer to Appendix A.

3.3 Models

We run the experiments on the following publicly available models that achieved competitive performance on SuperGLUE and Russian SuperGLUE benchmarks. **Models for English** include monolingual (en_bert_base) and multilingual base BERT (bert-multilingual) Devlin et al. (2019), RoBERTa-base Liu et al. (2019) (en_roberta_base), ALBERT-base Lan et al. (2019) (albert), and GPT-2-large Radford et al. (2019) (en_gpt2). **Models for Russian** involve multilingual BERT-base (bert-multilingual), 3 variants of ruGPT-3[4] (rugpt3-small, rugpt3-medium, and rugpt3-large), RuBERT-base (rubert) Kuratov and Arkhipov (2019), and Conversational RuBERT-base[5] (rubert-conversational) trained on social media data.

4   Results

Figure 1 demonstrates the results for Russian SuperGLUE (top) and SuperGLUE for English (bottom) based on the received Q, Tp, and M (see Sect. 3.1). These figures discover the models' relative positions for both languages. For English language GPT-2 seems to be the least optimal model, while RoBERTa is the most

---

[4] https://github.com/sberbank-ai/ru-gpts.
[5] https://huggingface.co/DeepPavlov/rubert-base-cased-conversational.

optimal one. One could apply Pareto rule to monolingual BERT and ALBERT models and find them roughly equivalent. For Russian, the most optimal model is RuBERT and RuGPT3-large is the least optimal one. Pareto rule gives an equivalence of multilingual BERT and RuGPT3-medium.

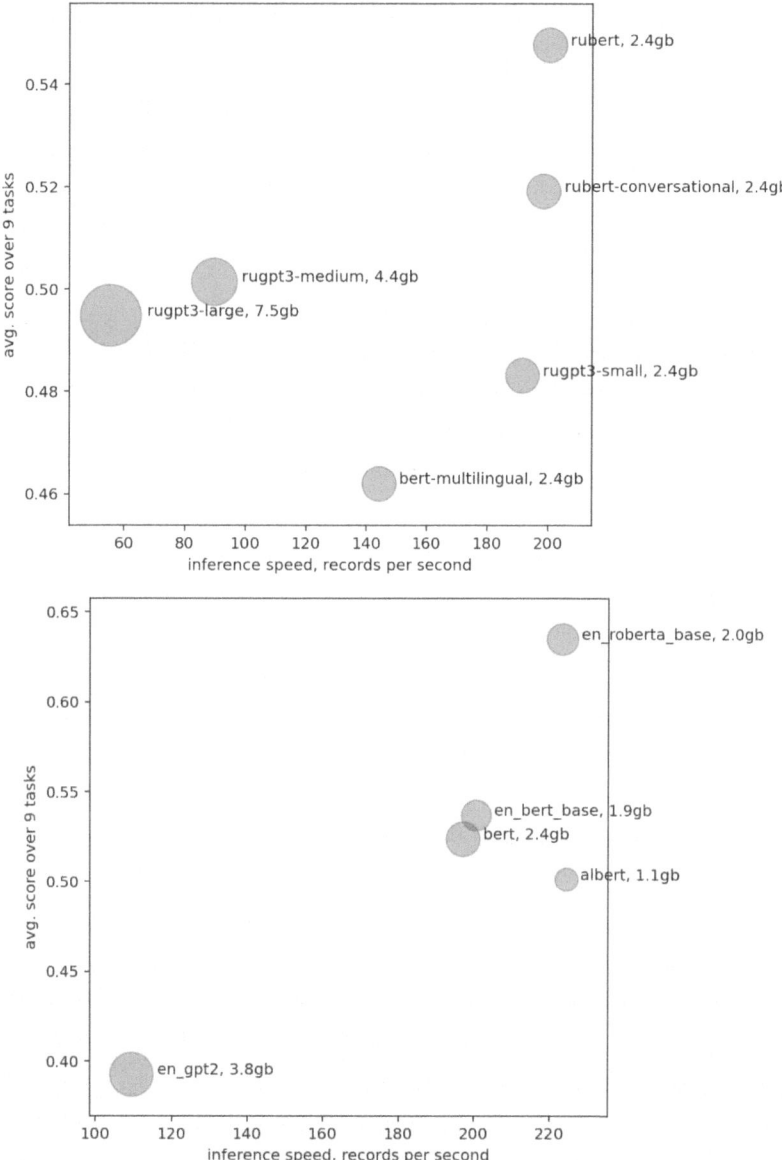

Fig. 1. Model evaluation on RussianSuperGLUE (top) and SuperGLUE (bottom). **X-axis** = Inference speed Tp (RPS). **Y-axis** = Task-specific performance Q. The memory footprint M is represented by the size of the circle.

The fitness metric F results are presented in Table 1. RoBERTa model had shown the best score for English, while RuBERT is the best fit among the tested models for Russian. The multilingual BERT model showed significantly different results on the two languages. We hypothesize that it attributes to the difference in the datasets in SuperGLUE and RussianSuperGLUE and the model's training data askew towards the English language. Overall, the evaluation results have revealed better models using task-specific quality, memory footprint, and inference speed.

Table 1. Fitness evaluation for the models in English and Russian. The models ordered to ease comparison between the languages.

English		Russian	
en_bert_base	5.05	rubert	**4.84**
bert-multilingual	4.79	bert-multilingual	3.30
en_roberta_base	**6.63**	rubert-conversational	4.59
albert	5.41	rugpt3-small	3.89
en_gpt2	1.95	rugpt3-medium	1.89
		rugpt3-large	1.24

5 Discussion

5.1 Detailed Metric Examination

Averaging the estimates of Q, Tp, and M is one of the main limitations of the proposed evaluation procedure. Averaging memory consumption M is less problematic, as it is relatively stable for any reasonable sample size. However, two other metrics require a more detailed investigation. Figure 2 compares the mean and maximum values of Q for different models. Each model was trained five times with different random seeds and was scored ten times, making fifty runs overall. The only exception was made to the largest model, rugpt3-large, fine-tuned only one time. Blue dots present evaluation for a single run, pale red dots show mean results for all runs, and full red dots show the maximum results. The ranking, achieved by maximum and mean scores, is identical.

Figure 3 compares averaged normalized inference speed for different task sets, adopted from RussianSuperGLUE. The normalization is done alongside the X-axis, thus, one can compare the models' ranking for different task sets. The model order remains mostly unchanged, while occasionally, top models exchange positions.

We conclude that our evaluation procedure is stable. Averaging the estimates of Q, Tp, and M does not introduce issues to the evaluation procedure and makes model comparison informative.

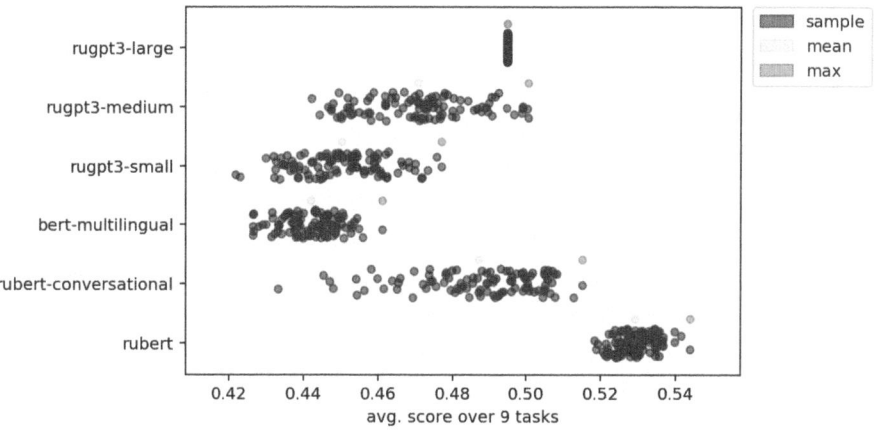

Fig. 2. Mean, maximum and averaged task-specific scores for the Russian SuperGLUE tasks.

5.2   Limitations

Our methodology inevitably has its limitations. First of all, it is fundamentally dependable on the length of the input sequences, once current Transformer language models have $O(n^2)$ in memory and computational time complexity. In the presented case, we use standardized datasets from SuperGLUE (for English) and RussianSuperGLUE. Although the datasets are fixed in the benchmarks they have different parameters. Table 2 shows average length in tokens for several datasets from RussianSuperGLUE. As one can see, the lowest value could be twice as small as the largest one. It is interesting to compare this statistics to Fig. 3. The seven lowest rows in it are single datasets listed in the mentioned table. We can see that the different models are perform instably on the datasets. We hypothesise that it is due to different lengths. Although once we compare the models on the combinations of the datasets (from two at once to all of them at once) the results became more and more stable.

Table 2. Statistics for selected datasets.

Dataset	RUSSE	TERRa	RCB	RWSD	MuSeRC	RuCoS	DaNetQA
Tokens per sample	12.11	18.46	13.63	14.93	19.76	20.55	21.02

Another important point is that we measure the highest memory consumption during the run. This decision allows us to compare the models in the worst conditions, thus providing the information important to run on a particular hardware piece with a limited memory. Although, the flaw of this method is that the average memory consumption could be lower and the evaluated values will be overestimated.

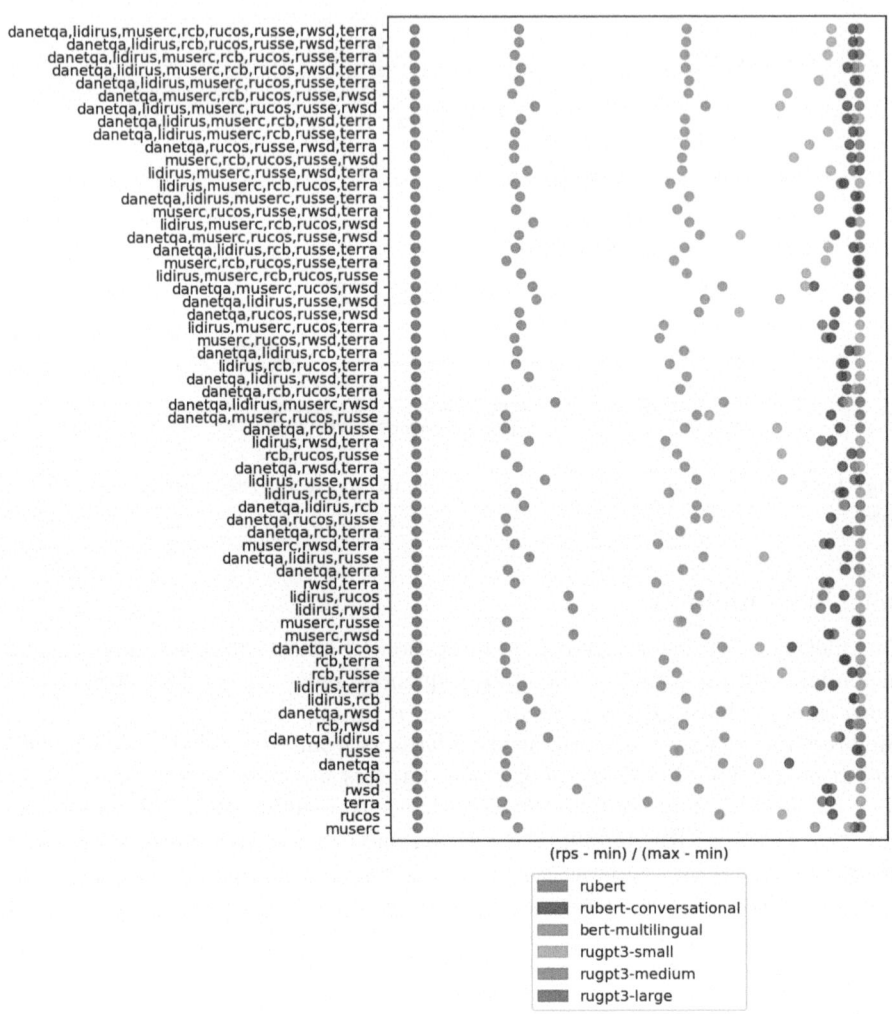

Fig. 3. Averaged inference speed for different combinations of the Russian SuperGLUE tasks.

5.3 Fitness Formulation

As stated in Eq. 1, there are three functions, namely f_Q, f_{Tp}, and f_M, which could affect the calculation. There are infinite possibilities for these functions, so we need to somehow limit our search with most standard and easy computable ones. We run a series of experiments with power function for all three, and additionally logarithm for f_M. We fix two of three functions to be identity ones, while changing the third one.

Since Q is measured in range $[0, 1]$ it is natural somehow increase its contribution to final fitness value. But we put an exponent in range of $[-1, 1]$ to evaluate the fitness behaviour with different Q contributions. The resulting plots for RussianSuperGLUE models are presented at Fig. 4. As one can see on the figure, almost all the models are keeping their relative positions, except for RuGPT-3. This model takes second place in range $[-0.6, 0.5]$ and even the first one at $[-1, -0.6]$.

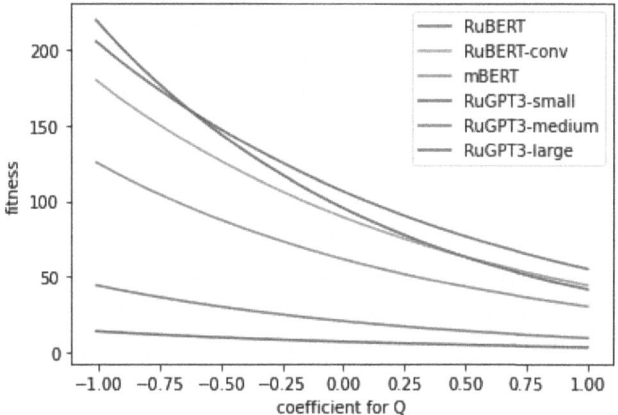

Fig. 4. f_Q function evaluation with different exponents.

Tp measures in natural numbers for the specific tasks, and cannot be less than 1. While aggregated it could be a real number, but still bigger than 1. For these numbers the ordering keeping the same for all the positive real exponents. The resulting plots for RussianSuperGLUE models are presented at Fig. 5. We keep only range $[0.75, 2.0]$ as the most informative.

As it stated in Sect. 3.1, M is measured in bytes, which makes it effectively much larger than other two metrics. To handle this fact we apply normalization, dividing M by 2^{30}, thus M is represented in Gebibytes for our power function evaluation. The resulting plots for RussianSuperGLUE models are presented at Fig. 6. We also provide results for $f_M(\cdot) = \log(\cdot)$ as in Eq. 2. We scale the achieved results by factor of 15 to match the other presented values. As one can see in all the variants the existing ordering keeps the same.

Given these experiments we could conclude that the only important factor for evaluation is f_Q. We preferred put it to identity due to on the one hand computation efficiency and on the other hand the quality measuring conventionality, as discussed in Sect. 3.1.

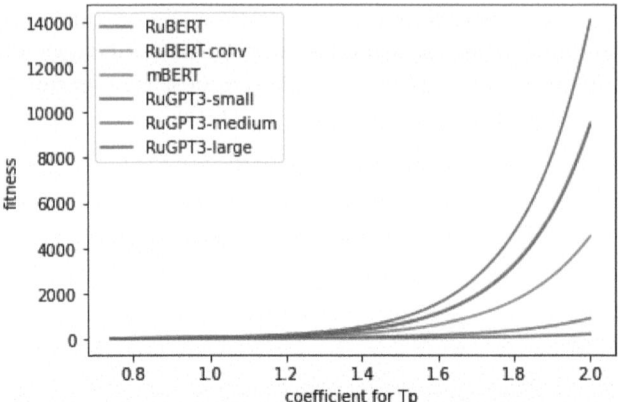

Fig. 5. f_{Tp} function evaluation with different exponents.

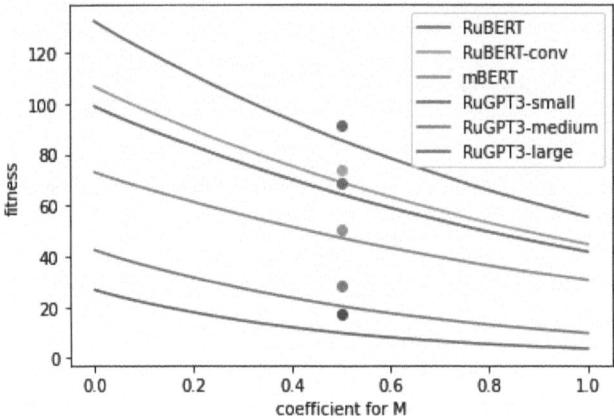

Fig. 6. f_M function evaluation with different exponents. The dots represent scaled values for logarithm function evaluation.

6   Conclusion

This work introduces the MOROCCO framework, which provides an assessment of language models' downstream quality combined with two computational efficiency metrics such as memory consumption and throughput during the inference stage. The proposed fitness metric allows to compose the GLUE-style leaderboards in a new way: to rank them so that the more high-precision, smaller and faster models are at the top, the accurate ones, but bigger and slower models are in the middle, and the most imprecise, largest and slowest ones are at the very bottom. Thus, to obtain a higher place on the leaderboard, researchers need to strive not for the score on the individual tasks, but also, develop optimal models in terms of their practical use. A similar conditional assessment of the results has been mainly adopted for image classification and QA tasks. We expand this idea by integrating MOROCCO with the canonical SuperGLUE leaderboards showing the applicability for two languages. The presented framework is also compatible with the `jiant` framework and transformer models, making it easily applicable to evaluate a wide range of popular architectures, both multilingual and monolingual. We hope that our framework can be utilized in other `jiant`-based projects to provide a better and more detailed evaluation. This paper aims to stimulate the research on a compromise evaluation of the overall performance of NLP-models, which could be an alternative to the existing dominant "bigger is better" trend and would consider the problems of overfitting, over-parametrization, data redundancy, and many others.

A fruitful direction for future work is to cooperate with NLP developers and enthusiasts to further search for the most optimal solutions, including organizing the competition of multilingual NLP models on existing benchmarks as a possible step. Another line of work includes extending the framework with other metrics such as time and memory use required for fine-tuning, the time needed to achieve the best quality, and robustness towards task-specific adversarial attacks.

A   Task Description

The experiments are run on a diverse set of 9 tasks[6] from the SuperGLUE benchmarks for each language (see Table 3): **Recognizing Textual Entailment** (RTE) task is aimed to capture textual entailment in a binary classification form; **Commitment Bank** belongs to the natural language inference (NLI) group of tasks type with a 3-way classification; **Diagnostic dataset** which is another test set for the RTE task annotated with various linguistic and semantic phenomena; **Words in Context** task is based on word sense disambiguation problem in a binary classification form; **Choice of Plausible Alternatives** is a binary classification task aimed at accessing commonsense causal reasoning; **Yes/No Questions** is a binary QA task for closed questions; **Multi-Sentence Reading Comprehension** is a task on multi-hop machine reading comprehension (MRC); **Reading Comprehension with Commonsense Reasoning** is an MRC task, where it is required to fill the masked gaps in the sentence with the best fitting entities from the given text paragraph; **Winograd Schema Challenge** is devoted to co-reference resolution in a binary classification form.

---

[6] SuperGLUE benchmark also includes an additional Winogender Schema Diagnostics task which is a dataset which we do not consider in the experiments since it is not included in Russian SuperGLUE.

Table 3. Datasets statistics. MCC stands for Matthews' Correlation Coefficient; Acc - Accuracy; EM - Exact Match. The size train/validation/test splits are provided in "Samples" columns.

Task Type	Task	SuperGLUE		Russian SuperGLUE		Metric
		Name	Samples	Name	Samples	
NLI	Recognizing Textual Entailment	RTE	2490/277/3000	TERRa	2616/307/3198	Acc
	Commitment Bank	CB	250/56/250	RCB	438/220/438	Avg. F1/Acc
NLI & diagnostics	Diagnostic	AX-b	0/0/1104	LiDiRus	0/0/1104	MCC
Common Sense	Words in Context	WiC	5428/638/1400	RUSSE	19845/8508/18892	Acc
	Choice of Plausible Alternatives	COPA	400/100/500	PARus	400/100/500	Acc
World Knowledge	Yes/No Questions	BoolQ	9427/3270/3245	DaNetQA	1749/821/805	Acc
Machine Reading	Multi-Sentence Reading Comprehension	MultiRC	456/83/166	MuSeRC	500/100/322	F1/EM
	Reading Comprehension with Commonsense Reasoning	ReCoRD	65709/7481/7484	RuCoS	72193/7577/7257	F1/EM
Reasoning	The Winograd Schema Challenge	WSC	554/104/146	RWSD	606/204/154	Acc

B   Standalone Run

The user needs to clone the project repository to their machine to run the framework locally. MOROCCO works with the Docker container engine and provides the corresponding code. We consider the following procedure for the evaluation: fine-tune a model for a specific task, build a Docker container with the model, run the container on the test data to get the outputs, collect the outputs for multiple runs and conduct the evaluation. The downstream performance can be received by submitting to the corresponding leaderboard.

For instance, the fine-tuning (training) the RuBERT model for RUSSE could be done with this command:

```
python main.py train rubert russe \
~/path/for/logs ~/data/RUSSE
--seed=3
```

Note that this run uses the fixed random seed which can be adjusted.

To infer the trained model for the specific task, run the following code snippet:

```
python main.py infer \
~/path/for/logs/rubert/ russe \
--batch-size=32
```

To build the Docker container with the trained model, run the following code snippet:

```
python main.py docker build \
~/path/for/logs/rubert/ russe \
rubert-russe
```

To infer the container with the model, storing its outputs, run the following code snippet:

```
docker run --gpus all \
--interactive --rm rubert-russe \
--batch-size 8 \
<~/data/RUSSE/val.jsonl \
>preds.jsonl
```

To evaluate the model by the task-specific metrics, submit your model predictions to the leaderboard or run the following code snippet on the validation set (preliminarily making the predictions):

```
python main.py eval russe \
preds.jsonl \
~/data/RUSSE/val.jsonl
```

Finally, to get the memory footprint and inference speed results, run the following code snippet:

```
for index in 01 02 03 04 05;
  do python main.py docker \
  bench rubert-russe ~/data \
  russe --input-size=2000 \
  --batch-size=32 \
  >~/benches/rubert/\
  russe/2000_32_$index.jl;
done
```

Submission to RussianSuperGLUE. To make a submission for the leaderboard, you need to push a Docker container to a Docker repository, login to Russian-SuperGLUE website, and add a link to the Docker container you created into the submission form.

References

Aleskerov, F., Bouyssou, D., Monjardet, B.: Utility Maximization, Choice and Preference, vol. 16. Springer, Heidelberg (2007). https://doi.org/10.1007/978-3-540-34183-3

Birch, A., Finch, A., Luong, M.T., Neubig, G., Oda, Y.: Findings of the second workshop on neural machine translation and generation. In: Proceedings of the 2nd Workshop on Neural Machine Translation and Generation, pp. 1–10 (2018)

Devlin, J., Chang, M.-W., Lee, K., Toutanova, K.: Bert: pre-training of deep bidirectional transformers for language understanding, pp. 4171–4186 (2019)

Ethayarajh, K., Jurafsky, D.: Utility is in the eye of the user: a critique of NLP leaderboards. arXiv preprint arXiv:2009.13888 (2020)

Hayashi, H., et al.: Findings of the third workshop on neural generation and translation. In: Proceedings of the 3rd Workshop on Neural Generation and Translation, pp. 1–14 (2019)

He, P., Liu, X., Gao, J., Chen, W.: Deberta: decoding-enhanced bert with disentangled attention (2021)

Heafield, K., et al.: Findings of the fourth workshop on neural generation and translation. In: Proceedings of the Fourth Workshop on Neural Generation and Translation, pp. 1–9 (2020)

Henderson, P., Jieru, H., Romoff, J., Brunskill, E., Jurafsky, D., Pineau, J.: Towards the systematic reporting of the energy and carbon footprints of machine learning. J. Mach. Learn. Res. 21(248), 1–43 (2020)

Hu, J., Ruder, S., Siddhant, A., Neubig, G., Firat, O., Johnson, M.: Xtreme: a massively multilingual multi-task benchmark for evaluating cross-lingual generalisation. In: International Conference on Machine Learning, pp. 4411–4421. PMLR (2020)

Kuratov, Y., Arkhipov, M.: Adaptation of deep bidirectional multilingual transformers for Russian language. arXiv preprint arXiv:1905.07213 (2019)

Lan, Z., Chen, M., Goodman, S., Gimpel, K., Sharma, P., Soricut, R.: Albert: a lite bert for self-supervised learning of language representations (2019)

Liang, Y., et al.: XGLUE: a new benchmark dataset for cross-lingual pre-training, understanding and generation (2020). http://arxiv.org/abs/2004.01401

Liu, Q., Kusner, M.J., Blunsom, P.: A survey on contextual embeddings (2020). http://arxiv.org/abs/2003.07278

Liu, Y., et al.: Roberta: a robustly optimized bert pretraining approach. arXiv preprint arXiv:1907.11692 (2019)

Ma, Z., et al.: Dynaboard: an evaluation-as-a-service platform for holistic next-generation benchmarking. arXiv preprint arXiv:2106.06052 (2021)

Min, S., et al.: Neurips 2020 efficientqa competition: systems, analyses and lessons learned (2021). http://arxiv.org/abs/2101.00133

Pires, T., Schlinger, E., Garrette, D.: How Multilingual is Multilingual BERT? pp. 4996–5001 (2019)

Pruksachatkun, Y., et al.: jiant: a software toolkit for research on general-purpose text understanding models. In: Proceedings of the 58th Annual Meeting of the Association for Computational Linguistics: System Demonstrations, pp. 109–117 (2020)

Radford, A., Jeffrey, W., Child, R., Luan, D., Amodei, D., Sutskever, I.: Language models are unsupervised multitask learners. OpenAI Blog 1(8), 9 (2019)

Rogers, A.: How the transformers broke NLP leaderboards (2019). https://hackingsemantics.xyz/2019/leaderboards/

Sanh, V., Debut, L., Chaumond, J., Wolf, T.: Distilbert, a distilled version of bert: smaller, faster, cheaper and lighter. In: The 5th Workshop on Energy Efficient Machine Learning and Cognitive Computing (2019)

Shavrina, T., et al.: Russiansuperglue: a Russian language understanding evaluation benchmark. In: Proceedings of the 2020 Conference on Empirical Methods in Natural Language Processing (EMNLP), pp. 4717–4726 (2020)

Shavrina, T., Malykh, V.: How not to lie with a benchmark: rearranging NLP learderboards. In: ICBINB@NeurIPS 2021 (2021)

Sun, C., Qiu, X., Xu, Y., Huang, X.: How to fine-tune BERT for text classification? In: Sun, M., Huang, X., Ji, H., Liu, Z., Liu, Y. (eds.) CCL 2019. LNCS (LNAI), vol. 11856, pp. 194–206. Springer, Cham (2019). https://doi.org/10.1007/978-3-030-32381-3_16

Tsai, H., Riesa, J., Johnson, M., Arivazhagan, N., Li, X., Archer, A.: Small and practical bert models for sequence labeling. In: EMNLP/IJCNLP (1) (2019)

van Rijsbergen, C.J.: Information Retrieval. Butterworth-Heinemann (1979)

Vaswani, A., et al.: Attention is All You Need, pp. 5998–6008 (2017)

Wang, A., et al.: Superglue: a stickier benchmark for general-purpose language understanding systems. In: Advances in Neural Information Processing Systems, vol. 32 (2019)

Wang, A., Singh, A., Michael, J., Hill, F., Levy, O., Bowman, S.: GLUE: a multi-task benchmark and analysis platform for natural language understanding, pp. 353–355 (2018). https://doi.org/10.18653/v1/W18-5446

Wang, A., Wolf, T.: Overview of the sustainlp 2020 shared task. In: Proceedings of SustaiNLP: Workshop on Simple and Efficient Natural Language Processing, pp. 174–178 (2020)

Wolf, T., et al.: Huggingface's transformers: state-of-the-art natural language processing. arXiv preprint arXiv:1910.03771 (2019)

Zellers, R., et al.: Defending against neural fake news (2020). http://arxiv.org/abs/1905.12616

Zhu, J., et al.: Incorporating bert into neural machine translation. In: International Conference on Learning Representations (2019)

Evaluating the Influence of Argumentation Markers on the Identification of Reasoning Models

Ivan S. Pimenov[1,3](✉) ⓘ and Natalia V. Salomatina[2,3] ⓘ

[1] Novosibirsk State University, 1 Pirogova, Novosibirsk, Russia
`pimenov.1330@yandex.ru`
[2] Sobolev Institute of Mathematics, 4 Koptuga, Novosibirsk, Russia
[3] Ershov Institute of Informatics Systems, 6 Acad. Lavrentjev pr., Novosibirsk, Russia

Abstract. The article focuses on evaluating the influence of argumentation markers on the identification of specific reasoning models with machine learning methods. The evaluation process consists of a sequence of classification experiments with different feature sets. The experiments cover the identification of arguments with three specific reasoning models: "Expert Opinion", "Example", and "Practical Reasoning". These models are characterized by 1) an active use in scientific articles (as evidenced by their high frequency in the employed corpus) and 2) reliance of their textual expression on typical words and phrases (markers). Each model corresponds to a separate subset of the overall dataset: 680 arguments for classifying the "Example" model, 386 for "Practical Reasoning", 172 for "Expert Opinion" (in each case, a half of the arguments employs the corresponding model, while the other half relies on any other model except for these three). The overall dataset contains 1975 arguments from 45 scientific articles in Russian language (on linguistics and computational technologies). The argumentation in these articles is annotated with the ArgNetBank Studio platform. Classification experiments employ machine learning methods of different types: multinomial naive Bayes, support vector machine, and multilayer perceptron. The feature sets differ by the inclusion or exclusion of discourse markers and persuasion modes indicators (expressions characterizing three argumentation aspects: logos, pathos, and ethos). The experiments show that the best improvement of identification scores (on average across all schemes and classifiers) corresponds to the representation of arguments with discourse markers (plus 10% for precision and 7% for F-measure over the lemmas baseline).

Keywords: Argument mining · reasoning models · machine learning · discourse markers · argumentation markers · scientific articles

1 Introduction

Automatic identification of arguments in texts becomes a prerequisite step for the diverse practical applications. These uses include, for instance, the analysis of texts persuasiveness (in tasks such as the assessment of scientific articles in the aspect of their conclusions

© The Author(s), under exclusive license to Springer Nature Switzerland AG 2024
J. Baixeries et al. (Eds.): DAMDID/RCDL 2023, CCIS 2086, pp. 282–297, 2024.
https://doi.org/10.1007/978-3-031-67826-4_21

justification). The identification of arguments also enables the detection of specific types of premises for analyzed theses (why authors of a scientific article apply a certain method to the task, or choose one algorithm over others, etc.).

The use of the supervised machine learning methods (ML) requires the availability of text corpora with annotated argumentation. One of the questions during the argumentation annotation concerns the inclusion of discourse markers within the boundaries of premises and conclusions (whether such expressions as "if…, then…" or "according to…" should be annotated along with the propositional content of statements or separated from it), particularly in case of corpora intended for ML methods training. Experiments in the presented research address the necessity of such an inclusion.

The article describes experiments in identifying the semantic types of arguments at the level of reasoning models (also called argumentation schemes). This task corresponds to the separate and final stage in the pipeline of extracting argumentation structures. The aim of the work is to evaluate the influence of argumentation markers on the identification of reasoning models with machine learning methods. As such, optimization of identification scores across feature sets lies beyond the scope of the article. We calculate the lower bounds for the schemes identification (baseline) without using Deep Learning models, as the available corpus does not contain enough arguments at the moment.

2 Related Works

The task of automatic identification of argumentation schemes (in the general framework of argumentation extraction) is first addressed in works [1] and [2]. The author of the first, Douglas Walton, suggests a six-stage approach to identifying arguments and reasoning models in their organization. The approach consists in first detecting arguments in a text, and then classifying them over the list of specific schemes.

The traditional approach to different stages of extracting arguments relies on employing different ML methods (DL, as of late). Vector representation of arguments for ML can characterize them in various aspects, particularly by usage of the argumentation markers. However, works with marker-based representation of arguments differ in evaluation of the markers influence on the extraction quality.

The article [3] provides an example of analyzing the influence of features of different types on the quality of extracting argumentative sentences. The authors demonstrate that the exclusive use of discourse markers (and analogical constructions for improving text coherency) without other feature types does not yield satisfying results: in the experiment, accuracy for the Maxent classifier reaches only 57.98%. In turn, combining markers and unigrams results in classification accuracy exceeding 70%. The work does not specify the classification accuracy for unigrams and bigrams in absence of markers (in case when discourse markers are excluded from annotated arguments).

The work [4] describes an experiment in context-based classification of text segments into argumentative and non-argumentative by using a BERT model and a fully connected neural network. The experiment employs a corpus of popular science articles in Russian language. The authors show that the explicit marking of argumentation markers results in a moderate increase of the classification precision (by 1%) and recall (by 5%).

The paper [5] describes a combined approach to identifying distinct elements of the argumentation structures. The authors employ discourse markers for the detection of argumentative connections (to check whether adjacent propositions in a text are related in the structure of reasoning). While the use of markers reaches the precision of 89%, the recall equals only 4% due to the low frequency of the markers in the dataset (and the need of supplementing them with other features). In turn, for identifying the exact argumentation schemes the authors employ a Naïve Bayes classifier with features of diverse types (unigrams, bigrams, part-of-speech tags, punctuation signs). The experiment covers the identification of two argumentation schemes ("Expert Opinion" and "Positive Consequences"), where the average precision across proposition types reaches respectively 87% and 80%, while the average recall equals 81% and 67%. The article does not contain a specific evaluation of the markers influence on identifying argumentation schemes.

The authors of [6] analyze the applicability of discourse markers for the automatic identification of argumentative relations in scientific papers (in biomedical domain). They employ a set of regular expressions for more than 100 discourse indicators (both separate words and compound phrases). However, the experiment shows that the use of discourse markers results in a decrease of identification quality from the baseline approach (based on the textual intersection processing). After analyzing the identification errors, the authors suggest that discourse markers in scientific articles do not necessarily organize the relevant reasoning, but instead frequently express the decorative (rhetoric) function in non-argumentative contexts.

The authors of [7] employ a multi-class SVM classifier for identifying argumentative roles of text segments in student essays, as well as for detecting arguments in support of a given thesis. They analyze the applicability of specific features from a composite set with various structural, lexical, syntactic, and contextual characteristics, as well as markers of different types (discourse and temporal markers, personal and possessive pronouns). The experiment demonstrates that the F1 value for marker-based classification of argumentative roles ranges from 26.5% to 73% (depending on the role). Addressing the task of detecting arguments in support of a given thesis, the authors arrive at conclusion that while the separate use of markers is less efficient than the separate use of lexical and syntactic characteristics, the combination of both feature types achieves the best results.

In [2], Feng and Hirst develop a similar method to [1] based on classifying arguments by their schematic structures (for the five most frequent models in their dataset). Their study focuses on automatic classification of arguments with five frequent schemes ("Example", "Cause to Effect", "Practical Reasoning", "Positive/Negative Consequences", "Verbal Classification"). The classification is approached as a separate step in the pipeline (with the assumption that arguments have been extracted on the previous step). The dataset contains 393 arguments overall (from 41 to 149 for a specific scheme). The feature set combines general features for all schemes (seven positional characteristics, such as the relative position of the conclusion and the premise, the length of the interval between them in the text) and scheme-specific features (which range from keywords and punctuation signs to the syntactic dependency relations). The authors employ the decision tree algorithm in two classification modes: one scheme against others and binary across scheme pairs. They demonstrate the significant dependence of the classification quality on the analyzed scheme: the best average accuracy reaches 90% for "Example" and "Practical Reasoning", but only 70% for "Cause to Effect" and 60% for "Positive

/ Negative Consequences" and "Verbal Classification" (least represented in the dataset, with 44 and 41 arguments). Features that are specific to particular schemes effectively correspond to markers of these schemes, but the article does not address the influence of these features.

The authors of [8] address the identification of "Expert Opinion" arguments in texts in Russian language. The identification employs lexical-grammatical patterns that are constructed by experts. These patterns correspond to specific combinations of discourse connectives, verbs and nouns of diverse semantic classes, as well as their integrating constructions (templates with variables). Constants in the templates correspond to markers. The precision of the identification reaches 86.5%.

Overall, the existing works in automatic identification of arguments with specific reasoning schemes focus prevalently on texts in English language. One existing work that addresses the task for texts in Russian language ([8]) limits the scope to just one type of reasoning schemes ("Expert Opinion"), and the identification of its arguments relies on expert patterns (which require extensive labor for construction and do not support the identification of arguments with other schemes). Another known work [9] addresses the automatic extraction of arguments at the level of their stance (supporting or attacking a given thesis).

3   Identification of Reasoning Models as a Classification Task

We employ a pipeline-based approach to extracting argumentation structures from texts and, in the present article, focus on the final stage: the identification of specific reasoning models. The input at this stage corresponds to arguments (sets of detected argumentative statements with established connections between them). We have addressed the preceding stages in our earlier works [10, 11].

Let $C = \{t_i\}$ denote a corpus of texts, $0 < i \leq I$, where I is the number of texts in the corpus. $A^i = \{a_j^i\}$ is the set of arguments in the text t_i ($0 < j \leq J$, J is the number of arguments in this text). An argument a_j^i consists of its forming statements (u_j^i) and their connecting reasoning scheme (sch_j^i): $a_j^i = \{u_j^i, sch_j^i\}$ ($u_j^i = \{\{p_{jk}^i\}, c_j^i\}$, where $\{p_{jk}^i\}$ is the set of premises ($k > 0$) in the argument a_j^i, while c_j^i is its conclusion). In this article we focus on three specific argumentation schemes: Sch = {Expert Opinion, Example, Practical Reasoning}. These schemes are characterized by 1) an active use in scientific articles of the chosen thematic areas (information technologies and linguistics), and 2) their frequent expression in texts with explicit markers of diverse types.

Figure 1 provides an example of an argumentation graph fragment with three arguments implementing the analyzed schemes. Arguments A33 and A29 support the same conclusion (S36) with different premises (S37 and S20) connected to this conclusion by different argumentation schemes. The statement S20 serves as a premise in A29 and as a conclusion in A17. The text of statements has been translated from Russian.

We address the task of binary classification for each $a_j^{t_i}$: whether the analyzed argument corresponds to the reasoning through the given argumentation scheme.

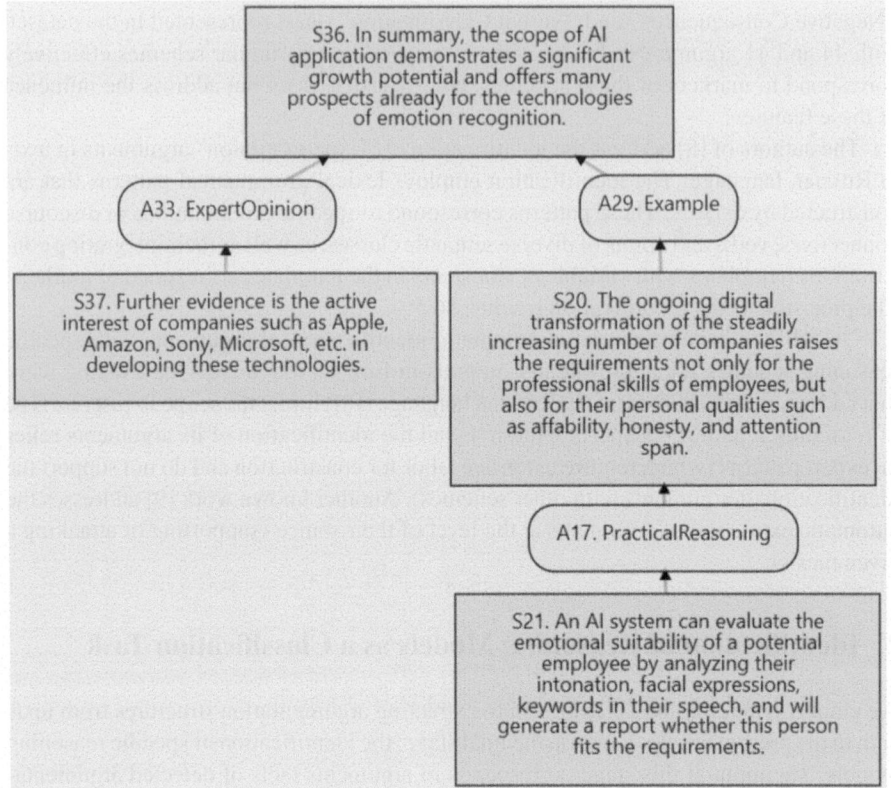

Fig. 1. An example of an argumentation graph fragment with the three analyzed schemes.

4 Methods for Identification of Reasoning Models

4.1 Classification by Machine Learning Algorithms

The base representation of an argument $\{\{p_{jk}^{i}\}, c_j^{i}\}$ (its textual expression) corresponds to a vector of lemmas containing at least one Cyrillic symbol. The selection of lemmas for the base feature set relies on the χ^2 criterion: this criterion enables the filtering of lemmas whose distribution across arguments does not provide an informative indication of argumentation schemes. We assign the χ^2 threshold in accordance with the empirical observations.

The experiments employ three classification algorithms belonging to different functional types and frequently used in Argument Mining research. These algorithms are SVM (support vector machine), MNB (multinomial Naïve Bayes), and MLP (multilayer perceptron, a basic neural network). We use program realizations of the algorithms from the Scikit-Learn library for machine learning in Python [12].

The SVM Algorithm. The training of the algorithm relies on the RBF kernel (Radial Basis Function). Two key parameters for this kernel are C (which regulates the balance between classification errors and simplicity of the training surface) and γ (which defines the influence of a single training example). We assign their values empirically: $C = 100$, $\gamma = 1/(n\_features * var)$, where n_features is the number of features, and var is the variance of the training matrix. The choice of the RBF over other kernel functions is based on preliminary experimental results: as the classification reached similar scores for RBF and other kernel types, we have chosen the former due to it being the default kernel function in the employed implementation of the algorithm.

The MLP Algorithm. The training of the neural network consists in assignation of weights for its constituting neurons by the backpropagation method. The network contains 100 neurons in the hidden layer and uses the logistic sigmoid function for its activation. Iterations continue until convergence (when the score or loss are not increasing by more than $1e-4$ for 10 iterations) or until reaching the maximum number of iterations (200). Additionally, an early stop check analyzes the validation data (10% of the training examples) to avoid overfitting. The regularization parameter α equals 0.0001.

The MNB Algorithm. To choose a class label for an input vector, the classifier evaluates the probability of this vector belonging to each of the possible classes with the Bayes' theorem. The Laplace smoothing parameter ($\alpha = 1$) balances the incompleteness of the training set by preventing the assignation of zero probability to the absent features.

4.2 Two Types of Argumentation Markers

The study investigates the influence of two types of argumentation markers on reasoning models identification. The first type denotes discourse markers that organize a text as the coherent unit (including the level of transitions between argument components, from premises to conclusions). The second type corresponds to indicators of three persuasion modes (aspects of the argumentative effect): persuasion by logical facts (logos), by emotional manipulation (pathos), or by appeals to a source of authority (ethos).

Discourse Markers. We extract markers from a corpus of texts with expert-annotated argumentation, and then expand their list with separate rhetoric markers from the RSTree-Bank resource (https://rstreebank.ru/), as well as with marker synonyms from various online synonym dictionaries. The resulting dictionary of discourse markers contains 407 words and word combinations. Some of the markers include punctuation signs, begin strictly with the capitalized letter (to indicate their position at the start of a sentence), or correspond to a shortened form. These specifications serve to improve the model identifications. Examples of such markers are "Tak," ("So,"), "Poetomu" ("That is why"), "napr." (a shortened version of "naprimer", which means "for example", corresponds to "e.g.").

Indicators of Ethos, Pathos, and Logos. According to the classic model by Aristotle, argumentation achieves a persuasive effect through three different aspects: intellectual (logos), emotional (pathos), and authoritative (ethos). These three aspects correspond to three different modes of influencing the audience. Textual expression of the modes relies on specific constructions with lexical units that correspond to each mode.

Identification of persuasion modes in argumentative statements presupposes the creation of dictionaries with indicators that mark these modes. In turn, specification of indicator forms needs to follow explicitly formulated criteria, such as given in [11].

Indicators of *pathos* correspond to at least one of the three following properties:

1) *Redundancy*: a language expression does not contribute to presentation of information (neither to the semantic content nor the structure), and its removal does not influence the statement neither in content nor in relations to other statements. Examples of such indicators are *"lish"* ("only", "merely"), *"dazhe"* ("even"), *"vovse"* (an emphatic particle with different contextual meanings, such as "at all" or "completely").
2) *Deontic modality*: an expression conveys a meaning of prescribing an action (*"trebuetsa"* ("it is required to"), *nuzhno* ("it is necessary to"), *"vazhno"* ("it is important to"), etc.).
3) Stylistic marking: an expression can be substituted by a stylistically neutral synonym without affecting the meaning of the statement (*"vpechatlyaushij"* ("impressive"); *"vydajushijsya"* ("remarkable")).

In the presented research, we employ a broad definition of ethos: justification of a claim through backing it with a source of authority. This source can correspond to a specific individual (a named scientist, an expert), or a group (such as a research team), or an impersonal agent (a popular opinion, uncertain informer), or an applicable example case (a precedent, a traditional practice). Consequently, indicators of *ethos* possess the property of *authorization*: they indicate a source of information (*"experty chitajut"* ("experts think that"), *"po mneniju avtora"* ("in the author's opinion")). This definition also includes bibliographic references, usually denoted by square or round brackets.

Finally, indicators of logos are expressions that organize the reasoning structure in its textual presentation. They contribute to *logical connectivity* of the text. Examples of logos indicators are *"esli..., to..."* (if ..., then...), *"vo-pervyh"*, *"vo-vtoryh"* ("first of all", "secondly").

To perform the automatic identification of indicators, we construct search patterns in form of regular expressions. An indicator may contain several elements (framed in square brackets in the pattern description) with specified lexical and/or grammatical properties of each. If an element of an indicator corresponds to several possible (alternative) constants, a delimiter "|" separates them in a list. Certain indicators permit insertions of arbitrary length (limited by the statement length), and the symbol "..." denotes these insertions. Below we present the examples of patterns for indicators of ethos (E_i), pathos (P_i), and logos (L_i).

E_i: [soglasno // PREP] [- // ADJF...(datv|loct)] [- // NOUN...(datv|loct)];
P_i: [(sovershenno|yavno|ochen') // -][- // (ADJF|ADJS|ADVB)];
L_i: [v // PREP] [(zaklyuchenie|itog|chastnost') // -]

The first example corresponds to an expression "according to...", where a source of authority is specified by a pair of an adjective and noun in a dative or locative case ("according to the new data", "according to the recent results", etc.). The second pattern specifies a combination of an emphatic particle ("completely", "obviously", "very") and an adjective or an adverb. The third example denotes a connective for accentuating the beginning of a new information block (in conclusion, in particular).

At present, the indicators dictionary contains 89 patterns: 32 for ethos, 39 for pathos, 18 for logos. These numbers correspond to separate patterns (sequences of several elements) without accounting for alternative constants within indicators.

The use of indicator dictionaries in identification of argumentation schemes enables the processing of persuasion modes characteristics that are implicitly expressed in these schemes. The article [11] describes a qualitative evaluation of persuasion modes weights in argumentation schemes by their comparative functional analysis. That study groups schemes by their functional similarity and then contrasts the similar schemes within each group. At the first level, schemes are separated into two groups based on scope of the expressed reasoning. Namely, argumentation can advance either by analyzing facts within the propositional content of presented statements or by appealing to external sources of authority (which are not directly commeasurable with the analyzed phenomena). Correspondingly, the first group contains argumentation schemes with dominant *logos*, while models in the second group rely on *ethos*. However, specific schemes might complement the main persuasion component with others at different intensity.

In particular, among the schemes with the prevalent logos, abstract causal models (such as "Cause to Effect" or "Correlation to Cause") rely more on the logical component than do practically-oriented schemes (such as "Practical Reasoning" or "Positive/Negative Consequences"). The latter models potentially convey a stronger complementary component of pathos (especially if an analysis of possible results accentuates their sentiment-based evaluation). Similarly, arguments from authority convey ethos most clearly by specifying an exact specialist (through the "Expert Opinion" model). The authoritativeness of the cited source decreases if an appeal addresses an impersonal agent (by the "Popular Opinion" scheme).

5 Classification Experiments with Different Features

5.1 Data Set with Argumentation Annotation

The experiment dataset contains 1975 arguments and 1809 argumentative statements extracted from 45 short scientific articles in Russian language with annotated argumentation. The articles range in length from 800 to 1500 words and belong to two research areas: information technologies (23 articles) and linguistics (22). Their texts have been downloaded from the freely accessible sources: online scientific library "CyberLeninka" and proceedings of the "Corpus Linguistics" conference. The expert annotation of argumentation uses tools of the ArgNetBank Studio web platform [13]. The annotated texts are available at the platform website [14]. The annotating process follows the Argument Interchange Format standard, an example of employing which for modelling argumentation in texts in Russian language is given in [15].

Two expert annotators (qualified both in linguistics and information technologies) perform the annotation of argumentation in texts. They follow a detailed annotation instruction formulated in advance. The annotation of a text consists in constructing an argumentation graph that is oriented, connected, acyclic, and rooted. The root node in each graph denotes the main thesis of the respective text. For each argument identified in a text, the annotators specify its constituents (premises and conclusions) and the semantic type of the argumentative connection between them (by indicating its argumentation scheme from Walton's compendium [16]).

For a quarter of the corpus texts (12 articles), both experts have constructed separate annotation versions. Double annotation of these texts enables the calculation of correspondence coefficients across all three levels of the argumentation structure (to ascertain the reliability of annotations). The average values across 12 texts are given below.

1) The average ratio of the number of argumentative statements, identified by both experts for the same text, to the sum of argumentative statements, identified in it by at least one, equals 78%.
2) The average ratio of corresponding connections between argument components (to the similar sum of all connections identified by both annotators) reaches 55%. This value serves as the lower bound: the same connections between same statements yet with different configurations (parallel or sequential) are considered non-corresponding.
3) The average percentage of matching argumentation schemes in connections equals 60%.

The resulting dataset contains only one argument-annotated version for each text. For texts with two annotation versions, the choice of a version follows the joint decision of the annotators. The corpus includes 330 arguments with the "Example" scheme, 193 with "Practical Reasoning", 86 with "Expert Opinion".

5.2 Construction of Training and Test Sets

For each of the three analyzed schemes, around 80% of its arguments constitute its training set (LS) for the classification, while the other 20% form the test set (TS). The exact percentage varies due to the principle of *text integrity* in dividing arguments between sets: arguments from the same text can belong either only to the LS or only to the TS (if an argument from a text belongs to the LS, all other arguments from this text can appear only in the LS, but not the TS, and vice versa). Additionally, the TS for its scheme contains an equal number of arguments from texts of both thematic fields (IT and linguistics). Negative classification examples in sets (arguments with other schemes) are extracted from the same texts as positive. They are selected at random, so that the number of negative examples equals the number of positive both in the TS and LS.

The resulting classification sets for each analyzed scheme contain the following numbers of arguments:

a) *Example*: 520 arguments in the LS (260 with "Example", 260 with other models) and 140 arguments in the TS (70 with "Example", 70 with other schemes).
b) *Practical Reasoning*: 306 arguments in the LS (153 arguments of both types) and 80 arguments in the TS (40 with "Practical Reasoning", 40 with other schemes).
c) *Expert Opinion*: 136 arguments in the LS (68 for "Expert Opinion", 68 others) and 36 arguments in the TS (18 with "Expert Opinion", 18 with other schemes).

The selection of lemmas for the vector representations of arguments relies on the χ^2 criterion (described in Sect. 4.1). The threshold values equal 10% for "Example" and "Practical Reasoning", and 20% for "Expert Opinion" (established empirically). The lemmatization of words in argument components (premises and conclusions) employs the PyMorphy2 library for Python.

5.3 Comparing Classification Results Across Feature Sets

We perform 10 experiments in binary classification for identifying each of the three chosen argumentation schemes. The experiments employ different feature sets. The construction of feature sets consists in different combinations of feature types in order to evaluate the influence of discourse markers and persuasion modes indicators on the identification of schemes. The number of features in each experiment depends on the employed feature type (407 discourse markers, 89 patterns of persuasion indicators) and the analyzed scheme (due to different χ^2 thresholds for the selection of lemmas: 444 lemmas for the "Example" scheme, 181 for "Expert Opinion", 325 for 'Practical Reasoning").

Before the experiments on different feature sets, we perform the preliminary tests to empirically assign the threshold values for the formal filtration of features (by the χ^2 criterion). The chosen threshold values improve the average F-measure across all classification algorithms and schemes by 6.3% (over the unfiltered lemmas). The following experiments (1–11) employ the formal filtration in all cases when feature sets include lemmas.

There are 4 different types of features: lemmas without positional specification (1), discourse markers (2), persuasion indicators (3), lemmas with positional specification (4), where types (1) and (4) are mutually exclusive. The number of possible feature types combinations can be calculated with the formula $M = 2 \times \sum_{k=1}^{n-1} C(n-1, k) - 3 = 14 - 3 = 11$, where $C(n-1, k)$ is the number of combinations of n − 1 elements taken k at a time. The deduction of the constant 3 is based on the mutual exclusivity of the feature types (1) and (4).

Experiment 1. Vector representations of arguments consist only of lemmas and do not include discourse markers nor persuasion modes indicators (described in Sect. 4.2). In effect, the initial experiment addresses the influence of thematic content of arguments on identification of each scheme.

Experiment 2. Vector representations of arguments contain only discourse markers (with the exclusion of ordinary lemmas). The classification results demonstrate the exclusive role of markers in indicating specific schemes.

Experiment 3. Vector representations of arguments contain only indicators of three persuasion modes (ethos, pathos, logos). This experiment addresses the distinguishing potential of persuasion modes indicators, which emphasize the form of arguments expression (by accentuating logical connections between facts, by invoking an emotional reaction in a reader, or by underlining authoritativeness of cited sources).

Experiment 4. Vector representation of arguments combines the features used in experiments 1 and 2. The new experiment aims at comparing the results from combining the features of different types with the quality scores for lemmas and markers when used separately (whether these scores improve, and if yes, to which degree).

Experiment 5. This experiment resembles the preceding one, but persuasion modes indicators replace discourse markers in vector representation of arguments. The comparison with the previous results will enable the evaluation of indicators efficiency in identifying schemes.

Experiment 6. Vector representations of arguments contain only persuasion modes indicators and discourse markers. The classification results will demonstrate how these two feature types strengthen or weaken each other in joint use and absence of lemmas.

Experiment 7. Vector representation of arguments combines all types of features separately employed in experiments 1, 2, and 3. The classification results demonstrate the efficiency of a multi-aspect argument modelling.

The next four experiments (8, 9, 10, 11) address the influence of the argumentative role-based distinguishing between lemmas (whether they appear in premises of an argument or the conclusion) on the identification of schemes. This is achieved by doubling the number of lemmas in the feature set (lemmas occurrence in premises and in conclusions are examined separately). The identification of roles of statements in arguments can employ machine learning methods in a similar classification task.

Experiments 8, 9, 10, 11. Each lemma in vector representation of arguments (as in experiments 1, 4, 5, 7) corresponds to two features: one specifies its occurrences in premises, another in conclusions.

The eleven experiments cover all possible combinations of the analyzed feature types. Table 1 provides the results of experiments 1–11, performed with three ML methods (described in Sect. 4.1), separately for each analyzed scheme. We employ the standard precision (P), recall (R) scores and F-measure (F) to evaluate the classification results. Experiments in table are denoted by letter "E" and their number, while supplementary labels specify the feature set composition ("L" for filtered lemmas, "M" for discourse markers, "P" for persuasion modes indicators, "R" for role specification of lemmas).

Comparing the Classification Results. The table shows a notable influence of a choice of a classifying algorithm on identification of schemes. For the "Example" model, the MNB algorithm achieves the best precision value across most of the experiments, while the best recall characterizes the SVM method. In turn, for "Expert Opinion" and "Practical Reasoning", MLP demonstrates the best precision and recall both. For all three schemes across feature sets, the basic neural network (MLP) surpasses two other algorithms in identification F-measure by 8.7% on average.

The experiments demonstrate the significant role of markers and indicators in characterizing argument components. In E2 and E3, feature sets contain only markers or indicators respectively, yet the separate classification scores (for "Expert Opinion", all scores) in these experiments are higher than in E1 (in representation of arguments by thematic lemmas) for 2 algorithms out of 3. The lesser efficiency of lemmas appears to be caused by thematic diversity of specialized articles in the corpus. Most of the represented themes correspond only to 2–3 articles: for example, 23 articles in IT cover a variety of subfields, such as image analysis, information security, artificial intelligence, machine translation, speech recognition, medical applications of IT, etc. As a result, thematic vocabularies differ significantly between articles. If all texts for a specific theme appear

Table 1. Classification quality scores for Experiments 1–11.

	Example								
	MNB			SVM			MLP		
	P	R	F	P	R	F	P	R	F
E1 (L)	0.64	0.26	**0.37**	0.52	0.43	**0.47**	0.55	0.49	**0.52**
E2 (M)	0.60	0.41	**0.49**	0.42	0.49	**0.45**	0.46	0.61	**0.53**
E3 (P)	0.67	0.29	**0.40**	0.71	0.21	**0.33**	0.61	0.24	**0.35**
E4 (LM)	0.65	0.29	**0.40**	0.57	0.49	**0.52**	0.53	0.43	**0.47**
E5 (LP)	0.74	0.24	**0.37**	0.60	0.30	**0.40**	0.52	0.39	**0.44**
E6(MP)	0.63	0.34	**0.44**	0.42	0.34	**0.38**	0.49	0.47	**0.48**
E7 (LMP)	0.63	0.31	**0.42**	0.42	0.34	**0.38**	0.50	0.53	**0.51**
E8 (LR)	0.82	0.26	**0.39**	0.54	0.44	**0.49**	0.58	0.44	**0.50**
E9 (LMR)	0.84	0.23	**0.36**	0.57	0.41	**0.48**	0.57	0.39	**0.46**
E10 (LPR)	0.78	0.30	**0.43**	0.55	0.41	**0.47**	0.65	0.51	**0.58**
E11 (All)	0.85	0.31	0.46	0.52	0.39	0.44	0.60	0.40	0.48
	Expert Opinion								
	MNB			SVM			MLP		
	P	R	F	P	R	F	P	R	F
E1 (L)	0.33	0.22	**0.27**	0.40	0.33	**0.36**	0.46	0.61	**0.52**
E2 (M)	0.53	0.50	**0.51**	0.65	0.72	**0.68**	0.50	0.50	**0.50**
E3 (P)	0.43	0.17	**0.24**	0.38	0.17	**0.23**	0.56	0.78	**0.65**
E4 (LM)	0.33	0.22	**0.27**	0.36	0.33	**0.28**	0.42	0.28	**0.33**
E5 (LP)	0.50	0.11	**0.18**	1.00	0.06	**0.11**	0.82	0.50	**0.62**
E6(MP)	0.64	0.50	**0.56**	0.64	0.50	**0.56**	0.64	0.39	**0.48**
E7 (LMP)	0.64	0.50	**0.56**	0.64	0.50	**0.56**	0.69	0.50	**0.58**
E8 (LR)	0.80	0.22	**0.35**	0.67	0.11	**0.19**	0.60	0.33	**0.43**
E9 (LMR)	0.75	0.17	**0.27**	0.67	0.11	**0.19**	0.50	0.22	**0.31**
E10 (LPR)	0.80	0.22	**0.35**	0.67	0.11	**0.19**	0.50	0.17	**0.25**
E11 (All)	0.80	0.22	**0.35**	0.67	0.11	**0.19**	0.50	0.17	**0.25**
	Practical Reasoning								
	MNB			SVM			MLP		
	P	R	F	P	R	F	P	R	F
E1 (L)	0.64	0.35	**0.45**	0.79	0.28	**0.41**	0.73	0.40	**0.52**
E2 (M)	0.52	0.28	**0.36**	0.45	0.50	**0.48**	0.50	0.53	**0.51**

(continued)

Table 1. (*continued*)

	Practical Reasoning								
	MNB			SVM			MLP		
	P	R	F	P	R	F	P	R	F
E3 (P)	0.70	0.40	**0.51**	0.63	0.30	**0.41**	0.57	0.30	**0.39**
E4 (LM)	0.73	0.20	**0.31**	0.77	0.25	**0.38**	0.62	0.33	**0.43**
E5 (LP)	0.45	0.23	**0.30**	0.57	0.30	**0.39**	0.53	0.23	**0.30**
E6(MP)	0.40	0.30	**0.34**	0.37	0.42	**0.40**	0.41	0.33	**0.36**
E7 (LMP)	0.40	0.30	**0.34**	0.37	0.42	**0.40**	0.44	0.38	**0.41**
E8 (LR)	0.52	0.33	**0.40**	0.77	0.25	**0.38**	0.67	0.45	**0.54**
E9 (LMR)	0.50	0.30	**0.37**	0.75	0.23	**0.35**	0.54	0.38	**0.44**
E10 (LPR)	0.48	0.33	**0.39**	0.69	0.23	**0.34**	0.57	0.33	**0.41**
E11 (All)	0.50	0.30	**0.37**	0.60	0.15	**0.24**	0.48	0.35	**0.41**

exclusively in either the LS or TS, identification scores decrease for vector representation of arguments by formally filtered lemmas. As a rule, persuasion modes indicators exhibit better precision, while discourse markers in most cases reach a better recall (for instance, as in E2 and E3 for the "Example" model).

The noted tendencies principally characterize the identification of the "Example" and "Expert Opinion" models. They apply to "Practical Reasoning" to a lesser extent, as the present incompleteness of its marker dictionary constrains its identification (the dictionary of markers for "Practical Reasoning" is at the stage of development). This difference in typical markers is caused by a specific functional trait of the "Practical Reasoning" scheme (a greater positional distance between connected statements in an argument). Namely, discourse markers specify connections mostly between adjacent statements in a text (due to their role as rhetoric connectives). The "Example" and "Expert Opinion" models operate with such adjacent statements (an example or an appeal to a source authority tends to directly succeed or precede the corresponding claim). However, in arguments based on "Practical Reasoning", the premises and conclusions can be presented in different sections of a text (for example, one paragraph specifies the aim of a research or a separate subtask, while another describes a choice of a method in accordance with the earlier formulation of the analyzed problem). In these cases, discourse markers might not suffice for identification of "Practical Reasoning" arguments, and the use of thematic lemmas for the vector representation becomes preferable.

The analysis of classification errors across the experiments reveals three prominent causes of incorrect scheme identification. One type of errors occurs due to innate overlapping of arguments in argumentation structures, where one statement can be used a support for proving another and then in turn be justified by a third statement (within another argument). As such, the intermediary statement might contain markers relevant to the scheme of either argument (but not the other), and these markers potentially mislead the classifier. Errors of the second type stem from the appearance of ambiguous markers that either can introduce several possible schemes or correspond in general to a specific scheme, yet in a particular context are used for implementing another. The third type of errors occurs for arguments without explicit markers (either of discourse or of a persuasion mode). In these cases, classification by the exclusive use of lemmas without relying on markers can exceed in quality scores the markers-focused classification (particularly for the "Practical Reasoning" scheme).

Not always justified is the initial assumption that the combined representation of arguments by lemmas and markers (or indicators) will result in a considerable improvement of the identification scores. The E4 for "Expert Opinion" provides an example of the scores, on the contrary, decreasing. The reasons for such decreases correspond to the two outlined causes: thematic diversity of specialized articles, insufficient size of the discourse markers dictionary. The MLP classifier (best across identification scores on average) demonstrates only a 4% increase in precision when extending lemmas with indicators. Its best recall (on average across schemes) corresponds to the exclusive use of markers (4.7% higher than with lemmas). If we employ the lemmas-based classification as a baseline for comparison (on average across all algorithms and schemes), the use of discourse markers yields a 10% increase in recall, while indicators provide a 2% increase in precision. The combination of lemmas, markers, and indicators improves the precision by 6%, yet the combination only of lemmas and indicators further increases the gain to 7.3%. In terms of F-measure, markers give a 7% increase over the lemma-based baseline, while the combination of all feature types provides a boost of 3%. Finally, the elaboration of lemmas position (in a premise or a conclusion) improves the classification precision in most of cases.

6 Conclusion

The presented research focuses on a sequence of experiments to evaluate the influence of argumentation markers of different types (discourse markers and persuasion modes indicators) on identification of specific argumentation schemes with machine learning methods. The analyzed schemes correspond to arguments from "Example", "Expert Opinion", and "Practical Reasoning". The binary classification employs algorithms of different functional types: SVM, MNB, and MLP. Experiments differ in feature sets for vector representation of arguments through inclusion or exclusion of various types of features (lemmas after formal filtration, discourse markers or persuasion modes indicators from the constructed dictionaries).

The experiments show that, when using markers from a dictionary of a sufficient size, the introduction of thematic lemmas into the feature set does not effectively influence the identification scores. Another observation concerns the importance of thematic homogeneity of the analyzed data (not only at the level of a general theme, but also in its sub-topics). In classifying arguments from articles in different thematic sub-fields of information technologies and linguistics, the introduction of lemmas into vectors might result in a decrease of identification scores on combined feature sets (with features of different types). As a rule, discourse markers improve the precision of identifying argumentation schemes, while persuasion modes indicators increase its recall. Precision can be further increased by specifying the positional properties of thematic lemmas (whether they occur in premises of an argument or its conclusion), yet such a specification requires the preliminary identification of statement roles in arguments.

Acknowledgements. The work was funded by Russian Science Foundation according to the research project no. 23-11-00261.

Disclosure of Interests. The authors have no competing interests to declare that are relevant to the content of this article.

References

1. Walton, D.: Argument mining by applying argumentation schemes. Stud. Log. **4**(1), 38–64 (2011)
2. Feng, V−W., Hirst, G.: Classifying arguments by scheme. In: Proceedings of the 49th Annual Meeting of the Association for Computational Linguistics: Human Language Technologies 2011, vol. 1, pp. 987–996. Association for Computational Linguistics (ACL) (2011)
3. Moens, M.-F., Boiy, E., Palau, R.M., Reed, C.: Automatic detection of arguments in legal texts. In: ICAIL 2007: Proceedings of the 11th International Conference on Artificial Intelligence and Law, pp. 225–230 (2007)
4. Sidorova, E., Akhmadeeva, I. Kononenko, I., Chagina, P.: Izvlechenie argumentatsii na osnove indikatornogo podhoda [An indicator-based approach to argumentation mining] [In Russian]. In: Proceedings of the 20th National Conference on Artificial Intelligence KII-2022, pp. 219–233 (2022)
5. Lawrence, J., Reed, C.: Combining argument mining techniques. In: Proceedings of the 2nd Workshop on Argumentation Mining, Denver, CO, pp. 127–136. Association for Computational Linguistics (2015)
6. Gao, Y., Gu, N., Lam, J., Hahnloser, R.: Do discourse indicators reflect the main ar-guments in scientific papers? In: Proceedings of the 9th Workshop on Argumentation Mining, pp. 34–50. Association for Computational Linguistics (ACL) (2022)
7. Stab, C., Gurevych, I.: Identifying argumentative discourse structures in persuasive essays. In: Proceedings of the 2014 Conference on Empirical Methods in Natural Language Processing (Doha), pp. 46–56 (2014)
8. Achmadeeva, I., Kononenko, I., Salomatina, N., Sidorova, E.: Indicator patterns as fea-tures for argument mining. In: International Multi-Conference on Engineering, Computer and Information Sciences (SIBIRCON) Proceedings, Novosibirsk, Russia, pp. 0886−0891 (2019)
9. Fishcheva, I., Kotelnikov, E.: Cross-lingual argumentation mining for russian texts. In: van der Aalst, W., et al. (eds.) AIST 2019. LNCS, vol. 11832, pp. 134−144. Springer, Cham (2019). https://doi.org/10.1007/978-3-030-37334-4_12
10. Salomatina, N., Sidorova, E., Pimenov, I.: Identification of argumentative sentences in Russian scientific and popular science texts. In: Journal of Physics: Conference Series, International Conference «Marchuk Scientific Readings 2021» (MSR-2021), Russia, Novosibirsk, vol. 2099, pp. 1–8 (2021)
11. Pimenov, I., Salomatina, N., Timofeeva, M.: The quantitative evaluation of the pathos to ethos ratio in scientific texts. In: Proceedings of the 2022 IEEE 23rd International Conference of Young Professionals in Electron Devices and Materials (EDM). Altai, Russia, pp. 312–317 (2022)

12. Scikit Learn Homepage. https://scikit-learn.org/stable/supervised_learning.html. Accessed 21 Aug 2023
13. Sidorova, E., Ahmadeeva, I., Zagorul'ko, Y.U., Seryj, A., Shestakov, V.: Research platform for the study of argumentation in popular science discourse. Ontol. Des. **10**(4(38)), 489–502 (2020)
14. ArgNetBank Studio Homepage. https://uniserv.iis.nsk.su/arg/. Accessed 21 Aug 2023
15. Pimenov, I.: Specifika argumentacionnogo annotirovaniya nauchnih I nauchno-populyarnih tekstov. In: Proceedings of the Corpora 2021 International Conference, Saint-Petersburg, Skifiya-Print, pp. 330–337 (2021)
16. Walton, D., Reed, C., Macagno, F.: Argumentation Schemes. Cambridge University Press, New York (2008)

Author Index

J. Baixeries et al. (Eds.): DAMDID/RCDL 2023, CCIS 2086, pp. 299–300, 2024.
https://doi.org/10.1007/978-3-031-67826-4

GPSR Compliance

The European Union's (EU) General Product Safety Regulation (GPSR) is a set of rules that requires consumer products to be safe and our obligations to ensure this.

If you have any concerns about our products, you can contact us on ProductSafety@springernature.com

In case Publisher is established outside the EU, the EU authorized representative is:

Springer Nature Customer Service Center GmbH
Europaplatz 3
69115 Heidelberg, Germany

The manufacturer's authorised representative in the EU is Springer
Nature Customer Service Centre GmbH, Europaplatz 3, 69115 Heidelberg,
Germany. If you have any concerns regarding our products, please
contact ProductSafety@springernature.com

Printed and bound by CPI Group (UK) Ltd, Croydon, CR0 4YY
29/04/2026
02099532-0004